FUNDAMENTALS
OF NEURAL NETWORKS

FUNDAMENTALS
OF NEURAL NETWORKS
ARCHITECTURES, ALGORITHMS, AND APPLICATIONS

Laurene Fausett

Florida Institute of Technology

Prentice Hall, Englewood Cliffs, NJ 07632

Library of Congress Cataloging-in-Publication Data

Fausett, Laurene V.
  Fundamentals of neural networks: architectures, algorithms,
and applications / by Laurene Fausett
    p. cm.
  Includes bibliological references and index.
  ISBN: 0-13-334186-0
  1. Neural networks (Computer science) I. Title.
QA76.87.F38 1994
006.3--dc20
                        93-8629
                         CIP

Acquisitions editor: **DON FOWLEY**
Editorial/production supervision
  and interior design: **RICHARD DeLORENZO**
Copy editor: **BRIAN BAKER**
Cover designer: **RICH DOMBROWSKI**
Production coordinators: **LINDA BEHRENS & DAVID DICKEY**
Editorial assistant: **JENNIFER KLEIN**
Supplements editor: **ALICE DWORKIN**

 © 1994 by Prentice-Hall, Inc.
A Paramount Communications Company
Englewood Cliffs, New Jersey 07632

The author and publisher of this book have used their best efforts in preparing this book. These efforts include the development, research, and testing of the theories and programs to determine their effectiveness. The author and publisher make no warranty of any kind, expressed or implied, with regard to these programs or the documentation contained in this book. The author and publisher shall not be liable in any event for incidental or consequential damages in connection with, or arising out of, the furnishing, performance, or use of these programs.

Printed in the United States of America

10 9 8 7 6 5 4 3 2

ISBN 0-13-334186-0

Prentice-Hall International (UK) Limited, London
Prentice-Hall of Australia Pty. Limited, Sydney
Prentice-Hall Canada Inc., Toronto
Prentice-Hall Hispanoamericana, S.A., Mexico
Prentice-Hall of India Private Limited, New Delhi
Prentice-Hall of Japan, Inc., Tokyo
Simon & Schuster Asia Pte. Ltd., Singapore
Editora Prentice-Hall do Brasil, Ltda., Rio de Janeiro

To my husband
Don
and our children
Ben, Beth, Bev, and Jenny

Contents

PREFACE   *xiii*

ACKNOWLEDGMENTS   *xv*

CHAPTER 1   **INTRODUCTION**   **1**

1.1   Why Neural Networks, and Why Now?   1

1.2   What Is a Neural Net?   3
         1.2.1   Artificial Neural Networks, 3
         1.2.2   Biological Neural Networks, 5

1.3   Where Are Neural Nets Being Used?   7
         1.3.1   Signal Processing, 7
         1.3.2   Control, 8
         1.3.3   Pattern Recognition, 8
         1.3.4   Medicine, 9
         1.3.5   Speech Production, 9
         1.3.6   Speech Recognition, 10
         1.3.7   Business, 11

1.4   How Are Neural Networks Used?   11
         1.4.1   Typical Architectures, 12
         1.4.2   Setting the Weights, 15
         1.4.3   Common Activation Functions, 17
         1.4.4   Summary of Notation, 20

1.5   Who Is Developing Neural Networks?   22
        1.5.1   The 1940s: The Beginning of Neural Nets, 22
        1.5.2   The 1950s and 1960s: The First Golden Age of
                Neural Networks, 23
        1.5.3   The 1970s: The Quiet Years, 24
        1.5.4   The 1980s: Renewed Enthusiasm, 25

1.6   When Neural Nets Began: the McCulloch-Pitts
        Neuron   26
        1.6.1   Architecture, 27
        1.6.2   Algorithm, 28
        1.6.3   Applications, 30

1.7   Suggestions for Further Study   35
        1.7.1   Readings, 35
        1.7.2   Exercises, 37

CHAPTER 2   SIMPLE NEURAL NETS FOR PATTERN
CLASSIFICATION   39

2.1   General Discussion   39
        2.1.1   Architecture, 40
        2.1.2   Biases and Thresholds, 41
        2.1.3   Linear Separability, 43
        2.1.4   Data Representation, 48

2.2   Hebb Net   48
        2.2.1   Algorithm, 49
        2.2.2   Application, 50

2.3   Perceptron   59
        2.3.1   Architecture, 60
        2.3.2   Algorithm, 61
        2.3.3   Application, 62
        2.3.4   Perceptron Learning Rule Convergence Theorem, 76

2.4   Adaline   80
        2.4.1   Architecture, 81
        2.4.2   Algorithm, 81
        2.4.3   Applications, 82
        2.4.4   Derivations, 86
        2.4.5   Madaline, 88

2.5   Suggestions for Further Study   96
        2.5.1   Readings, 96
        2.5.2   Exercises, 97
        2.5.3   Projects, 100

CHAPTER 3   PATTERN ASSOCIATION   101

3.1   Training Algorithms for Pattern Association   103
       3.1.1   Hebb Rule for Pattern Association, 103
       3.1.2   Delta Rule for Pattern Association, 106

3.2   Heteroassociative Memory Neural Network   108
       3.2.1   Architecture, 108
       3.2.2   Application, 108

3.3   Autoassociative Net   121
       3.3.1   Architecture, 121
       3.3.2   Algorithm, 122
       3.3.3   Application, 122
       3.3.4   Storage Capacity, 125

3.4   Iterative Autoassociative Net   129
       3.4.1   Recurrent Linear Autoassociator, 130
       3.4.2   Brain-State-in-a-Box, 131
       3.4.3   Autoassociator With Threshold Function, 132
       3.4.4   Discrete Hopfield Net, 135

3.5   Bidirectional Associative Memory (BAM)   140
       3.5.1   Architecture, 141
       3.5.2   Algorithm, 141
       3.5.3   Application, 144
       3.5.4   Analysis, 148

3.6   Suggestions for Further Study   149
       3.6.1   Readings, 149
       3.6.2   Exercises, 150
       3.6.3   Projects, 152

CHAPTER 4   NEURAL NETWORKS BASED ON COMPETITION   156

4.1   Fixed-Weight Competitive Nets   158
       4.1.1   Maxnet, 158
       4.1.2   Mexican Hat, 160
       4.1.3   Hamming Net, 164

4.2   Kohonen Self-Organizing Maps   169
       4.2.1   Architecture, 169
       4.2.2   Algorithm, 170
       4.2.3   Application, 172

4.3   Learning Vector Quantization   187
       4.3.1   Architecture, 187
       4.3.2   Algorithm, 188
       4.3.3   Application, 189
       4.3.4   Variations, 192

4.4   Counterpropagation 195
         4.4.1 Full Counterpropagation, 196
         4.4.2 Forward-Only Counterpropagation, 206

4.5   Suggestions For Further Study 211
         4.5.1 Readings, 211
         4.5.2 Exercises, 211
         4.5.3 Projects, 214

CHAPTER 5   ADAPTIVE RESONANCE THEORY   218

5.1   Introduction 218
         5.1.1 Motivation, 218
         5.1.2 Basic Architecture, 219
         5.1.3 Basic Operation, 220

5.2   ART1 222
         5.2.1 Architecture, 222
         5.2.2 Algorithm, 225
         5.2.3 Applications, 229
         5.2.4 Analysis, 243

5.3   ART2 246
         5.3.1 Architecture, 247
         5.3.2 Algorithm, 250
         5.3.3 Applications, 257
         5.3.4 Analysis, 276

5.4   Suggestions for Further Study 283
         5.4.1 Readings, 283
         5.4.2 Exercises, 284
         5.4.3 Projects, 287

CHAPTER 6   BACKPROPAGATION NEURAL NET   289

6.1   Standard Backpropagation 289
         6.1.1 Architecture, 290
         6.1.2 Algorithm, 290
         6.1.3 Applications, 300

6.2   Variations 305
         6.2.1 Alternative Weight Update Procedures, 305
         6.2.2 Alternative Activation Functions, 309
         6.2.3 Strictly Local Backpropagation, 316
         6.2.4 Number of Hidden Layers, 320

6.3   Theoretical Results 324
         6.3.1 Derivation of Learning Rules, 324
         *6.3.2 Multilayer Neural Nets as Universal Approximators,
                 328*

6.4   Suggestions for Further Study   330
       6.4.1   *Readings, 330*
       6.4.2   *Exercises, 330*
       6.4.3   *Projects, 332*

CHAPTER 7   A SAMPLER OF OTHER NEURAL NETS   334

7.1   Fixed Weight Nets for Constrained Optimization   335
       7.1.1   *Boltzmann Machine, 338*
       7.1.2   *Continuous Hopfield Net, 348*
       7.1.3   *Gaussian Machine, 357*
       7.1.4   *Cauchy Machine, 359*

7.2   A Few More Nets that Learn   362
       7.2.1   *Modified Hebbian Learning, 362*
       7.2.2   *Boltzmann Machine with Learning, 367*
       7.2.3   *Simple Recurrent Net, 372*
       7.2.4   *Backpropagation in Time, 377*
       7.2.5   *Backpropagation Training for Fully Recurrent Nets, 384*

7.3   Adaptive Architectures   385
       7.3.1   *Probabilistic Neural Net, 385*
       7.3.2   *Cascade Correlation, 390*

7.4   Neocognitron   398
       7.4.1   *Architecture, 399*
       7.4.2   *Algorithm, 407*

7.5   Suggestions for Further Study   418
       7.5.1   *Readings, 418*
       7.5.2   *Exercises, 418*
       7.5.3   *Project, 420*

GLOSSARY   422

REFERENCES   437

INDEX   449

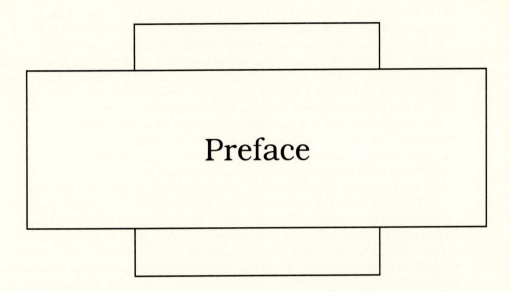

Preface

There has been a resurgence of interest in artificial neural networks over the last few years, as researchers from diverse backgrounds have produced a firm theoretical foundation and demonstrated numerous applications of this rich field of study. However, the interdisciplinary nature of neural networks complicates the development of a comprehensive, but introductory, treatise on the subject. Neural networks are useful tools for solving many types of problems. These problems may be characterized as mapping (including pattern association and pattern classification), clustering, and constrained optimization. There are several neural networks available for each type of problem. In order to use these tools effectively it is important to understand the characteristics (strengths and limitations) of each.

This book presents a wide variety of standard neural networks, with diagrams of the architecture, detailed statements of the training algorithm, and several examples of the application for each net. In keeping with our intent to show neural networks in a fair but objective light, typical results of simple experiments are included (rather than the best possible). The emphasis is on computational characteristics, rather than psychological interpretations. To illustrate the similarities and differences among the neural networks discussed, similar examples are used wherever it is appropriate.

Fundamentals of Neural Networks has been written for students and for researchers in academia, industry, and government who are interested in using neural networks. It has been developed both as a textbook for a one semester, or two quarter, Introduction to Neural Networks course at Florida Institute of Technology, and as a resource book for researchers. Our course has been developed jointly by neural networks researchers from applied mathematics, computer science, and computer and electrical engineering. Our students are seniors, or graduate students, in science and engineering; many work in local industry.

It is assumed that the reader is familiar with calculus and some vector-matrix notation and operations. The mathematical treatment has been kept at a minimal level, consistent with the primary aims of clarity and correctness. Derivations, theorems and proofs are included when they serve to illustrate the important features of a particular neural network. For example, the mathematical derivation of the backpropagation training algorithm makes clear the correct order of the operations. The level of mathematical sophistication increases somewhat in the later chapters, as is appropriate for the networks presented in chapters 5, 6, and 7. However, derivations and proofs (when included) are presented at the end of a section or chapter, so that they can be skipped without loss of continuity.

The order of presentation of the topics was chosen to reflect increasing complexity of the networks. The material in each chapter is largely independent, so that the chapters (after the first chapter) may be used in almost any order desired. The McCulloch-Pitts neuron discussed at the end of Chapter 1 provides a simple example of an early neural net. Single layer nets for pattern classification and pattern association, covered in chapters 2 and 3, are two of the earliest applications of neural networks with adaptive weights. More complex networks, discussed in later chapters, are also used for these types of problems, as well as for more general mapping problems. Chapter 6, backpropagation, can logically follow chapter 2, although the networks in chapters 3–5 are somewhat simpler in structure. Chapters 4 and 5 treat networks for clustering problems (and mapping networks that are based on these clustering networks). Chapter 7 presents a few of the most widely used of the many other neural networks, including two for constrained optimization problems.

Algorithms, rather than computer codes, are provided to encourage the reader to develop a thorough understanding of the mechanisms of training and applying the neural network, rather than fostering the more superficial familiarity that sometimes results from using completely developed software packages. For many applications, the formulation of the problem for solution by a neural network (and choice of an appropriate network) requires the detailed understanding of the networks that comes from performing both hand calculations and developing computer codes for extremely simple examples.

Acknowledgments

Many people have helped to make this book a reality. I can only mention a few of them here.

I have benefited either directly or indirectly from short courses on neural networks taught by Harold Szu, Robert Hecht-Nielsen, Steven Rogers, Bernard Widrow, and Tony Martinez.

My thanks go also to my colleagues for stimulating discussions and encouragement, especially Harold K. Brown, Barry Grossman, Fred Ham, Demetrios Lainiotis, Moti Schneider, Nazif Tepedelenlioglu, and Mike Thursby.

My students have assisted in the development of this book in many ways; several of the exampls are based on student work. Joe Vandeville, Alan Lindsay, and Francisco Gomez performed the computations for many of the examples in Chapter 2. John Karp provided the results for Example 4.8. Judith Lipofsky did Examples 4.9 and 4.10. Fred Parker obtained the results shown in Examples 4.12 and 4.13. Joseph Oslakovic performed the computations for several of the examples in Chapter 5. Laurie Walker assisted in the development of the backpropagation program for several of the examples in Chapter 6; Ti-Cheng Shih did the computations for Example 6.5; Abdallah Said developed the logarithmic activation function used in Examples 6.7 and 6.8. Todd Kovach, Robin Schumann, and Hong-wei Du assisted with the Boltzmann machine and Hopfield net examples in Chapter 7; Ki-suck Yoo provided Example 7.8.

Several of the network architecture diagrams are adapted from the original publications as referenced in the text. The spanning tree test data (Figures 4.11, 4.12, 5.11, and 5.12) are used with permission from Springer-Verlag. The illustrations of modified Hebbian learning have been adapted from the original publications: Figure 7.10 has been adapted from Hertz, Krogh, Palmer, Introduction to the Theory of Neural Computation, @ 1991 by Addison-Wesley Publishing Company, Inc. Figure 7.11 has been adapted and reprinted from Neural Networks, Vol. 5, Xu, Oja, and Suen, Modified Hebbian Learning for Curve and Surface Fitting, pp. 441–457, 1992 with permission from Pergamon Press Ltd, Headington Hill Hall, Oxford 0X3 0BW, UK. Several of the figures for the neocognitron are adapted from (Fukushima, et al., 1983); they are used with permission of IEEE. The diagrams of the ART2 architecture are used with permission of the Optical Society of America, and Carpenter and Grossberg. The diagrams of the simple recurrent net for learning a context sensitive grammar (Servan-Schreiber, et al., 1989) are used with the permission of the authors.

The preparation of the manuscript and software for the examples has been greatly facilitated by the use of a Macintosh IIci furnished by Apple Computers under the AppleSeed project. I thank Maurice Kurtz for making it available to me.

I appreciate the constructive and encouraging comments of the manuscript reviewers: Stanley Ahalt, The Ohio State University; Peter Anderson, Rochester Institute of Technology; and Nirmal Bose, Penn State University.

I would like to thank the Prentice-Hall editorial staff, and especially Rick DeLorenzo, for their diligent efforts to produce an accurate and attractive product within the inevitable time and budget constraints.

But first, last, and always, I would like to thank my husband and colleague, Don Fausett for introducing me to neural networks, and for his patience, encouragement, and advice when asked, during the writing of this book (as well as other times).

of the net) and the method of training the net are discussed further in Section 1.4. A detailed consideration of these ideas for specific nets, together with simple examples of an application of each net, is the focus of the following chapters.

1.2.2 Biological Neural Networks

The extent to which a neural network models a particular biological neural system varies. For some researchers, this is a primary concern; for others, the ability of the net to perform useful tasks (such as approximate or represent a function) is more important than the biological plausibility of the net. Although our interest lies almost exclusively in the computational capabilities of neural networks, we shall present a brief discussion of some features of biological neurons that may help to clarify the most important characteristics of artificial neural networks. In addition to being the original inspiration for artificial nets, biological neural systems suggest features that have distinct computational advantages.

There is a close analogy between the structure of a biological neuron (i.e., a brain or nerve cell) and the processing element (or artificial neuron) presented in the rest of this book. In fact, the structure of an individual neuron varies much less from species to species than does the organization of the system of which the neuron is an element.

A biological neuron has three types of components that are of particular interest in understanding an artificial neuron: its *dendrites, soma,* and *axon.* The many dendrites receive signals from other neurons. The signals are electric impulses that are transmitted across a synaptic gap by means of a chemical process. The action of the chemical transmitter modifies the incoming signal (typically, by scaling the frequency of the signals that are received) in a manner similar to the action of the weights in an artificial neural network.

The soma, or cell body, sums the incoming signals. When sufficient input is received, the cell fires; that is, it transmits a signal over its axon to other cells. It is often supposed that a cell either fires or doesn't at any instant of time, so that transmitted signals can be treated as binary. However, the frequency of firing varies and can be viewed as a signal of either greater or lesser magnitude. This corresponds to looking at discrete time steps and summing all activity (signals received or signals sent) at a particular point in time.

The transmission of the signal from a particular neuron is accomplished by an action potential resulting from differential concentrations of ions on either side of the neuron's axon sheath (the brain's "white matter"). The ions most directly involved are potassium, sodium, and chloride.

A generic biological neuron is illustrated in Figure 1.3, together with axons from two other neurons (from which the illustrated neuron could receive signals) and dendrites for two other neurons (to which the original neuron would send signals). Several key features of the processing elements of artificial neural networks are suggested by the properties of biological neurons, viz., that:

1. The processing element receives many signals.
2. Signals may be modified by a weight at the receiving synapse.
3. The processing element sums the weighted inputs.
4. Under appropriate circumstances (sufficient input), the neuron transmits a single output.
5. The output from a particular neuron may go to many other neurons (the axon branches).

Other features of artificial neural networks that are suggested by biological neurons are:

6. Information processing is local (although other means of transmission, such as the action of hormones, may suggest means of overall process control).
7. Memory is distributed:
   a. Long-term memory resides in the neurons' synapses or weights.
   b. Short-term memory corresponds to the signals sent by the neurons.
8. A synapse's strength may be modified by experience.
9. Neurotransmitters for synapses may be excitatory or inhibitory.

Yet another important characteristic that artificial neural networks share with biological neural systems is *fault tolerance*. Biological neural systems are fault tolerant in two respects. First, we are able to recognize many input signals that are somewhat different from any signal we have seen before. An example of this is our ability to recognize a person in a picture we have not seen before or to recognize a person after a long period of time.

Second, we are able to tolerate damage to the neural system itself. Humans are born with as many as 100 billion neurons. Most of these are in the brain, and most are not replaced when they die [Johnson & Brown, 1988]. In spite of our continuous loss of neurons, we continue to learn. Even in cases of traumatic neural

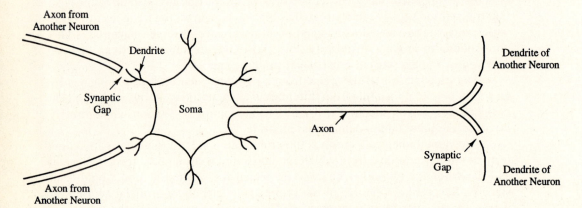

Figure 1.3 Biological neuron.

loss, other neurons can sometimes be trained to take over the functions of the damaged cells. In a similar manner, artificial neural networks can be designed to be insensitive to small damage to the network, and the network can be retrained in cases of significant damage (e.g., loss of data and some connections).

Even for uses of artificial neural networks that are not intended primarily to model biological neural systems, attempts to achieve biological plausibility may lead to improved computational features. One example is the use of a planar array of neurons, as is found in the neurons of the visual cortex, for Kohonen's self-organizing maps (see Chapter 4). The topological nature of these maps has computational advantages, even in applications where the structure of the output units is not itself significant.

Other researchers have found that computationally optimal groupings of artificial neurons correspond to biological bundles of neurons [Rogers & Kabrisky, 1989]. Separating the action of a backpropagation net into smaller pieces to make it more local (and therefore, perhaps more biologically plausible) also allows improvement in computational power (cf. Section 6.2.3) [D. Fausett, 1990].

1.3 WHERE ARE NEURAL NETS BEING USED?

The study of neural networks is an extremely interdisciplinary field, both in its development and in its application. A brief sampling of some of the areas in which neural networks are currently being applied suggests the breadth of their applicability. The examples range from commercial successes to areas of active research that show promise for the future.

1.3.1 Signal Processing

There are many applications of neural networks in the general area of signal processing. One of the first commercial applications was (and still is) to suppress noise on a telephone line. The neural net used for this purpose is a form of ADALINE. (We discuss ADALINES in Chapter 2.) The need for adaptive echo cancelers has become more pressing with the development of transcontinental satellite links for long-distance telephone circuits. The two-way round-trip time delay for the radio transmission is on the order of half a second. The switching involved in conventional echo suppression is very disruptive with path delays of this length. Even in the case of wire-based telephone transmission, the repeater amplifiers introduce echoes in the signal.

The adaptive noise cancellation idea is quite simple. At the end of a long-distance line, the incoming signal is applied to both the telephone system component (called the hybrid) and the adaptive filter (the ADALINE type of neural net). The difference between the output of the hybrid and the output of the ADALINE is the error, which is used to adjust the weights on the ADALINE. The ADALINE is trained to remove the noise (echo) from the hybrid's output signal. (See Widrow and Stearns, 1985, for a more detailed discussion.)

1.3.2 Control

The difficulties involved in backing up a trailer are obvious to anyone who has either attempted or watched a novice attempt this maneuver. However, a driver with experience accomplishes the feat with remarkable ease. As an example of the application of neural networks to control problems, consider the task of training a neural "truck backer-upper" to provide steering directions to a trailer truck attempting to back up to a loading dock [Nguyen & Widrow, 1989; Miller, Sutton, & Werbos, 1990]. Information is available describing the position of the cab of the truck, the position of the rear of the trailer, the (fixed) position of the loading dock, and the angles that the truck and the trailer make with the loading dock. The neural net is able to learn how to steer the truck in order for the trailer to reach the dock, starting with the truck and trailer in any initial configuration that allows enough clearance for a solution to be possible. To make the problem more challenging, the truck is allowed only to back up.

The neural net solution to this problem uses two modules. The first (called the *emulator*) learns to compute the new position of the truck, given its current position and the steering angle. The truck moves a fixed distance at each time step. This module learns the "feel" of how a trailer truck responds to various steering signals, in much the same way as a driver learns the behavior of such a rig. The emulator has several hidden units and is trained using backpropagation (which is the subject of Chapter 6).

The second module is the *controller*. After the emulator is trained, the controller learns to give the correct series of steering signals to the truck so that the trailer arrives at the dock with its back parallel to the dock. At each time step, the controller gives a steering signal and the emulator determines the new position of the truck and trailer. This process continues until either the trailer reaches the dock or the rig jackknifes. The error is then determined and the weights on the controller are adjusted.

As with a driver, performance improves with practice, and the neural controller learns to provide a series of steering signals that direct the truck and trailer to the dock, regardless of the starting position (as long as a solution is possible). Initially, the truck may be facing toward the dock, may be facing away from the dock, or may be at any angle in between. Similarly, the angle between the truck and the trailer may have an initial value short of that in a jack-knife situation. The training process for the controller is similar to the recurrent backpropagation described in Chapter 7.

1.3.3 Pattern Recognition

Many interesting problems fall into the general area of pattern recognition. One specific area in which many neural network applications have been developed is the automatic recognition of handwritten characters (digits or letters). The large variation in sizes, positions, and styles of writing make this a difficult problem for traditional techniques. It is a good example, however, of the type of information processing that humans can perform relatively easily.

General-purpose multilayer neural nets, such as the backpropagation net (a multilayer net trained by backpropagation) described in Chapter 6, have been used for recognizing handwritten zip codes [Le Cun et al., 1990]. Even when an application is based on a standard training algorithm, it is quite common to customize the architecture to improve the performance of the application. This backpropagation net has several hidden layers, but the pattern of connections from one layer to the next is quite localized.

An alternative approach to the problem of recognizing handwritten characters is the ''neocognitron'' described in Chapter 7. This net has several layers, each with a highly structured pattern of connections from the previous layer and to the subsequent layer. However, its training is a layer-by-layer process, specialized for just such an application.

1.3.4 Medicine

One of many examples of the application of neural networks to medicine was developed by Anderson et al. in the mid-1980s [Anderson, 1986; Anderson, Golden, and Murphy, 1986]. It has been called the ''Instant Physician'' [Hecht-Nielsen, 1990]. The idea behind this application is to train an autoassociative memory neural network (the ''Brain-State-in-a-Box,'' described in Section 3.4.2) to store a large number of medical records, each of which includes information on symptoms, diagnosis, and treatment for a particular case. After training, the net can be presented with input consisting of a set of symptoms; it will then find the full stored pattern that represents the ''best'' diagnosis and treatment.

The net performs surprisingly well, given its simple structure. When a particular set of symptoms occurs frequently in the training set, together with a unique diagnosis and treatment, the net will usually give the same diagnosis and treatment. In cases where there are ambiguities in the training data, the net will give the most common diagnosis and treatment. In novel situations, the net will prescribe a treatment corresponding to the symptom(s) it has seen before, regardless of the other symptoms that are present.

1.3.5 Speech Production

Learning to read English text aloud is a difficult task, because the correct phonetic pronunciation of a letter depends on the context in which the letter appears. A traditional approach to the problem would typically involve constructing a set of rules for the standard pronunciation of various groups of letters, together with a look-up table for the exceptions.

One of the most widely known examples of a neural network approach to the problem of speech production is NETtalk [Sejnowski and Rosenberg, 1986], a multilayer neural net (i.e., a net with hidden units) similar to those described in Chapter 6. In contrast to the need to construct rules and look-up tables for the exceptions, NETtalk's only requirement is a set of examples of the written input, together with the correct pronunciation for it. The written input includes both the letter that is currently being spoken and three letters before and after it (to provide a context). Additional symbols are used to indicate the end of a word or punctuation. The net is trained using the 1,000 most common English words. After training, the net can read new words with very few errors; the errors that it does make are slight mispronunciations, and the intelligibility of the speech is quite good.

It is interesting that there are several fairly distinct stages to the response of the net as training progresses. The net learns quite quickly to distinguish vowels from consonants; however, it uses the same vowel for all vowels and the same consonant for all consonants at this first stage. The result is a babbling sound. The second stage of learning corresponds to the net recognizing the boundaries between words; this produces a pseudoword type of response. After as few as 10 passes through the training data, the text is intelligible. Thus, the response of the net as training progresses is similar to the development of speech in small children.

1.3.6 Speech Recognition

Progress is being made in the difficult area of speaker-independent recognition of speech. A number of useful systems now have a limited vocabulary or grammar or require retraining for different speakers. Several types of neural networks have been used for speech recognition, including multilayer nets (see Chapter 6) or multilayer nets with recurrent connections (see Section 7.2). Lippmann (1989) summarizes the characteristics of many of these nets.

One net that is of particular interest, both because of its level of development toward a practical system and because of its design, was developed by Kohonen using the self-organizing map (Chapter 4). He calls his net a "phonetic typewriter." The output units for a self-organizing map are arranged in a two-dimensional array (rectangular or hexagonal). The input to the net is based on short segments (a few milliseconds long) of the speech waveform. As the net groups similar inputs, the clusters that are formed are positioned so that different examples of the same phoneme occur on output units that are close together in the output array.

After the speech input signals are mapped to the phoneme regions (which has been done without telling the net what a phoneme is), the output units can be connected to the appropriate typewriter key to construct the phonetic typewriter. Because the correspondence between phonemes and written letters is very regular in Finnish (for which the net was developed), the spelling is often correct. See Kohonen (1988) for a more extensive description.

1.3.7 Business

Neural networks are being applied in a number of business settings [Harston, 1990]. We mention only one of many examples here, the mortgage assessment work by Nestor, Inc. [Collins, Ghosh, & Scofield, 1988a, 1988b].

Although it may be thought that the rules which form the basis for mortgage underwriting are well understood, it is difficult to specify completely the process by which experts make decisions in marginal cases. In addition, there is a large financial reward for even a small reduction in the number of mortgages that become delinquent. The basic idea behind the neural network approach to mortgage risk assessment is to use past experience to train the net to provide more consistent and reliable evaluation of mortgage applications.

Using data from several experienced mortgage evaluators, neural nets were trained to screen mortgage applicants for mortgage origination underwriting and mortgage insurance underwriting. The purpose in each of these is to determine whether the applicant should be given a loan. The decisions in the second kind of underwriting are more difficult, because only those applicants assessed as higher risks are processed for mortgage insurance. The training input includes information on the applicant's years of employment, number of dependents, current income, etc., as well as features related to the mortgage itself, such as the loan-to-value ratio, and characteristics of the property, such as its appraised value. The target output from the net is an ''accept'' or ''reject'' response.

In both kinds of underwriting, the neural networks achieved a high level of agreement with the human experts. When disagreement did occur, the case was often a marginal one where the experts would also disagree. Using an independent measure of the quality of the mortgages certified, the neural network consistently made better judgments than the experts. In effect, the net learned to form a consensus from the experience of all of the experts whose actions had formed the basis for its training.

A second neural net was trained to evaluate the risk of default on a loan, based on data from a data base consisting of 111,080 applications, 109,072 of which had no history of delinquency. A total of 4,000 training samples were selected from the data base. Although delinquency can result from many causes that are not reflected in the information available on a loan application, the predictions the net was able to make produced a 12% reduction in delinquencies.

1.4 HOW ARE NEURAL NETWORKS USED?

Let us now consider some of the fundamental features of how neural networks operate. Detailed discussions of these ideas for a number of specific nets are presented in the remaining chapters. The building blocks of our examination here are the network architectures and the methods of setting the weights (training).

We also illustrate several typical activation functions and conclude the section with a summary of the notation we shall use throughout the rest of the text.

1.4.1 Typical Architectures

Often, it is convenient to visualize neurons as arranged in layers. Typically, neurons in the same layer behave in the same manner. Key factors in determining the behavior of a neuron are its activation function and the pattern of weighted connections over which it sends and receives signals. Within each layer, neurons usually have the same activation function and the same pattern of connections to other neurons. To be more specific, in many neural networks, the neurons within a layer are either fully interconnected or not interconnected at all. If any neuron in a layer (for instance, the layer of hidden units) is connected to a neuron in another layer (say, the output layer), then each hidden unit is connected to every output neuron.

The arrangement of neurons into layers and the connection patterns within and between layers is called the *net architecture*. Many neural nets have an input layer in which the activation of each unit is equal to an external input signal. The net illustrated in Figure 1.2 consists of input units, output units, and one hidden unit (a unit that is neither an input unit nor an output unit).

Neural nets are often classified as single layer or multilayer. In determining the number of layers, the input units are not counted as a layer, because they perform no computation. Equivalently, the number of layers in the net can be defined to be the number of layers of weighted interconnect links between the slabs of neurons. This view is motivated by the fact that the weights in a net contain extremely important information. The net shown in Figure 1.2 has two layers of weights.

The single-layer and multilayer nets illustrated in Figures 1.4 and 1.5 are examples of *feedforward* nets—nets in which the signals flow from the input units to the output units, in a forward direction. The fully interconnected competitive net in Figure 1.6 is an example of a *recurrent* net, in which there are closed-loop signal paths from a unit back to itself.

Single-Layer Net

A single-layer net has one layer of connection weights. Often, the units can be distinguished as input units, which receive signals from the outside world, and output units, from which the response of the net can be read. In the typical single-layer net shown in Figure 1.4, the input units are fully connected to output units but are not connected to other input units, and the output units are not connected to other output units. By contrast, the Hopfield net architecture, shown in Figure 3.7, is an example of a single-layer net in which all units function as both input and output units.

For pattern classification, each output unit corresponds to a particular cat-

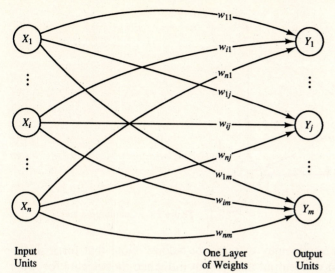

Input
Units

One Layer
of Weights

Output
Units

Figure 1.4 A single-layer neural net.

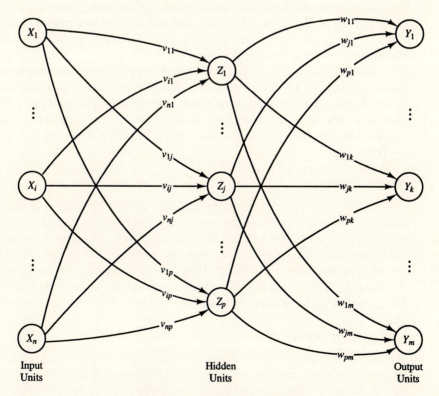

Input
Units

Hidden
Units

Output
Units

Figure 1.5 A multilayer neural net.

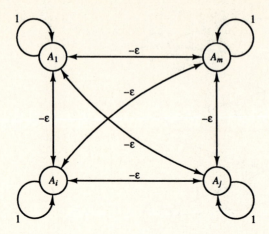

Figure 1.6 Competitive layer.

egory to which an input vector may or may not belong. Note that for a single-layer net, the weights for one output unit do not influence the weights for other output units. For pattern association, the same architecture can be used, but now the overall pattern of output signals gives the response pattern associated with the input signal that caused it to be produced. These two examples illustrate the fact that the same type of net can be used for different problems, depending on the interpretation of the response of the net.

On the other hand, more complicated mapping problems may require a multilayer network. The characteristics of the problems for which a single-layer net is satisfactory are considered in Chapters 2 and 3. The problems that require multilayer nets may still represent a classification or association of patterns; the type of problem influences the choice of architecture, but does not uniquely determine it.

Multilayer net

A multilayer net is a net with one or more layers (or levels) of nodes (the so-called hidden units) between the input units and the output units. Typically, there is a layer of weights between two adjacent levels of units (input, hidden, or output). Multilayer nets can solve more complicated problems than can single-layer nets, but training may be more difficult. However, in some cases, training may be more successful, because it is possible to solve a problem that a single-layer net cannot be trained to perform correctly at all.

Competitive layer

A competitive layer forms a part of a large number of neural networks. Several examples of these nets are discussed in Chapters 4 and 5. Typically, the interconnections between neurons in the competitive layer are not shown in the architecture diagrams for such nets. An example of the architecture for a competitive layer is given in Figure 1.6; the competitive interconnections have weights of $-\epsilon$. The operation of a winner-take-all competition, MAXNET [Lippman, 1987], is described in Section 4.1.1.

1.4.2 Setting the Weights

In addition to the architecture, the method of setting the values of the weights (training) is an important distinguishing characteristic of different neural nets. For convenience, we shall distinguish two types of training—supervised and unsupervised—for a neural network; in addition, there are nets whose weights are fixed without an iterative training process.

Many of the tasks that neural nets can be trained to perform fall into the areas of mapping, clustering, and constrained optimization. Pattern classification and pattern association may be considered special forms of the more general problem of mapping input vectors or patterns to the specified output vectors or patterns.

There is some ambiguity in the labeling of training methods as supervised or unsupervised, and some authors find a third category, self-supervised training, useful. However, in general, there is a useful correspondence between the type of training that is appropriate and the type of problem we wish to solve. We summarize here the basic characteristics of supervised and unsupervised training and the types of problems for which each, as well as the fixed-weight nets, is typically used.

Supervised training

In perhaps the most typical neural net setting, training is accomplished by presenting a sequence of training vectors, or patterns, each with an associated target output vector. The weights are then adjusted according to a learning algorithm. This process is known as *supervised training*.

Some of the simplest (and historically earliest) neural nets are designed to perform pattern classification, i.e., to classify an input vector as either belonging or not belonging to a given category. In this type of neural net, the output is a bivalent element, say, either 1 (if the input vector belongs to the category) or -1 (if it does not belong). In the next chapter, we consider several simple single-layer nets that were designed or typically used for pattern classification. These nets are trained using a supervised algorithm. The characteristics of a classification problem that determines whether a single-layer net is adequate are considered in Chapter 2 also. For more difficult classification problems, a multilayer net, such as that trained by backpropagation (presented in Chapter 6) may be better.

Pattern association is another special form of a mapping problem, one in which the desired output is not just a "yes" or "no," but rather a pattern. A neural net that is trained to associate a set of input vectors with a corresponding set of output vectors is called an *associative memory*. If the desired output vector is the same as the input vector, the net is an *autoassociative memory*; if the output target vector is different from the input vector, the net is a *heteroassociative memory*. After training, an associative memory can recall a stored pattern when it is given an input vector that is sufficiently similar to a vector it has learned. Associative memory neural nets, both feedforward and recurrent, are discussed in Chapter 3.

Multilayer neural nets can be trained to perform a nonlinear mapping from an *n*-dimensional space of input vectors (*n*-tuples) to an *m*-dimensional output space—i.e., the output vectors are *m*-tuples.

The single-layer nets in Chapter 2 (pattern classification nets) and Chapter 3 (pattern association nets) use supervised training (the Hebb rule or the delta rule). Backpropagation (the generalized delta rule) is used to train the multilayer nets in Chapter 6. Other forms of supervised learning are used for some of the nets in Chapter 4 (learning vector quantization and counterpropagation) and Chapter 7. Each learning algorithm will be described in detail, along with a description of the net for which it is used.

Unsupervised training

Self-organizing neural nets group similar input vectors together without the use of training data to specify what a typical member of each group looks like or to which group each vector belongs. A sequence of input vectors is provided, but no target vectors are specified. The net modifies the weights so that the most similar input vectors are assigned to the same output (or cluster) unit. The neural net will produce an exemplar (representative) vector for each cluster formed. Self-organizing nets are described in Chapters 4 (Kohonen self-organizing maps) and Chapter 5 (adaptive resonance theory).

Unsupervised learning is also used for other tasks, in addition to clustering. Examples are included in Chapter 7.

Fixed-weight nets

Still other types of neural nets can solve constrained optimization problems. Such nets may work well for problems that can cause difficulty for traditional techniques, such as problems with conflicting constraints (i.e., not all constraints can be satisfied simultaneously). Often, in such cases, a nearly optimal solution (which the net can find) is satisfactory. When these nets are designed, the weights are set to represent the constraints and the quantity to be maximized or minimized. The Boltzmann machine (without learning) and the continuous Hopfield net (Chapter 7) can be used for constrained optimization problems.

Fixed weights are also used in contrast-enhancing nets (see Section 4.1).

1.4.3 Common Activation Functions

As mentioned before, the basic operation of an artificial neuron involves summing its weighted input signal and applying an output, or activation, function. For the input units, this function is the identity function (see Figure 1.7). Typically, the same activation function is used for all neurons in any particular layer of a neural net, although this is not required. In most cases, a nonlinear activation function is used. In order to achieve the advantages of multilayer nets, compared with the limited capabilities of single-layer nets, nonlinear functions are required (since the results of feeding a signal through two or more layers of linear processing elements—i.e., elements with linear activation functions—are no different from what can be obtained using a single layer).

(*i*) Identity function:

$$f(x) = x \qquad \text{for all } x.$$

Single-layer nets often use a step function to convert the net input, which is a continuously valued variable, to an output unit that is a binary (1 or 0) or bipolar (1 or -1) signal (see Figure 1.8). The use of a *threshold* in this regard is discussed in Section 2.1.2. The binary step function is also known as the threshold function or Heaviside function.

(*ii*) Binary step function (with threshold θ):

$$f(x) = \begin{cases} 1 & \text{if } x \geq \theta \\ 0 & \text{if } x < \theta \end{cases}$$

Sigmoid functions (*S*-shaped curves) are useful activation functions. The logistic function and the hyperbolic tangent functions are the most common. They are especially advantageous for use in neural nets trained by backpropagation, because the simple relationship between the value of the function at a point and the value of the derivative at that point reduces the computational burden during training.

The logistic function, a sigmoid function with range from 0 to 1, is often

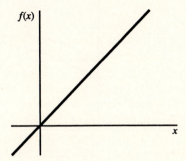

Figure 1.7   Identity function.

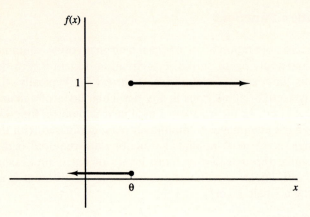

Figure 1.8 Binary step function.

used as the activation function for neural nets in which the desired output values either are binary or are in the interval between 0 and 1. To emphasize the range of the function, we will call it the *binary sigmoid*; it is also called the *logistic sigmoid*. This function is illustrated in Figure 1.9 for two values of the *steepness parameter* σ.

(***iii***) Binary sigmoid:

$$f(x) = \frac{1}{1 + \exp(-\sigma x)}.$$

$$f'(x) = \sigma f(x) \, [1 - f(x)].$$

As is shown in Section 6.2.3, the logistic sigmoid function can be scaled to have any range of values that is appropriate for a given problem. The most common range is from −1 to 1; we call this sigmoid the *bipolar sigmoid*. It is illustrated in Figure 1.10 for σ = 1.

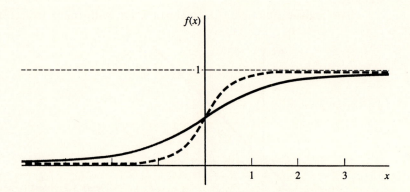

Figure 1.9 Binary sigmoid. Steepness parameters σ = 1 and σ = 3.

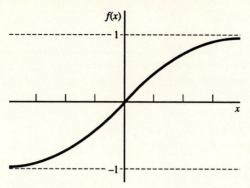

Figure 1.10   Bipolar sigmoid.

(*iv*) Bipolar sigmoid:

$$g(x) = 2f(x) - 1 = \frac{2}{1 + \exp(-\sigma x)} - 1$$

$$= \frac{1 - \exp(-\sigma x)}{1 + \exp(-\sigma x)}.$$

$$g'(x) = \frac{\sigma}{2}[1 + g(x)][1 - g(x)].$$

The bipolar sigmoid is closely related to the hyperbolic tangent function, which is also often used as the activation function when the desired range of output values is between -1 and 1. We illustrate the correspondence between the two for $\sigma = 1$. We have

$$g(x) = \frac{1 - \exp(-x)}{1 + \exp(-x)}.$$

The hyperbolic tangent is

$$h(x) = \frac{\exp(x) - \exp(-x)}{\exp(x) + \exp(-x)}$$

$$= \frac{1 - \exp(-2x)}{1 + \exp(-2x)}.$$

The derivative of the hyperbolic tangent is

$$h'(x) = [1 + h(x)][1 - h(x)].$$

For binary data (rather than continuously valued data in the range from 0 to 1), it is usually preferable to convert to bipolar form and use the bipolar sigmoid or hyperbolic tangent. A more extensive discussion of the choice of activation functions and different forms of sigmoid functions is given in Section 6.2.2.

1.4.4 Summary of Notation

The following notation will be used throughout the discussions of specific neural nets, unless indicated otherwise for a particular net (appropriate values for the parameter depend on the particular neural net model being used and will be discussed further for each model):

x_i, y_j Activations of units X_i, Y_j, respectively:
          For input units X_i,

$$x_i = \text{input signal};$$

          for other units Y_j,

$$y_j = f(y\_in_j).$$

w_{ij} Weight on connection from unit X_i to unit Y_j:
          Beware: Some authors use the opposite convention, with w_{ij} denoting the weight from unit Y_j to unit X_i.

b_j Bias on unit Y_j:
          A bias acts like a weight on a connection from a unit with a constant activation of 1 (see Figure 1.11).

y_in_j Net input to unit Y_j:

$$y\_in_j = b_j + \sum_i x_i w_{ij}$$

W Weight matrix:

$$W = \{w_{ij}\}.$$

$\mathbf{w}_{\cdot j}$ Vector of weights:

$$\mathbf{w}_{\cdot j} = (w_{1j}, w_{2j}, \ldots, w_{nj})^T.$$

          This is the jth column of the weight matrix.

$\| \mathbf{x} \|$ Norm or magnitude of vector $\mathbf{x}$.

θ_j Threshold for activation of neuron Y_j:
          A step activation function sets the activation of a neuron to 1 whenever its net input is greater than the specified threshold value θ_j; otherwise its activation is 0 (see Figure 1.8).

$\mathbf{s}$ Training input vector:

$$\mathbf{s} = (s_1, \ldots, s_i, \ldots, s_n).$$

$\mathbf{t}$ Training (or target) output vector:

$$\mathbf{t} = (t_1, \ldots, t_j, \ldots, t_m).$$

$\mathbf{x}$ Input vector (for the net to classify or respond to):

$$\mathbf{x} = (x_1, \ldots, x_i, \ldots, x_n).$$

Δw_{ij} Change in w_{ij}:

$$\Delta w_{ij} = [w_{ij} \text{(new)} - w_{ij} \text{(old)}].$$

α Learning rate:
> The learning rate is used to control the amount of weight adjustment at each step of training.

Matrix multiplication method for calculating net input

If the connection weights for a neural net are stored in a matrix $\mathbf{W} = (w_{ij})$, the net input to unit Y_j (with no bias on unit j) is simply the dot product of the vectors $\mathbf{x} = (x_1, \dots, x_i, \dots, x_n)$ and $\mathbf{w}_{\cdot j}$ (the jth column of the weight matrix):

$$y\_in_j = \mathbf{x} \cdot \mathbf{w}_{\cdot j}$$

$$= \sum_{i=1}^{n} x_i w_{ij}.$$

Bias

A bias can be included by adding a component $x_0 = 1$ to the vector $\mathbf{x}$, i.e., $\mathbf{x} = (1, x_1, \dots, x_i, \dots, x_n)$. The bias is treated exactly like any other weight, i.e., $w_{0j} = b_j$. The net input to unit Y_j is given by

$$y\_in_j = \mathbf{x} \cdot \mathbf{w}_{\cdot j}$$

$$= \sum_{i=0}^{n} x_i w_{ij}$$

$$= w_{0j} + \sum_{i=1}^{n} x_i w_{ij}$$

$$= b_j + \sum_{i=1}^{n} x_i w_{ij}.$$

The relation between a bias and a threshold is considered in Section 2.1.2.

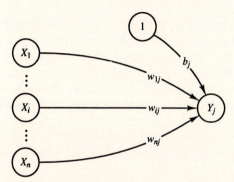

Figure 1.11 Neuron with a bias.

1.5 WHO IS DEVELOPING NEURAL NETWORKS?

This section presents a very brief summary of the history of neural networks, in terms of the development of architectures and algorithms that are widely used today. Results of a primarily biological nature are not included, due to space constraints. They have, however, served as the inspiration for a number of networks that are applicable to problems beyond the original ones studied. The history of neural networks shows the interplay among biological experimentation, modeling, and computer simulation/hardware implementation. Thus, the field is strongly interdisciplinary.

1.5.1 The 1940s: The Beginning of Neural Nets

McCulloch-Pitts neurons

Warren McCulloch and Walter Pitts designed what are generally regarded as the first neural networks [McCulloch & Pitts, 1943]. These researchers recognized that combining many simple neurons into neural systems was the source of increased computational power. The weights on a McCulloch-Pitts neuron are set so that the neuron performs a particular simple logic function, with different neurons performing different functions. The neurons can be arranged into a net to produce any output that can be represented as a combination of logic functions. The flow of information through the net assumes a unit time step for a signal to travel from one neuron to the next. This time delay allows the net to model some physiological processes, such as the perception of hot and cold.

   The idea of a threshold such that if the net input to a neuron is greater than the threshold then the unit fires is one feature of a McCulloch-Pitts neuron that is used in many artificial neurons today. However, McCulloch-Pitts neurons are used most widely as logic circuits [Anderson & Rosenfeld, 1988].

   McCulloch and Pitts subsequent work [Pitts & McCulloch, 1947] addressed issues that are still important research areas today, such as translation and rotation invariant pattern recognition.

Hebb learning

Donald Hebb, a psychologist at McGill University, designed the first learning law for artificial neural networks [Hebb, 1949]. His premise was that if two neurons were active simultaneously, then the strength of the connection between them should be increased. Refinements were subsequently made to this rather general statement to allow computer simulations [Rochester, Holland, Haibt & Duda, 1956]. The idea is closely related to the correlation matrix learning developed by Kohonen (1972) and Anderson (1972) among others. An expanded form of Hebb learning [McClelland & Rumelhart, 1988] in which units that are simultaneously off also reinforce the weight on the connection between them will be presented in Chapters 2 and 3.

1.5.2 The 1950s and 1960s: The First Golden Age of Neural Networks

Although today neural networks are often viewed as an alternative to (or complement of) traditional computing, it is interesting to note that John von Neumann, the "father of modern computing," was keenly interested in modeling the brain [von Neumann, 1958]. Johnson and Brown (1988) and Anderson and Rosenfeld (1988) discuss the interaction between von Neumann and early neural network researchers such as Warren McCulloch, and present further indication of von Neumann's views of the directions in which computers would develop.

Perceptrons

Together with several other researchers [Block, 1962; Minsky & Papert, 1988 (originally published 1969)], Frank Rosenblatt (1958, 1959, 1962) introduced and developed a large class of artificial neural networks called *perceptrons*. The most typical perceptron consisted of an input layer (the retina) connected by paths with fixed weights to associator neurons; the weights on the connection paths were adjustable. The perceptron learning rule uses an iterative weight adjustment that is more powerful than the Hebb rule. Perceptron learning can be proved to converge to the correct weights if there are weights that will solve the problem at hand (i.e., allow the net to reproduce correctly all of the training input and target output pairs). Rosenblatt's 1962 work describes many types of perceptrons. Like the neurons developed by McCulloch and Pitts and by Hebb, perceptrons use a threshold output function.

The early successes with perceptrons led to enthusiastic claims. However, the mathematical proof of the convergence of iterative learning under suitable assumptions was followed by a demonstration of the limitations regarding what the perceptron type of net can learn [Minsky & Papert, 1969].

ADALINE

Bernard Widrow and his student, Marcian (Ted) Hoff [Widrow & Hoff, 1960], developed a learning rule (which usually either bears their names, or is designated the least mean squares or delta rule) that is closely related to the perceptron learning rule. The perceptron rule adjusts the connection weights to a unit whenever the response of the unit is incorrect. (The response indicates a classification of the input pattern.) The delta rule adjusts the weights to reduce the difference between the net input to the output unit and the desired output. This results in the smallest mean squared error. The similarity of models developed in psychology by Rosenblatt to those developed in electrical engineering by Widrow and Hoff is evidence of the interdisciplinary nature of neural networks. The difference in learning rules, although slight, leads to an improved ability of the net to generalize (i.e., respond to input that is similar, but not identical, to that on which it was trained). The Widrow-Hoff learning rule for a single-layer network is a precursor of the backpropagation rule for multilayer nets.

Work by Widrow and his students is sometimes reported as neural network research, sometimes as adaptive linear systems. The name ADALINE, interpreted as either ADAptive LInear NEuron or ADAptive LINEar system, is often given to these nets. There have been many interesting applications of ADALINES, from neural networks for adaptive antenna systems [Widrow, Mantey, Griffiths, & Goode, 1967] to rotation-invariant pattern recognition to a variety of control problems, such as broom balancing and backing up a truck [Widrow, 1987; Tolat & Widrow, 1988; Nguyen & Widrow, 1989]. MADALINES are multilayer extensions of ADALINES [Widrow & Hoff, 1960; Widrow & Lehr, 1990].

1.5.3 The 1970s: The Quiet Years

In spite of Minsky and Papert's demonstration of the limitations of perceptrons (i.e., single-layer nets), research on neural networks continued. Many of the current leaders in the field began to publish their work during the 1970s. (Widrow, of course, had started somewhat earlier and is still active.)

Kohonen

The early work of Teuvo Kohonen (1972), of Helsinki University of Technology, dealt with associative memory neural nets. His more recent work [Kohonen, 1982] has been the development of self-organizing feature maps that use a topological structure for the cluster units. These nets have been applied to speech recognition (for Finnish and Japanese words) [Kohonen, Torkkola, Shozakai, Kangas, & Venta, 1987; Kohonen, 1988], the solution of the "Traveling Salesman Problem" [Angeniol, Vaubois, & Le Texier, 1988], and musical composition [Kohonen, 1989b].

Anderson

James Anderson, of Brown University, also started his research in neural networks with associative memory nets [Anderson, 1968, 1972]. He developed these ideas into his "Brain-State-in-a-Box" [Anderson, Silverstein, Ritz, & Jones, 1977], which truncates the linear output of earlier models to prevent the output from becoming too large as the net iterates to find a stable solution (or memory). Among the areas of application for these nets are medical diagnosis and learning multiplication tables. Anderson and Rosenfeld (1988) and Anderson, Pellionisz, and Rosenfeld (1990) are collections of fundamental papers on neural network research. The introductions to each are especially useful.

Grossberg

Stephen Grossberg, together with his many colleagues and coauthors, has had an extremely prolific and productive career. Klimasauskas (1989) lists 146 publications by Grossberg from 1967 to 1988. His work, which is very mathematical and very biological, is widely known [Grossberg, 1976, 1980, 1982, 1987, 1988]. Grossberg is director of the Center for Adaptive Systems at Boston University.

Carpenter

Together with Stephen Grossberg, Gail Carpenter has developed a theory of self-organizing neural networks called *adaptive resonance theory* [Carpenter & Grossberg, 1985, 1987a, 1987b, 1990]. Adaptive resonance theory nets for binary input patterns (ART1) and for continuously valued inputs (ART2) will be examined in Chapter 5.

1.5.4 The 1980s: Renewed Enthusiasm

Backpropagation

Two of the reasons for the "quiet years" of the 1970s were the failure of single-layer perceptrons to be able to solve such simple problems (mappings) as the XOR function and the lack of a general method of training a multilayer net. A method for propagating information about errors at the output units back to the hidden units had been discovered in the previous decade [Werbos, 1974], but had not gained wide publicity. This method was also discovered independently by David Parker (1985) and by LeCun (1986) before it became widely known. It is very similar to yet an earlier algorithm in optimal control theory [Bryson & Ho, 1969]. Parker's work came to the attention of the Parallel Distributed Processing Group led by psychologists David Rumelhart, of the University of California at San Diego, and James McClelland, of Carneigie-Mellon University, who refined and publicized it [Rumelhart, Hinton, & Williams, 1986a, 1986b; McClelland & Rumelhart, 1988].

Hopfield nets

Another key player in the increased visibility of and respect for neural nets is Nobel prize winner (in physics) John Hopfield, of the California Institute of Technology. Together with David Tank, a researcher at AT&T, Hopfield has developed a number of neural networks based on fixed weights and adaptive activations [Hopfield, 1982, 1984; Hopfield & Tank, 1985, 1986; Tank & Hopfield, 1987]. These nets can serve as associative memory nets and can be used to solve constraint satisfaction problems such as the "Traveling Salesman Problem." An article in *Scientific American* [Tank & Hopfield, 1987] helped to draw popular attention to neural nets, as did the message of a Nobel prize–winning physicist that, in order to make machines that can do what humans do, we need to study human cognition.

Neocognitron

Kunihiko Fukushima and his colleagues at NHK Laboratories in Tokyo have developed a series of specialized neural nets for character recognition. One example of such a net, called a *neocognitron,* is described in Chapter 7. An earlier self-organizing network, called the cognitron [Fukushima, 1975], failed to recognize position- or rotation-distorted characters. This deficiency was corrected in the neocognitron [Fukushima, 1988; Fukushima, Miyake, & Ito, 1983].

Boltzmann machine

A number of researchers have been involved in the development of nondeterministic neural nets, that is, nets in which weights or activations are changed on the basis of a probability density function [Kirkpatrick, Gelatt, & Vecchi, 1983; Geman & Geman, 1984; Ackley, Hinton, & Sejnowski, 1985; Szu & Hartley, 1987]. These nets incorporate such classical ideas as simulated annealing and Bayesian decision theory.

Hardware implementation

Another reason for renewed interest in neural networks (in addition to solving the problem of how to train a multilayer net) is improved computational capabilities. Optical neural nets [Farhat, Psaltis, Prata, & Paek, 1985] and VLSI implementations [Sivilatti, Mahowald, & Mead, 1987] are being developed.

Carver Mead, of California Institute of Technology, who also studies motion detection, is the coinventor of software to design microchips. He is also cofounder of Synaptics, Inc., a leader in the study of neural circuitry.

Nobel laureate Leon Cooper, of Brown University, introduced one of the first multilayer nets, the *reduced coulomb energy network*. Cooper is chairman of Nestor, the first public neural network company [Johnson & Brown, 1988], and the holder of several patents for information-processing systems [Klimasauskas, 1989].

Robert Hecht-Nielsen and Todd Gutschow developed several digital neurocomputers at TRW, Inc., during 1983–85. Funding was provided by the Defense Advanced Research Projects Agency (DARPA) [Hecht-Nielsen, 1990]. DARPA (1988) is a valuable summary of the state of the art in artificial neural networks (especially with regard to successful applications). To quote from the preface to his book, *Neurocomputing,* Hecht-Nielsen is "an industrialist, an adjunct academic, and a philanthropist without financial portfolio" [Hecht-Nielsen, 1990]. The founder of HNC, Inc., he is also a professor at the University of California, San Diego, and the developer of the counterpropagation network.

1.6 WHEN NEURAL NETS BEGAN: THE McCULLOCH-PITTS NEURON

The McCulloch-Pitts neuron is perhaps the earliest artificial neuron [McCulloch & Pitts, 1943]. It displays several important features found in many neural networks. The requirements for McCulloch-Pitts neurons may be summarized as follows:

1. The activation of a McCulloch-Pitts neuron is binary. That is, at any time step, the neuron either fires (has an activation of 1) or does not fire (has an activation of 0).

2. McCulloch-Pitts neurons are connected by directed, weighted paths.

3. A connection path is excitatory if the weight on the path is positive; otherwise it is inhibitory. All excitatory connections into a particular neuron have the same weights.

4. Each neuron has a fixed threshold such that if the net input to the neuron is greater than the threshold, the neuron fires.

5. The threshold is set so that inhibition is absolute. That is, any nonzero inhibitory input will prevent the neuron from firing.

6. It takes one time step for a signal to pass over one connection link.

The simple example of a McCulloch-Pitts neuron shown in Figure 1.12 illustrates several of these requirements. The connection from X_1 to Y is excitatory, as is the connection from X_2 to Y. These excitatory connections have the same (positive) weight because they are going into the same unit.

The threshold for unit Y is 4; for the values of the excitatory and inhibitory weights shown, this is the only integer value of the threshold that will allow Y to fire sometimes, but will prevent it from firing if it receives a nonzero signal over the inhibitory connection.

It takes one time step for the signals to pass from the X units to Y; the activation of Y at time t is determined by the activations of X_1, X_2, and X_3 at the previous time, $t - 1$. The use of discrete time steps enables a network of McCulloch-Pitts neurons to model physiological phenomena in which there is a time delay; such an example is given in Section 1.6.3.

1.6.1 Architecture

In general, a McCulloch-Pitts neuron Y may receive signals from any number of other neurons. Each connection path is either excitatory, with weight $w > 0$, or inhibitory, with weight $-p$ ($p > 0$). For convenience, in Figure 1.13, we assume there are n units, $X_1, \ldots, X_n$, which send excitatory signals to unit Y, and m units, $X_{n+1}, \ldots, X_{n+m}$, which send inhibitory signals. The activation function for unit Y is

$$f(y\_in) = \begin{cases} 1 & \text{if } y\_in \geq \theta \\ 0 & \text{if } y\_in < \theta \end{cases}$$

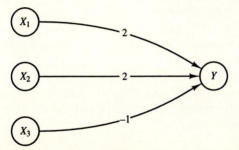

Figure 1.12 A simple McCulloch-Pitts neuron Y.

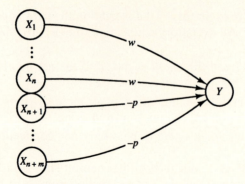

Figure 1.13 Architecture of a McCulloch-Pitts neuron Y.

where y_in is the total input signal received and θ is the threshold. The condition that inhibition is absolute requires that θ for the activation function satisfy the inequality

$$\theta > nw - p.$$

Y will fire if it receives k or more excitatory inputs and no inhibitory inputs, where

$$kw \geq \theta > (k - 1)w.$$

Although all excitatory weights coming into any particular unit must be the same, the weights coming into one unit, say, Y_1, do not have to be the same as the weights coming into another unit, say Y_2.

1.6.2 Algorithm

The weights for a McCulloch-Pitts neuron are set, together with the threshold for the neuron's activation function, so that the neuron will perform a simple logic function. Since analysis, rather than a training algorithm, is used to determine the values of the weights and threshold, several examples of simple McCulloch-Pitts neurons are presented in this section. Using these simple neurons as building blocks, we can model any function or phenomenon that can be represented as a logic function. In Section 1.6.3, an example is given of how several of these simple neurons can be combined to model an interesting physiological phenomenon.

Simple networks of McCulloch-Pitts neurons, each with a threshold of 2, are shown in Figures 1.14–1.17. The activation of unit X_i at time t is denoted $x_i(t)$. The activation of a neuron X_i at time t is determined by the activations, at time $t - 1$, of the neurons from which it receives signals.

Logic functions will be used as simple examples for a number of neural nets. The binary form of the functions for AND, OR, and AND NOT are defined here for reference. Each of these functions acts on two input values, denoted x_1 and x_2, and produces a single output value y.

AND

The AND function gives the response "true" if both input values are "true"; otherwise the response is "false." If we represent "true" by the value 1, and "false" by 0, this gives the following four training input, target output pairs:

x_1	x_2	$\rightarrow$	y
1	1		1
1	0		0
0	1		0
0	0		0

Example 1.1 A McCulloch-Pitts Neuron for the AND Function

The network in Figure 1.14 performs the mapping of the logical AND function. The threshold on unit Y is 2.

OR

The OR function gives the response "true" if either of the input values is "true"; otherwise the response is "false." This is the "inclusive or," since both input values may be "true" and the response is still "true." Representing "true" as 1, and "false" as 0, we have the following four training input, target output pairs:

x_1	x_2	$\rightarrow$	y
1	1		1
1	0		1
0	1		1
0	0		0

Example 1.2 A McCulloch-Pitts Neuron for the OR Function

The network in Figure 1.15 performs the logical OR function. The threshold on unit Y is 2.

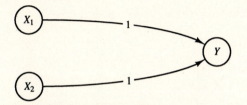

Figure 1.14 A McCulloch-Pitts neuron to perform the logical AND function.

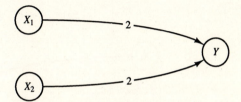

Figure 1.15 A McCulloch-Pitts neuron to perform the logical OR function.

AND NOT

The AND NOT function is an example of a logic function that is not symmetric in its treatment of the two input values. The response is "true" if the first input value, x_1, is "true" and the second input value, x_2, is "false"; otherwise the response is "false." Using a binary representation of the logical input and response values, the four training input, target output pairs are:

x_1	x_2	$\rightarrow$	y
1	1		0
1	0		1
0	1		0
0	0		0

Example 1.3 A McCulloch-Pitts Neuron for the AND NOT Function

The net in Figure 1.16 performs the function x_1 AND NOT x_2. In other words, neuron Y fires at time t if and only if unit X_1 fires at time $t - 1$ and unit X_2 does not fire at time $t - 1$. The threshold for unit Y is 2.

1.6.3 Applications

XOR

The XOR (exclusive or) function gives the response "true" if exactly one of the input values is "true"; otherwise the response is "false." Using a binary representation, the four training input, target output pairs are:

x_1	x_2	$\rightarrow$	y
1	1		0
1	0		1
0	1		1
0	0		0

Example 1.4 A McCulloch-Pitts Net for the XOR Function

The network in Figure 1.17 performs the logical XOR function. XOR can be expressed as

$$x_1 \text{ XOR } x_2 \leftrightarrow (x_1 \text{ AND NOT } x_2) \text{ OR } (x_2 \text{ AND NOT } x_1).$$

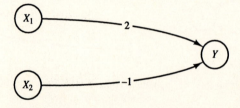

Figure 1.16   A McCulloch-Pitts neuron to perform the logical AND NOT function.

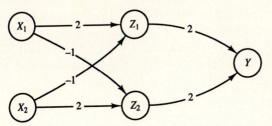

Figure 1.17 A McCulloch-Pitts neural net to perform the logical XOR function.

Thus, $y = x_1$ XOR x_2 is found by a two-layer net. The first layer forms

$$z_1 = x_1 \text{ AND NOT } x_2$$

and

$$z_2 = x_2 \text{ AND NOT } x_1.$$

The second layer consists of

$$y = z_1 \text{ OR } z_2.$$

Units Z_1, Z_2, and Y each have a threshold of 2.

Hot and cold

Example 1.5 Modeling the Perception of Hot and Cold with a McCulloch-Pitts Net

It is a well-known and interesting physiological phenomenon that if a cold stimulus is applied to a person's skin for a very short period of time, the person will perceive heat. However, if the same stimulus is applied for a longer period, the person will perceive cold. The use of discrete time steps enables the network of McCulloch-Pitts neurons shown in Figure 1.18 to model this phenomenon. The example is an elaboration of one originally presented by McCulloch and Pitts [1943]. The model is designed to give only the first perception of heat or cold that is received by the perceptor units.

In the figure, neurons X_1 and X_2 represent receptors for heat and cold, respectively, and neurons Y_1 and Y_2 are the counterpart perceptors. Neurons Z_1

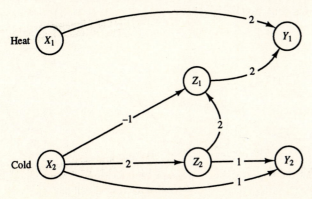

Figure 1.18 A network of McCulloch-Pitts neurons to model the perception of heat and cold.

and Z_2 are auxiliary units needed for the problem. Each neuron has a threshold of 2, i.e., it fires (sets its activation to 1) if the net input it receives is ≥ 2. Input to the system will be (1,0) if heat is applied and (0,1) if cold is applied. The desired response of the system is that cold is perceived if a cold stimulus is applied for two time steps, i.e.,

$$y_2(t) = x_2(t - 2) \text{ AND } x_2(t - 1).$$

The activation of unit Y_2 at time t is $y_2(t)$; $y_2(t) = 1$ if cold is perceived, and $y_2(t) = 0$ if cold is not perceived.

In order to model the physical phenomenon described, it is also required that heat be perceived if either a hot stimulus is applied or a cold stimulus is applied briefly (for one time step) and then removed. This condition is expressed as

$$y_1(t) = \{x_1(t - 1)\} \text{ OR } \{x_2(t - 3) \text{ AND NOT } x_2(t - 2)\}.$$

To see that the net shown in Figure 1.18 does in fact represent the two logical statements required, consider first the neurons that determine the response of Y_1 at time t (illustrated in Figure 1.19). The figure shows that

$$y_1(t) = x_1(t - 1) \text{ OR } z_1(t - 1).$$

Now consider the neurons (illustrated in Figure 1.20) that determine the response of unit Z_1 at time $t - 1$. This figure shows that

$$z_1(t - 1) = z_2(t - 2) \text{ AND NOT } x_2(t - 2).$$

Finally, the response of unit Z_2 at time $t - 2$ is simply the value of X_2 at the previous time step:

$$z_2(t - 2) = x_2(t - 3).$$

Substituting in the preceding expression for $y_1(t)$ gives

$$y_1(t) = \{x_1(t - 1)\} \text{ OR } \{x_2(t - 3) \text{ AND NOT } x_2(t - 2)\}.$$

The analysis for the response of neuron Y_2 at time t proceeds in a similar manner. Figure 1.21 shows that $y_2(t) = z_2(t - 1) \text{ AND } x_2(t - 1)$. However, as

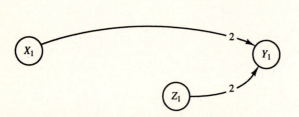

Figure 1.19   The neurons that determine the response of unit Y_1.

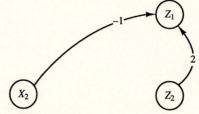

Figure 1.20   The neurons that determine the response of unit Z_1.

before, $z_2(t - 1) = x_2(t - 2)$; substituting in the expression for $y_2(t)$ then gives

$$y_2(t) = x_2(t - 2) \text{ AND } x_2(t - 1),$$

as required.

It is also informative to trace the flow of activations through the net, starting with the presentation of a stimulus at $t = 0$. Case 1, a cold stimulus applied for one time step and then removed, is illustrated in Figure 1.22. Case 2, a cold stimulus applied for two time steps, is illustrated in Figure 1.23, and case 3, a hot stimulus applied for one time step, is illustrated in Figure 1.24. In each case, only the activations that are known at a particular time step are indicated. The weights on the connections are as in Figure 1.18.

Case 1: A cold stimulus applied for one time step.

The activations that are known at $t = 0$ are shown in Figure 1.22(a).

The activations that are known at $t = 1$ are shown in Figure 1.22(b). The activations of the input units are both 0, signifying that the cold stimulus presented at $t = 0$ was removed after one time step. The activations of Z_1 and Z_2 are based on the activations of X_2 at $t = 0$.

The activations that are known at $t = 2$ are shown in Figure 1.22(c). Note that the activations of the input units are not specified, since their value at $t = 2$ does not determine the first response of the net to the situation being modeled. Although the responses of the perceptor units are determined, no perception of hot or cold has reached them yet.

The activations that are known at $t = 3$ are shown in Figure 1.22(d). A perception of heat is indicated by the fact that unit Y_1 has an activation of 1 and unit Y_2 has an activation of 0.

Case 2: A cold stimulus applied for two time steps.

The activations that are known at $t = 0$ are shown in Figure 1.23(a), and those that are known at $t = 1$ are shown in Figure 1.23(b).

The activations that are known at $t = 2$ are shown in Figure 1.23(c). Note that the activations of the input units are not specified, since the first response of the net to the cold stimulus being applied for two time steps is not influenced by whether or not the stimulus is removed after the two steps. Although the responses of the auxiliary units Z_1 and Z_2 are indicated, the responses of the perceptor units are determined by the activations of all of the units at $t = 1$.

Case 3: A hot stimulus applied for one time step.

The activations that are known at $t = 0$ are shown in Figure 1.24(a).

The activations that are known at $t = 1$ are shown in Figure 1.24(b). Unit Y_1 fires because it has received a signal from X_1. Y_2 does not fire because it requires input signals from both X_2 and Z_2 in order to fire, and X_2 had an activation of 0 at $t = 0$.

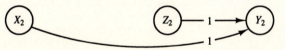

Figure 1.21   The neurons that determine the response of unit Y_2.

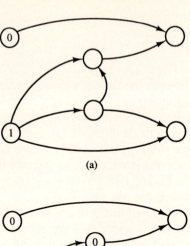

(a)

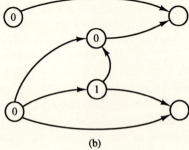

(b)

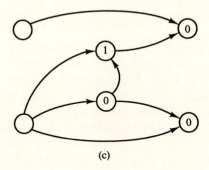

(c)

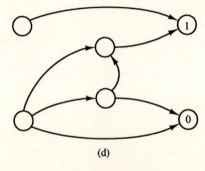

(d)

Figure 1.22 A cold stimulus applied for one time step. Activations at (a) $t = 0$, (b) $t = 1$, (c) $t = 2$, and (d) $t = 3$.

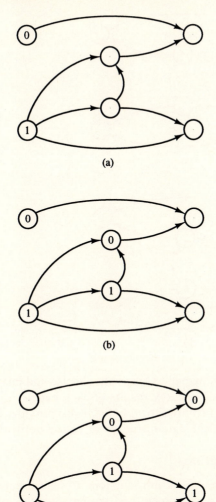

(a)

(b)

(c)

Figure 1.23 A cold stimulus applied for two time steps. Activations at (a) $t = 0$, (b) $t = 1$, and (c) $t = 2$.

1.7 SUGGESTIONS FOR FURTHER STUDY

1.7.1 Readings

Many of the applications and historical developments we have summarized in this chapter are described in more detail in two collections of original research:

- *Neurocomputing: Foundations of Research* [Anderson & Rosenfeld, 1988].
- *Neurocomputing 2: Directions for Research* [Anderson, Pellionisz & Rosenfeld, 1990].

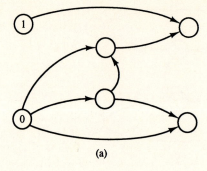

(a)

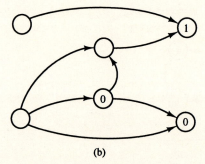

(b)

Figure 1.24 A hot stimulus applied for one time step. Activations at (a) $t = 0$ and (b) $t = 1$.

These contain useful papers, along with concise and insightful introductions explaining the significance and key results of each paper.

The *DARPA Neural Network Study* (1988) also provides descriptions of both the theoretical and practical state of the art of neural networks that year.

Nontechnical introductions

Two very readable nontechnical introductions to neural networks, with an emphasis on the historical development and the personalities of the leaders in the field, are:

- *Cognizers* [Johnson & Brown, 1988].
- *Apprentices of Wonder: Inside the Neural Network Revolution* [Allman, 1989].

Applications

Among the books dealing with neural networks for particular types of applications are:

- *Neural Networks for Signal Processing* [Kosko, 1992b].
- *Neural Networks for Control* [Miller, Sutton, & Werbos, 1990].

- *Simulated Annealing and Boltzmann Machines* [Aarts & Korst, 1989].
- *Adaptive Pattern Recognition and Neural Networks* [Pao, 1989].
- *Adaptive Signal Processing* [Widrow & Sterns, 1985].

History

The history of neural networks is a combination of progress in experimental work with biological neural systems, computer modeling of biological neural systems, the development of mathematical models and their applications to problems in a wide variety of fields, and hardware implementation of these models. In addition to the collections of original papers already mentioned, in which the introductions to each paper provide historical perspectives, *Embodiments of Mind* [McCulloch, 1988] is a wonderful selection of some of McCulloch's essays. *Perceptrons* [Minsky & Papert, 1988] also places the development of neural networks into historical context.

Biological neural networks

Introduction to Neural and Cognitive Modeling [Levine, 1991] provides extensive information on the history of neural networks from a mathematical and psychological perspective. For additional writings from a biological point of view, see *Neuroscience and Connectionist Theory* [Gluck & Rumelhart, 1990] and *Neural and Brain Modeling* [MacGregor, 1987].

1.7.2 Exercises

1.1 Consider the neural network of McCulloch-Pitts neurons shown in Figure 1.25. Each neuron (other than the input neurons, N_1 and N_2) has a threshold of 2.

    a. Define the response of neuron N_5 at time t in terms of the activations of the input neurons, N_1 and N_2, at the appropriate time.

    b. Show the activation of each neuron that results from an input signal of $N_1 = 1$, $N_2 = 0$ at $t = 0$.

1.2 There are at least two ways to express XOR in terms of other simple logic functions that can be represented by McCulloch-Pitts neurons. One such example is presented in Section 1.4. Find another representation and the net for it. How do the two nets (yours and the one in Section 1.4) compare?

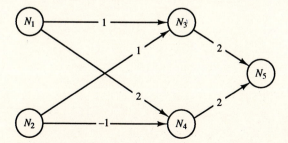

Figure 1.25 Neural network for Exercise 1.1.

1.3 In the McCulloch-Pitts model of the perception of heat and cold, a cold stimulus applied at times $t - 2$ and $t - 1$ is perceived as cold at time t. Can you modify the net to require the cold stimulus to be applied for three time steps before cold is felt?

1.4 In the hot and cold model, consider what is felt after the first perception. (That is, if the first perception of hot or cold is at time t, what is felt at time $t + 1$?) State clearly any further assumptions as to what happens to the inputs (stimuli) that may be necessary or relevant.

1.5 Design a McCulloch-Pitts net to model the perception of simple musical patterns. Use three input units to correspond to the three pitches, "do," "re," and "mi." Assume that only one pitch is presented at any time. Use two output units to correspond to the perception of an "upscale segment" and a "downscale segment"—specifically,

a. the pattern of inputs "do" at time $t = 1$, "re" at $t = 2$, and "mi" at $t = 3$ should elicit a positive response from the "upscale segment" unit;

b. the pattern of inputs "mi" at time $t = 1$, "re" at $t = 2$, and "do" at $t = 3$ should elicit a positive response from the "downscale segment" unit;

c. any other pattern of inputs should generate no response.

You may wish to elaborate on this example, allowing for more than one input unit to be "on" at any instant of time, designing output units to detect chords, etc.

CHAPTER 2

Simple Neural Nets
for Pattern Classification

2.1 GENERAL DISCUSSION

One of the simplest tasks that neural nets can be trained to perform is pattern classification. In pattern classification problems, each input vector (pattern) belongs, or does not belong, to a particular class or category. For a neural net approach, we assume we have a set of training patterns for which the correct classification is known. In the simplest case, we consider the question of membership in a single class. The output unit represents membership in the class with a response of 1; a response of -1 (or 0 if binary representation is used) indicates that the pattern is not a member of the class. For the single-layer nets described in this chapter, extension to the more general case in which each pattern may or may not belong to any of several classes is immediate. In such case, there is an output unit for each class. Pattern classification is one type of pattern recognition; the associative recall of patterns (discussed in Chapter 3) is another.

Pattern classification problems arise in many areas. In 1963, Donald Specht (a student of Widrow) used neural networks to detect heart abnormalities with EKG types of data as input (46 measurements). The output was "normal" or "abnormal," with an "on" response signifying normal [Specht, 1967; Caudill & Butler, 1990]. In the early 1960s, Minsky and Papert used neural nets to classify input patterns as convex or not convex and connected or not connected [Minsky & Papert, 1988]. There are many other examples of pattern classification problems being solved by neural networks, both the simple nets described in this chapter, other early nets not discussed here, and multilayer nets (especially the backpropagation nets described in Chapter 6).

In this chapter, we shall discuss three methods of training a simple single-layer neural net for pattern classification: the Hebb rule, the perceptron learning rule, and the delta rule (used by Widrow in his ADALINE neural net). First, however, we discuss some issues that are common to all single-layer nets that perform pattern classification. We conclude the chapter with some comparisons of the nets discussed and an extension to a multilayer net, MADALINE.

Many real-world examples need more sophisticated architecture and training rules because the conditions for a single-layer net to be adequate (see Section 2.1.3) are not met. However, if the conditions are met approximately, the results may be sufficiently accurate. Also, insight can be gained from the more simple nets, since the meaning of the weights may be easier to interpret.

2.1.1 Architecture

The basic architecture of the simplest possible neural networks that perform pattern classification consists of a layer of input units (as many units as the patterns to be classified have components) and a single output unit. Most neural nets we shall consider in this chapter use the single-layer architecture shown in Figure 2.1. This allows classification of vectors, which are n-tuples, but considers membership in only one category.

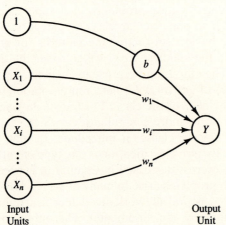

Input
Units

Output
Unit

Figure 2.1 Single-layer net for pattern classification.

An example of a net that classifies the input into several categories is considered in Section 2.3.3. This net is a simple extension of the nets that perform a single classification. The MADALINE net considered in Section 2.4.5 is a multilayer extension of the single-layer ADALINE net.

2.1.2 Biases and Thresholds

A bias acts exactly as a weight on a connection from a unit whose activation is always 1. Increasing the bias increases the net input to the unit. If a bias is included, the activation function is typically taken to be

$$f(\text{net}) = \begin{cases} 1 & \text{if net} \geq 0; \\ -1 & \text{if net} < 0; \end{cases}$$

where

$$\text{net} = b + \sum_i x_i w_i.$$

Some authors do not use a bias weight, but instead use a fixed threshold θ for the activation function. In that case,

$$f(\text{net}) = \begin{cases} 1 & \text{if net} \geq \theta; \\ -1 & \text{if net} < \theta; \end{cases}$$

where

$$\text{net} = \sum_i x_i w_i.$$

However, as the next example will demonstrate, this is essentially equivalent to the use of an adjustable bias.

Example 2.1 The role of a bias or a threshold

In this example and in the next section, we consider the separation of the input space into regions where the response of the net is positive and regions where the response is negative. To facilitate a graphical display of the relationships, we illustrate the ideas for an input with two components while the output is a scalar (i.e., it has only one component). The architecture of these examples is given in Figure 2.2.

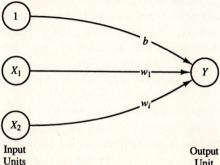

Input
Units

Output
Unit

Figure 2.2 Single-layer neural network for logic functions.

The boundary between the values of x_1 and x_2 for which the net gives a positive response and the values for which it gives a negative response is the separating line

$$b + x_1 w_1 + x_2 w_2 = 0,$$

or (assuming that $w_2 \neq 0$),

$$x_2 = -\frac{w_1}{w_2} x_1 - \frac{b}{w_2}.$$

The requirement for a positive response from the output unit is that the net input it receives, namely, $b + x_1 w_1 + x_2 w_2$, be greater than 0. During training, values of w_1, w_2, and b are determined so that the net will have the correct response for the training data.

If one thinks in terms of a threshold, the requirement for a positive response from the output unit is that the net input it receives, namely, $x_1 w_1 + x_2 w_2$, be greater than the threshold. This gives the equation of the line separating positive from negative output as

$$x_1 w_1 + x_2 w_2 = \theta,$$

or (assuming that $w_2 \neq 0$),

$$x_2 = -\frac{w_1}{w_2} x_1 + \frac{\theta}{w_2}.$$

During training, values of w_1 and w_2 are determined so that the net will have the correct response for the training data. In this case, the separating line cannot pass through the origin, but a line can be found that passes arbitrarily close to the origin.

The form of the separating line found by using an adjustable bias and the form obtained by using a fixed threshold illustrate that there is no advantage to including both a bias and a nonzero threshold for a neuron that uses the step function as its activation function. On the other hand, including neither a bias nor a threshold is equivalent to requiring the separating line (or plane or hyperplane for inputs with more components) to pass through the origin. This may or may not be appropriate for a particular problem.

As an illustration of a pseudopsychological analogy to the use of a bias, consider a simple (artificial) neural net in which the activation of the neuron corresponds to a person's action, "Go to the ball game." Each input signal corresponds to some factor influencing the decision to "go" or "not go" (other possible activities, the weather conditions, information about who is pitching, etc.). The weights on these input signals correspond to the importance the person places on each factor. (Of course, the weights may change with time, but methods for modifying them are not considered in this illustration.) A bias could represent a general inclination to "go" or "not go," based on past experiences. Thus, the bias would be modifiable, but the signal to it would not correspond to information about the specific game in question or activities competing for the person's time.

The threshold for this "decision neuron" indicates the total net input necessary to cause the person to "go," i.e., for the decision neuron to fire. The threshold would be different for different people; however, for the sake of this simple example, it should be thought of as a quantity that remains fixed for each individual. Since it is the relative values of the weights, rather than their actual magnitudes, that determine the response of the neuron, the model can cover all possibilities using either the fixed threshold or the adjustable bias.

2.1.3 Linear Separability

For each of the nets in this chapter, the intent is to train the net (i.e., adaptively determine its weights) so that it will respond with the desired classification when presented with an input pattern that it was trained on or when presented with one that is sufficiently similar to one of the training patterns. Before discussing the particular nets (which is to say, the particular styles of training), it is useful to discuss some issues common to all of the nets. For a particular output unit, the desired response is a "yes" if the input pattern is a member of its class and a "no" if it is not. A "yes" response is represented by an output signal of 1, a "no" by an output signal of -1 (for bipolar signals). Since we want one of two responses, the activation (or transfer or output) function is taken to be a step function. The value of the function is 1 if the net input is positive and -1 if the net input is negative. Since the net input to the output unit is

$$y\_in = b + \sum_i x_i w_i,$$

it is easy to see that the boundary between the region where $y\_in > 0$ and the region where $y\_in < 0$, which we call the *decision boundary*, is determined by the relation

$$b + \sum_i x_i w_i = 0.$$

Depending on the number of input units in the network, this equation represents a line, a plane, or a hyperplane.

If there are weights (and a bias) so that all of the training input vectors for which the correct response is $+1$ lie on one side of the decision boundary and all of the training input vectors for which the correct response is -1 lie on the other side of the decision boundary, we say that the problem is "linearly separable." Minsky and Papert [1988] showed that a single-layer net can learn only linearly separable problems. Furthermore, it is easy to extend this result to show that multilayer nets with linear activation functions are no more powerful than single-layer nets (since the composition of linear functions is linear).

It is convenient, if the input vectors are ordered pairs (or at most ordered triples), to graph the input training vectors and indicate the desired response by the appropriate symbol ("$+$" or "$-$"). The analysis also extends easily to nets with more input units; however, the graphical display is not as convenient. The region where y is positive is separated from the region where it is negative by the line

$$x_2 = -\frac{w_1}{w_2} x_1 - \frac{b}{w_2}.$$

These two regions are often called *decision regions* for the net. Notice in the following examples that there are many different lines that will serve to separate the input points that have different target values. However, for any particular line, there are also many choices of w_1, w_2, and b that give exactly the same line. The choice of the sign for b determines which side of the separating line corresponds to a $+1$ response and which side to a -1 response.

There are four different bipolar input patterns we can use to train a net with two input units. However, there are two possible responses for each input pattern, so there are 2^4 different functions that we might be able to train a very simple net to perform. Several of these functions are familiar from elementary logic, and we will use them for illustrations, for convenience. The first question we consider is, For this very simple net, do weights exist so that the net will have the desired output for each of the training input vectors?

Example 2.2 Response regions for the AND function

The AND function (for bipolar inputs and target) is defined as follows:

INPUT (x_1, x_2)	OUTPUT (t)
$(1, 1)$	$+1$
$(1, -1)$	-1
$(-1, 1)$	-1
$(-1, -1)$	-1

The desired responses can be illustrated as shown in Figure 2.3. One possible decision boundary for this function is shown in Figure 2.4.

An example of weights that would give the decision boundary illustrated in the figure, namely, the separating line

$$x_2 = -x_1 + 1,$$

is

$$b = -1,$$
$$w_1 = 1,$$
$$w_2 = 1.$$

The choice of sign for b is determined by the requirement that

$$b + x_1 w_1 + x_2 w_2 < 0$$

where $x_1 = 0$ and $x_2 = 0$. (Any point that is not on the decision boundary can be used to determine which side of the boundary is positive and which is negative; the origin is particularly convenient to use when it is not on the boundary.)

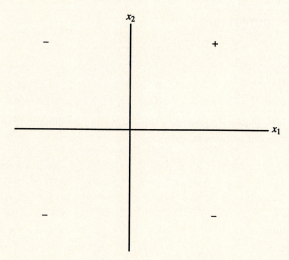

Figure 2.3 Desired response for the logic function AND (for bipolar inputs).

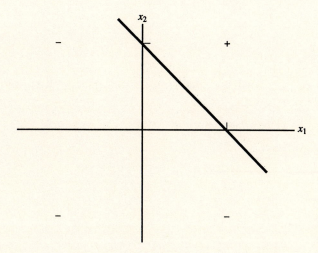

Figure 2.4 The logic function AND, showing a possible decision boundary.

Example 2.3 Response regions for the OR function

The logical OR function (for bipolar inputs and target) is defined as follows:

INPUT (x_1, x_2)	OUTPUT (t)
$(1, 1)$	$+1$
$(1, -1)$	$+1$
$(-1, 1)$	$+1$
$(-1, -1)$	-1

The weights must be chosen to provide a separating line, as illustrated in Figure 2.5. One example of suitable weights is

$$b = 1,$$
$$w_1 = 1,$$
$$w_2 = 1,$$

giving the separating line

$$x_2 = -x_1 - 1.$$

The choice of sign for b is determined by the requirement that

$$b + x_1 w_1 + x_2 w_2 > 0$$

where $x_1 = 0$ and $x_2 = 0$.

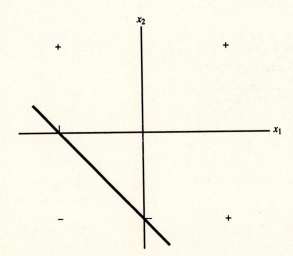

Figure 2.5 The logic function OR, showing a possible decision boundary.

The preceding two mappings (which can each be solved by a single-layer neural net) illustrate graphically the concept of linearly separable input. The input points to be classified positive can be separated from the input points to be classified negative by a straight line. The equations of the decision boundaries are not unique. We will return to these examples to illustrate each of the learning rules in this chapter.

Note that if a bias weight were not included in these examples, the decision boundary would be forced to go through the origin. In many cases (including Examples 2.2 and 2.3), that would change a problem that could be solved (i.e., learned, or one for which weights exist) into a problem that could not be solved.

Not all simple two-input, single-output mappings can be solved by a single-layer net (even with a bias included), as is illustrated in Example 2.4.

Example 2.4 Response regions for the XOR function

The desired response of this net is as follows:

INPUT (x_1, x_2)	OUTPUT (t)
(1, 1)	-1
(1, -1)	$+1$
(-1, 1)	$+1$
(-1, -1)	-1

It is easy to see that no single straight line can separate the points for which a positive response is desired from those for which a negative response is desired.

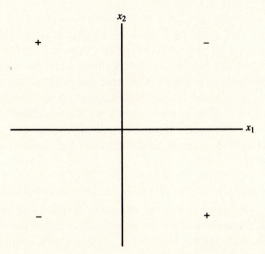

Figure 2.6 Desired response for the logic function XOR.

2.1.4 Data Representation

The previous examples show the use of a bipolar (values 1 and −1) representation of the training data, rather than the binary representation used for the McCulloch-Pitts neurons in Chapter 1. Many early neural network models used binary representation, although in most cases it can be modified to bipolar form. The form of the data may change the problem from one that can be solved by a simple neural net to one that cannot, as is illustrated in Examples 2.5–2.7 for the Hebb rule. Binary representation is also not as good as bipolar if we want the net to generalize (i.e., respond to input data similar, but not identical to, training data). Using bipolar input, missing data can be distinguished from mistaken data. Missing values can be represented by "0" and mistakes by reversing the input value from +1 to −1, or vice versa. We shall discuss some of the issues relating to the choice of binary versus bipolar representation further as they apply to particular neural nets. In general, bipolar representation is preferable.

The remainder of this chapter focuses on three methods of training single-layer neural nets that are useful for pattern classification: the Hebb rule, the perceptron learning rule, and the delta rule (or least mean squares). The Hebb rule, or correlational learning, is extremely simple but limited (even for linearly separable problems); the training algorithms for the perceptron and for ADALINE (adaptive linear neuron, trained by the delta rule) are closely related. Both are iterative techniques that are guaranteed to converge under suitable circumstances. The generalization of an ADALINE to a multilayer net (MADALINE) also will be examined.

2.2 HEBB NET

The earliest and simplest learning rule for a neural net is generally known as the Hebb rule. Hebb proposed that learning occurs by modification of the synapse strengths (weights) in a manner such that if two interconnected neurons are both "on" at the same time, then the weight between those neurons should be increased. The original statement only talks about neurons firing at the same time (and does not say anything about reinforcing neurons that do not fire at the same time). However, a stronger form of learning occurs if we also increase the weights if both neurons are "off" at the same time. We use this extended Hebb rule [McClelland & Rumelhart, 1988] because of its improved computational power and shall refer to it as the Hebb rule.

We shall refer to a single-layer (feedforward) neural net trained using the (extended) Hebb rule as a *Hebb net*. The Hebb rule is also used for training other specific nets that are discussed later. Since we are considering a single-layer net, one of the interconnected neurons will be an input unit and one an output unit (since no input units are connected to each other, nor are any output units interconnected). If data are represented in bipolar form, it is easy to express the desired weight update as

$$w_i(\text{new}) = w_i(\text{old}) + x_i y.$$

If the data are binary, this formula does not distinguish between a training pair in which an input unit is "on" and the target value is "off" and a training pair in which both the input unit and the target value are "off." Examples 2.5 and 2.6 (in Section 2.2.2) illustrate the extreme limitations of the Hebb rule for binary data. Example 2.7 shows the improved performance achieved by using bipolar representation for both the input and target values.

2.2.1 Algorithm

Step 0.   Initialize all weights:

$$w_i = 0 \qquad (i = 1 \text{ to } n).$$

Step 1.   For each input training vector and target output pair, $\mathbf{s} : t$, do steps 2–4.

    Step 2.   Set activations for input units:

$$x_i = s_i \qquad (i = 1 \text{ to } n).$$

    Step 3.   Set activation for output unit:

$$y = t.$$

    Step 4.   Adjust the weights for

$$w_i(\text{new}) = w_i(\text{old}) + x_i y \qquad (i = 1 \text{ to } n).$$

    Adjust the bias:

$$b(\text{new}) = b(\text{old}) + y.$$

Note that the bias is adjusted exactly like a weight from a "unit" whose output signal is always 1. The weight update can also be expressed in vector form as

$$\mathbf{w}(\text{new}) = \mathbf{w}(\text{old}) + \mathbf{x}y.$$

This is often written in terms of the weight change, $\Delta\mathbf{w}$, as

$$\Delta\mathbf{w} = \mathbf{x}y$$

and

$$\mathbf{w}(\text{new}) = \mathbf{w}(\text{old}) + \Delta\mathbf{w}.$$

There are several methods of implementing the Hebb rule for learning. The foregoing algorithm requires only one pass through the training set; other equivalent methods of finding the weights are described in Section 3.1.1, where the Hebb rule for pattern association (in which the target is a vector) is presented.

2.2.2 Application

Bias types of inputs are not explicitly used in the original formulation of Hebb learning. However, they are included in the examples in this section (shown as a third input component that is always 1) because without them, the problems discussed cannot be solved.

Logic functions

Example 2.5 A Hebb net for the AND function: binary inputs and targets

INPUT			TARGET
$(x_1$	x_2	1)	
(1	1	1)	1
(1	0	1)	0
(0	1	1)	0
(0	0	1)	0

For each training input: target, the weight change is the product of the input vector and the target value, i.e.,

$$\Delta w_1 = x_1 t, \qquad \Delta w_2 = x_2 t, \qquad \Delta b = t.$$

The new weights are the sum of the previous weights and the weight change. Only one iteration through the training vectors is required. The weight updates for the first input are as follows:

INPUT			TARGET	WEIGHT CHANGES			WEIGHTS		
$(x_1$	x_2	1)		$(\Delta w_1$	Δw_2	$\Delta b)$	$(w_1$	w_2	$b)$
							(0	0	0)
(1	1	1)	1	(1	1	1)	(1	1	1)

The separating line (see Section 2.1.3) becomes

$$x_2 = -x_1 - 1.$$

The graph, presented in Figure 2.7, shows that the response of the net will now be correct for the first input pattern. Presenting the second, third, and fourth training inputs shows that because the target value is 0, no learning occurs. Thus, using binary target values prevents the net from learning any pattern for which the target is "off":

INPUT			TARGET	WEIGHT CHANGES			WEIGHTS		
$(x_1$	x_2	1)		$(\Delta w_1$	Δw_2	$\Delta b)$	$(w_1$	w_2	$b)$
(1	0	1)	0	(0	0	0)	(1	1	1)
(0	1	1)	0	(0	0	0)	(1	1	1)
(0	0	1)	0	(0	0	0)	(1	1	1)

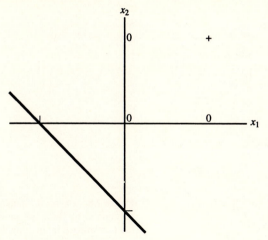

Figure 2.7 Decision boundary for binary AND function using Hebb rule after first training pair.

Example 2.6 A Hebb net for the AND function: binary inputs, bipolar targets

INPUT			TARGET
$(x_1$	x_2	1)	
(1	1	1)	1
(1	0	1)	-1
(0	1	1)	-1
(0	0	1)	-1

Presenting the first input, including a value of 1 for the third component, yields the following:

INPUT			TARGET	WEIGHT CHANGES			WEIGHTS		
$(x_1$	x_2	1)		$(\Delta w_1$	Δw_2	$\Delta b)$	$(w_1$	w_2	$b)$
							(0	0	0)
(1	1	1)	1	(1	1	1)	(1	1	1)

The separating line becomes

$$x_2 = -x_1 - 1.$$

Figure 2.8 shows that the response of the net will now be correct for the first input pattern.

Presenting the second, third, and fourth training patterns shows that learning continues for each of these patterns (since the target value is now -1, rather than 0, as in Example 2.5).

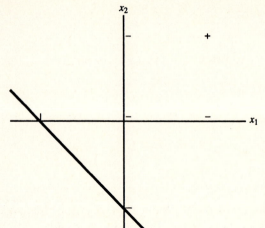

Figure 2.8 Decision boundary for AND function using Hebb rule after first training pair (binary inputs, bipolar targets).

INPUT			TARGET	WEIGHT CHANGES			WEIGHTS		
$(x_1$	x_2	1)		$(\Delta w_1$	Δw_2	$\Delta b)$	$(w_1$	w_2	$b)$
(1	0	1)	-1	$(-1$	0	$-1)$	(0	1	0)
(0	1	1)	-1	(0	-1	$-1)$	(0	0	$-1)$
(0	0	1)	-1	(0	0	$-1)$	(0	0	$-2)$

However, these weights do not provide the correct response for the first input pattern.

The choice of training patterns can play a significant role in determining which problems can be solved using the Hebb rule. The next example shows that the AND function can be solved if we modify its representation to express the inputs as well as the targets in bipolar form. Bipolar representation of the inputs and targets allows modification of a weight when the input unit and the target value are both "on" at the same time and when they are both "off" at the same time. The algorithm is the same as that just given, except that now all units will learn whenever there is an error in the output.

Example 2.7 A Hebb net for the AND function: bipolar inputs and targets

INPUT			TARGET
$(x_1$	x_2	1)	
(1	1	1)	1
(1	-1	1)	-1
$(-1$	1	1)	-1
$(-1$	-1	1)	-1

Presenting the first input, including a value of 1 for the third component, yields the following:

INPUT	TARGET	WEIGHT CHANGES	WEIGHTS
$(x_1 \quad x_2 \quad 1)$		$(\Delta w_1 \quad \Delta w_2 \quad \Delta b)$	$(w_1 \quad w_2 \quad b)$
			$(0 \quad 0 \quad 0)$
$(1 \quad 1 \quad 1)$	1	$(1 \quad 1 \quad 1)$	$(1 \quad 1 \quad 1)$

The separating line becomes

$$x_2 = -x_1 - 1.$$

The graph in Figure 2.9 shows that the response of the net will now be correct for the first input point (and also, by the way, for the input point $(-1, -1)$). Presenting the second input vector and target results in the following situation:

INPUT	TARGET	WEIGHT CHANGES	WEIGHTS
$(x_1 \quad x_2 \quad 1)$		$(\Delta w_1 \quad \Delta w_2 \quad \Delta b)$	$(w_1 \quad w_2 \quad b)$
			$(1 \quad 1 \quad 1)$
$(1 \quad -1 \quad 1)$	-1	$(-1 \quad 1 \quad -1)$	$(0 \quad 2 \quad 0)$

The separating line becomes

$$x_2 = 0.$$

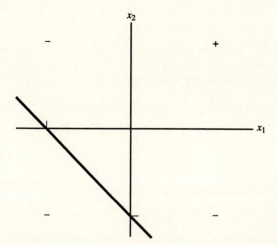

Figure 2.9 Decision boundary for the AND function using Hebb rule after first training pair (bipolar inputs and targets).

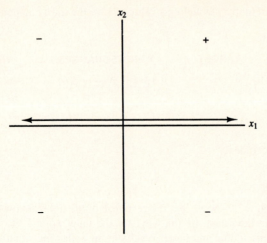

Figure 2.10 Decision boundary for bipolar AND function using Hebb rule after second training pattern (boundary is x_1-axis).

The graph in Figure 2.10 shows that the response of the net will now be correct for the first two input points, $(1, 1)$ and $(1, -1)$, and also, incidentally, for the input point $(-1, -1)$. Presenting the third input vector and target yields the following:

INPUT	TARGET	WEIGHT CHANGES	WEIGHTS
$(x_1 \quad x_2 \quad 1)$		$(\Delta w_1 \quad \Delta w_2 \quad \Delta b)$	$(w_1 \quad w_2 \quad b)$
			$(0 \quad 2 \quad 0)$
$(-1 \quad 1 \quad 1)$	-1	$(1 \quad -1 \quad -1)$	$(1 \quad 1 \quad -1)$

The separating line becomes

$$x_2 = -x_1 + 1.$$

The graph in Figure 2.11 shows that the response of the net will now be correct for the first three input points (and also, by the way, for the input point $(-1, -1)$). Presenting the last point, we obtain the following:

INPUT	TARGET	WEIGHT CHANGES	WEIGHTS
$(x_1 \quad x_2 \quad 1)$		$(\Delta w_1 \quad \Delta w_2 \quad \Delta b)$	$(w_1 \quad w_2 \quad b)$
			$(1 \quad 1 \quad -1)$
$(-1 \quad -1 \quad 1)$	-1	$(1 \quad 1 \quad -1)$	$(2 \quad 2 \quad -2)$

Even though the weights have changed, the separating line is still

$$x_2 = -x_1 + 1,$$

so the graph of the decision regions (the positive response and the negative response) remains as in Figure 2.11.

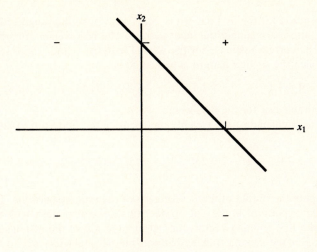

Figure 2.11 Decision boundary for bipolar AND function using Hebb rule after third training pattern.

Character recognition

Example 2.8 A Hebb net to classify two-dimensional input patterns (representing letters)

A simple example of using the Hebb rule for character recognition involves training the net to distinguish between the pattern "X" and the pattern "O". The patterns can be represented as

```
        # . . . #              . # # # .
        . # . # .              # . . . #
        . . # . .      and     # . . . #
        . # . # .              # . . . #
        # . . . #              . # # # .

         Pattern 1             Pattern 2
```

To treat this example as a pattern classification problem with one output class, we will designate that class "X" and take the pattern "O" to be an example of output that is not "X."

The first thing we need to do is to convert the patterns to input vectors. That is easy to do by assigning each # the value 1 and each "." the value −1. To convert from the two-dimensional pattern to an input vector, we simply concatenate the rows, i.e., the second row of the pattern comes after the first row, the third row follows, ect. Pattern 1 then becomes

$$1 \ -1 \ -1 \ -1 \ 1, \ -1 \ 1 \ -1 \ 1 \ -1, \ -1 \ -1 \ 1 \ -1 \ -1, \ -1 \ 1 \ -1 \ 1 \ -1,$$
$$1 \ -1 \ -1 \ -1 \ 1,$$

and pattern 2 becomes

$$-1 \ 1 \ 1 \ 1 \ -1, \ 1 \ -1 \ -1 \ -1 \ 1, \ 1 \ -1 \ -1 \ -1 \ 1, \ 1 \ -1 \ -1 \ -1 \ 1, \ -1 \ 1 \ 1 \ 1 \ -1,$$

where a comma denotes the termination of a line of the original matrix. For computer simulations, the program can be written so that the vector is read in from the two-dimensional format.

The correct response for the first pattern is "on," or $+1$, so the weights after presenting the first pattern are simply the input pattern. The bias weight after presenting this is $+1$. The correct response for the second pattern is "off," or -1, so the weight change when the second pattern is presented is

$$1 \; -1 \; -1 \; -1 \; 1, \; -1 \; 1 \; 1 \; 1 \; -1, \; -1 \; 1 \; 1 \; 1 \; -1, \; -1 \; 1 \; 1 \; 1 \; -1, \; 1 \; -1 \; -1 \; -1 \; 1.$$

In addition, the weight change for the bias weight is -1.

Adding the weight change to the weights representing the first pattern gives the final weights:

$$2 \; -2 \; -2 \; -2 \; 2, \; -2 \; 2 \; 0 \; 2 \; -2, \; -2 \; 0 \; 2 \; 0 \; -2, \; -2 \; 2 \; 0 \; 2 \; -2, \; 2 \; -2 \; -2 \; -2 \; 2.$$

The bias weight is 0.

Now, we compute the output of the net for each of the training patterns. The net input (for any input pattern) is the dot product of the input pattern with the weight vector. For the first training vector, the net input is 42, so the response is positive, as desired. For the second training pattern, the net input is -42, so the response is clearly negative, also as desired.

However, the net can also give reasonable responses to input patterns that are similar, but not identical, to the training patterns. There are two types of changes that can be made to one of the input patterns that will generate a new input pattern for which it is reasonable to expect a response. The first type of change is usually referred to as "mistakes in the data." In this case, one or more components of the input vector (corresponding to one or more pixels in the original pattern) have had their sign reversed, changing a 1 to a -1, or vice versa. The second type of change is called "missing data." In this situation, one or more of the components of the input vector have the value 0, rather than 1 or -1. In general, a net can handle more missing components than wrong components; in other words, with input data, "It's better not to guess."

Other simple examples

Example 2.9 Limitations of Hebb rule training for binary patterns

This example shows that the Hebb rule may fail, even if the problem is linearly separable (and even if 0 is not the target).

Consider the following input and target output pairs:

1	1	1	$\rightarrow$	1
1	1	0	$\rightarrow$	0
1	0	1	$\rightarrow$	0
0	1	1	$\rightarrow$	0

It is easy to see that the Hebb rule cannot learn any pattern for which the target is 0. So we must at least convert the targets to $+1$ and -1. Now consider

1	1	1	$\rightarrow$	1
1	1	0	$\rightarrow$	-1
1	0	1	$\rightarrow$	-1
0	1	1	$\rightarrow$	-1

Figure 2.12 shows that the problem is now solvable, i.e., the input points classified in one class (with target value $+1$) are linearly separable from those not in the class (with target value -1). The figure also shows that a nonzero bias will be necessary, since the separating plane does not pass through the origin. The plane pictured is $x_1 + x_2 + x_3 + (-2.5) = 0$, i.e., a weight vector of (1 1 1) and a bias of -2.5.

The weights (and bias) are found by taking the sum of the weight changes that occur at each stage of the algorithm. The weight change is simply the input pattern (augmented by the fourth component, the input to the bias weight, which is always 1) multiplied by the target value for the pattern. We obtain:

Weight change for first input pattern:	1	1	1	1
Weight change for second input pattern:	-1	-1	0	-1
Weight change for third input pattern:	-1	0	-1	-1
Weight change for fourth input pattern:	0	-1	-1	-1
Final weights (and bias)	-1	-1	-1	-2

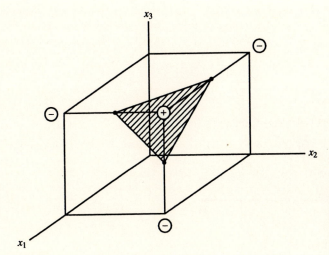

Figure 2.12 Linear separation of binary training inputs.

It is easy to see that these weights do not produce the correct output for the first pattern.

Example 2.10 Limitation of Hebb rule training for bipolar patterns

Examples 2.5, 2.6, and 2.7 show that even if the representation of the vectors does not change the problem from unsolvable to solvable, it can affect whether the Hebb rule works. In this example, we consider the same problem as in Example 2.9, but with the input points (and target classifications) in bipolar form. Accordingly, we have the following arrangement of values:

INPUT				TARGET	WEIGHT CHANGE				WEIGHT			
$(x_1$	x_2	x_3	1)		$(\Delta w_1$	Δw_2	Δw_2	$\Delta b)$	$(w_1$	w_2	w_3	$b)$
									(0	0	0	0)
(1	1	1	1)	1	(1	1	1	1)	(1	1	1	1)
(1	1	-1	1)	-1	$(-1$	-1	1	$-1)$	(0	0	2	0)
(1	-1	1	1)	-1	$(-1$	1	-1	$-1)$	$(-1$	1	1	$-1)$
$(-1$	1	1	1)	-1	(1	-1	-1	$-1)$	(0	0	0	$-2)$

Again, it is clear that the weights do not give the correct output for the first input pattern.

Figure 2.13 shows that the input points are linearly separable; one posssible plane, $x_1 + x_2 + x_3 + (-2) = 0$, to perform the separation is shown. This plane corresponds to a weight vector of (1 1 1) and a bias of -2.

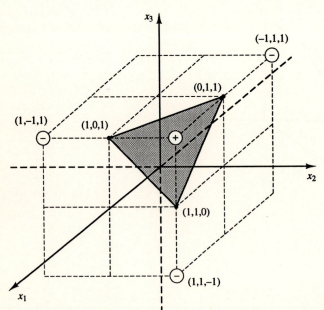

Figure 2.13   Linear separation of bipolar training inputs.

2.3 PERCEPTRON

Perceptrons had perhaps the most far-reaching impact of any of the early neural nets. The perceptron learning rule is a more powerful learning rule than the Hebb rule. Under suitable assumptions, its iterative learning procedure can be proved to converge to the correct weights, i.e., the weights that allow the net to produce the correct output value for each of the training input patterns. Not too surprisingly, one of the necessary assumptions is that such weights exist.

A number of different types of perceptrons are described in Rosenblatt (1962) and in Minsky and Papert (1969, 1988). Although some perceptrons were self-organizing, most were trained. Typically, the original perceptrons had three layers of neurons—sensory units, associator units, and a response unit—forming an approximate model of a retina. One particular simple perceptron [Block, 1962] used binary activations for the sensory and associator units and an activation of $+1$, 0, or -1 for the response unit. The sensory units were connected to the associator units by connections with fixed weights having values of $+1$, 0, or -1, assigned at random.

The activation function for each associator unit was the binary step function with an arbitrary, but fixed, threshold. Thus, the signal sent from the associator units to the output unit was a binary (0 or 1) signal. The output of the perceptron is $y = f(y\_in)$, where the activation function is

$$f(y\_in) = \begin{cases} 1 & \text{if } y\_in > \theta \\ 0 & \text{if } -\theta \le y\_in \le \theta \\ -1 & \text{if } y\_in < -\theta \end{cases}$$

The weights from the associator units to the response (or output) unit were adjusted by the perceptron learning rule. For each training input, the net would calculate the response of the output unit. Then the net would determine whether an error occurred for this pattern (by comparing the calculated output with the target value). The net did not distinguish between an error in which the calculated output was zero and the target -1, as opposed to an error in which the calculated output was $+1$ and the target -1. In either of these cases, the sign of the error denotes that the weights should be changed in the direction indicated by the target value. However, only the weights on the connections from units that sent a non-zero signal to the output unit would be adjusted (since only these signals contributed to the error). If an error occurred for a particular training input pattern, the weights would be changed according to the formula

$$w_i(\text{new}) = w_i(\text{old}) + \alpha t x_i,$$

where the target value t is $+1$ or -1 and α is the learning rate. If an error did not occur, the weights would not be changed.

Training would continue until no error occurred. The perceptron learning rule convergence theorem states that if weights exist to allow the net to respond correctly to all training patterns, then the rule's procedure for adjusting the weights will find values such that the net does respond correctly to all training patterns (i.e., the net solves the problem or learns the classification). Moreover, the net will find these weights in a finite number of training steps. We will consider a proof of this theorem in Section 2.3.4, since it helps clarify which aspects of the many variations on perceptron learning are significant.

2.3.1 Architecture

Simple perceptron for pattern classification

The output from the associator units in the original simple perceptron was a binary vector; that vector is treated as the input signal to the output unit in the sections that follow. As the proof of the perceptron learning rule convergence theorem given in Section 2.3.4 illustrates, the assumption of a binary input is not necessary. Since only the weights from the associator units to the output unit could be adjusted, we limit our consideration to the single-layer portion of the net, shown in Figure 2.14. Thus, the associator units function like input units, and the architecture is as given in the figure.

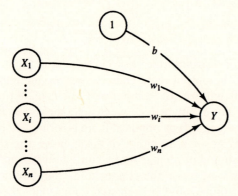

Figure 2.14 Perceptron to perform single classification.

The goal of the net is to classify each input pattern as belonging, or not belonging, to a particular class. Belonging is signified by the output unit giving a response of $+1$; not belonging is indicated by a response of -1. The net is trained to perform this classification by the iterative technique described earlier and given in the algorithm that follows.

2.3.2 Algorithm

The algorithm given here is suitable for either binary or bipolar input vectors (n-tuples), with a bipolar target, fixed θ, and adjustable bias. The threshold θ does not play the same role as in the step function illustrated in Section 2.1.2; thus, both a bias and a threshold are needed. The role of the threshold is discussed following the presentation of the algorithm. The algorithm is not particularly sensitive to the initial values of the weights or the value of the learning rate.

Step 0.   Initialize weights and bias.
              (For simplicity, set weights and bias to zero.)
              Set learning rate α ($0 < \alpha \leq 1$).
              (For simplicity, α can be set to 1.)

Step 1.   While stopping condition is false, do Steps 2–6.

   Step 2.   For each training pair s:t, do Steps 3–5.

      Step 3.   Set activations of input units:

$$x_i = s_i.$$

      Step 4.   Compute response of output unit:

$$y\_in = b + \sum_i x_i w_i;$$

$$y = \begin{cases} 1 & \text{if } y\_in > \theta \\ 0 & \text{if } -\theta \leq y\_in \leq \theta \\ -1 & \text{if } y\_in < -\theta \end{cases}$$

      Step 5.   Update weights and bias if an error occurred for this pattern.
                    If $y \neq t$,

$$w_i(\text{new}) = w_i(\text{old}) + \alpha t x_i,$$

$$b(\text{new}) = b(\text{old}) + \alpha t.$$

              else

$$w_i(\text{new}) = w_i(\text{old}),$$

$$b(\text{new}) = b(\text{old}).$$

   Step 6.   Test stopping condition:
                If no weights changed in Step 2, stop; else, continue.

Note that only weights connecting active input units ($x_i \neq 0$) are updated. Also, weights are updated only for patterns that do not produce the correct value of y. This means that as more training patterns produce the correct response, less learning occurs. This is in contrast to the training of the ADALINE units described in Section 2.4, in which learning is based on the difference between y_in and t.

The threshold on the activation function for the response unit is a fixed, non-negative value θ. The form of the activation function for the output unit (response unit) is such that there is an "undecided" band (of fixed width determined by θ) separating the region of positive response from that of negative response. Thus, the previous analysis of the interchangeability of bias and threshold does not apply, because changing θ would change the width of the band, not just the position.

Note that instead of one separating line, we have a line separating the region of positive response from the region of zero response, namely, the line bounding the inequality

$$w_1x_1 + w_2x_2 + b > \theta,$$

and a line separating the region of zero response from the region of negative response, namely, the line bounding the inequality

$$w_1x_1 + w_2x_2 + b < -\theta.$$

2.3.3 Application

Logic functions

Example 2.11 A Perceptron for the AND function: binary inputs, bipolar targets

Let us consider again the AND function with binary input and bipolar target, now using the perceptron learning rule. The training data are as given in Example 2.6 for the Hebb rule. An adjustable bias is included, since it is necessary if a single-layer net is to be able to solve this problem. For simplicity, we take $\alpha = 1$ and set the initial weights and bias to 0, as indicated. However, to illustrate the role of the threshold, we take $\theta = .2$.

The weight change is $\Delta\mathbf{w} = t(x_1, x_2, 1)$ if an error has occurred and zero otherwise. Presenting the first input, we have:

INPUT			NET	OUT	TARGET	WEIGHT CHANGES			WEIGHTS		
$(x_1$	x_2	1)							$(w_1$	w_2	$b)$
									(0	0	0)
(1	1	1)	0	0	1	(1	1	1)	(1	1	1)

The separating lines become

$$x_1 + x_2 + 1 = .2$$

and

$$x_1 + x_2 + 1 = -.2.$$

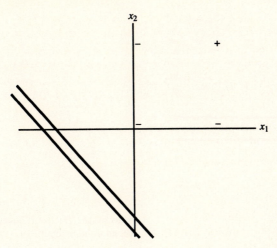

Figure 2.15 Decision boundary for logic function AND after first training input.

The graph in Figure 2.15 shows that the response of the net will now be correct for the first input pattern.

Presenting the second input yields the following:

INPUT			NET	OUT	TARGET	WEIGHT CHANGES			WEIGHTS		
$(x_1$	x_2	1)							$(w_1$	w_2	$b)$
									(1	1	1)
(1	0	1)	2	1	-1	$(-1$	0	$-1)$	(0	1	0)

The separating lines become

$$x_2 = .2$$

and

$$x_2 = -.2$$

The graph in Figure 2.16 shows that the response of the net will now (still) be correct for the first input point.

For the third input, we have:

INPUT			NET	OUT	TARGET	WEIGHT CHANGES			WEIGHTS		
$(x_1$	x_2	1)							$(w_1$	w_2	$b)$
									(0	1	0)
(0	1	1)	1	1	-1	(0	-1	$-1)$	(0	0	$-1)$

Since the components of the input patterns are nonnegative and the components of the weight vector are nonpositive, the response of the net will be negative (or zero).

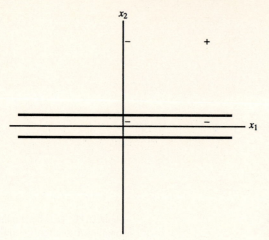

Figure 2.16 Decision boundary after second training input.

To complete the first epoch of training, we present the fourth training pattern:

	INPUT		NET	OUT	TARGET	WEIGHT CHANGES			WEIGHTS		
$(x_1$	x_2	1)							$(w_1$	w_2	$b)$
									(0	0	$-1)$
(0	0	1)	-1	-1	-1	(0	0	0)	(0	0	$-1)$

The response for all of the input patterns is negative for the weights derived; but since the response for input pattern (1, 1) is not correct, we are not finished.

The second epoch of training yields the following weight updates for the first input:

	INPUT		NET	OUT	TARGET	WEIGHT CHANGES			WEIGHTS		
$(x_1$	x_2	1)							$(w_1$	w_2	$b)$
									(0	0	$-1)$
(1	1	1)	-1	-1	1	(1	1	1)	(1	1	0)

The separating lines become

$$x_1 + x_2 = .2$$

and

$$x_1 + x_2 = -.2.$$

The graph in Figure 2.17 shows that the response of the net will now be correct for (1, 1).

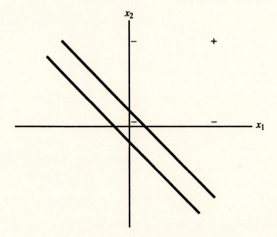

Figure 2.17 Boundary after first training input of second epoch.

For the second input in the second epoch, we have:

	INPUT		NET	OUT	TARGET	WEIGHT CHANGES			WEIGHTS		
$(x_1$	x_2	1)							$(w_1$	w_2	$b)$
									(1	1	0)
(1	0	1)	1	1	-1	(·-1	0	$-1)$	(0	1	$-1)$

The separating lines become

$$x_2 - 1 = .2$$

and

$$x_2 - 1 = -.2.$$

The graph in Figure 2.18 shows that the response of the net will now be correct (negative) for the input points (1, 0) and (0, 0); the response for input points (0, 1) and (1, 1) will be 0, since the net input would be 0, which is between $-.2$ and $.2$ ($\theta = .2$).

In the second epoch, the third input yields:

	INPUT		NET	OUT	TARGET	WEIGHT CHANGES			WEIGHTS		
$(x_1$	x_2	1)							$(w_1$	w_2	$b)$
									(0	1	$-1)$
(0	1	1)	0	0	-1	(0	-1	$-1)$	(0	0	$-2)$

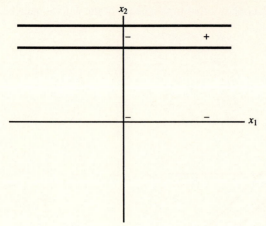

Figure 2.18 Boundary after second input of second epoch.

Again, the response will be negative for all of the input.

To complete the second epoch of training, we present the fourth training pattern:

INPUT			NET	OUT	TARGET	WEIGHTS CHANGE			WEIGHTS		
$(x_1$	x_2	1)							$(w_1$	w_2	$b)$
									(0	0	$-2)$
(0	0	1)	-2	-1	-1	(0	0	0)	(0	0	$-2)$

The results for the third epoch are:

INPUT			NET	OUT	TARGET	WEIGHT CHANGES			WEIGHTS		
$(x_1$	x_2	1)							$(w_1$	w_2	$b)$
									(0	0	$-2)$
(1	1	1)	-2	-1	1	(1	1	1)	(1	1	$-1)$
(1	0	1)	0	0	-1	$(-1$	0	$-1)$	(0	1	$-2)$
(0	1	1)	-1	-1	-1	(0	0	0)	(0	1	$-2)$
(0	0	1)	-2	-1	-1	(0	0	0)	(0	1	$-2)$

The results for the fourth epoch are:

(1	1	1)	-1	-1	1	(1	1	1)	(1	2	$-1)$
(1	0	1)	0	0	-1	$(-1$	0	$-1)$	(0	2	$-2)$
(0	1	1)	0	0	-1	(0	-1	$-1)$	(0	1	$-3)$
(0	0	1)	-3	-1	-1	(0	0	0)	(0	1	$-3)$

For the fifth epoch, we have

(1	1	1)	−2	−1	1	(1	1	1)	(1	2	−2)
(1	0	1)	−1	−1	−1	(0	0	0)	(1	2	−2)
(0	1	1)	0	0	−1	(0	−1	−1)	(1	1	−3)
(0	0	1)	−3	−1	−1	(0	0	0)	(1	1	−3)

and for the sixth epoch,

(1	1	1)	−1	−1	1	(1	1	1)	(2	2	−2)
(1	0	1)	0	0	−1	(−1	0	−1)	(1	2	−3)
(0	1	1)	−1	−1	−1	(0	0	0)	(1	2	−3)
(0	0	1)	−3	−1	−1	(0	0	0)	(1	2	−3)

The results for the seventh epoch are:

(1	1	1)	0	0	1	(1	1	1)	(2	3	−2)
(1	0	1)	0	0	−1	(−1	0	−1)	(1	3	−3)
(0	1	1)	0	0	−1	(0	−1	−1)	(1	2	−4)
(0	0	1)	−4	−1	−1	(0	0	0)	(1	2	−4)

The eighth epoch yields

(1	1	1)	−1	−1	1	(1	1	1)	(2	3	−3)
(1	0	1)	−1	−1	−1	(0	0	0)	(2	3	−3)
(0	1	1)	0	0	−1	(0	−1	−1)	(2	2	−4)
(0	0	1)	−4	−1	−1	(0	0	0)	(2	2	−4)

and the ninth

(1	1	1)	0	0	1	(1	1	1)	(3	3	−3)
(1	0	1)	0	0	−1	(−1	0	−1)	(2	3	−4)
(0	1	1)	−1	−1	−1	(0	0	0)	(2	3	−4)
(0	0	1)	−4	−1	−1	(0	0	0)	(2	3	−4)

Finally, the results for the tenth epoch are:

(1	1	1)	1	1	1	(0	0	0)	(2	3	−4)
(1	0	1)	−2	−1	−1	(0	0	0)	(2	3	−4)
(0	1	1)	−1	−1	−1	(0	0	0)	(2	3	−4)
(0	0	1)	−4	−1	−1	(0	0	0)	(2	3	−4)

Thus, the positive response is given by all points such that

$$2x_1 + 3x_2 - 4 > .2,$$

with boundary line

$$x_2 = -\frac{2}{3}x_1 + \frac{7}{5},$$

and the negative response is given by all points such that

$$2x_1 + 3x_2 - 4 < -.2,$$

with boundary line

$$x_2 = -\frac{2}{3}x_1 + \frac{19}{15}$$

(see Figure 2.19.)

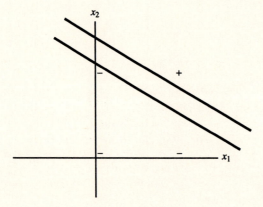

Figure 2.19 Final decision boundaries for AND function in perceptron learning.

Since the proof of the perceptron learning rule convergence theorem (Section 2.3.4) shows that binary input is not required, and in previous examples bipolar input was often preferable, we consider again the previous example, but with bipolar inputs, an adjustable bias, and $\theta = 0$. This variation provides the most direct comparison with Widrow-Hoff learning (an ADALINE net), which we consider in the next section. Note that it is not necessary to modify the training set so that all patterns are mapped to $+1$ (as is done in the proof of the perceptron learning rule convergence theorem); the weight adjustment is $t\mathbf{x}$ whenever the response of the net to input vector $\mathbf{x}$ is incorrect. The target value is still bipolar.

Example 2.12 A Perceptron for the AND function: bipolar inputs and targets

The training process for bipolar input, $\alpha = 1$, and threshold and initial weights = 0 is:

INPUT			NET	OUT	TARGET	WEIGHT CHANGES			WEIGHTS		
$(x_1$	x_2	1)							$(w_1$	w_2	$b)$
									(0	0	0)
(1	1	1)	0	0	1	(1	1	1)	(1	1	1)
(1	−1	1)	1	1	−1	(−1	1	−1)	(0	2	0)
(−1	1	1)	2	1	−1	(1	−1	−1)	(1	1	−1)
(−1	−1	1)	−3	−1	−1	(0	0	0)	(1	1	−1)

In the second epoch of training, we have:

(1	1	1)	1	1	1	(0	0	0)	(1	1	−1)
(1	−1	1)	−1	−1	−1	(0	0	0)	(1	1	−1)
(−1	1	1)	−1	−1	−1	(0	0	0)	(1	1	−1)
(−1	−1	1)	−3	−1	−1	(0	0	0)	(1	1	−1)

Since all the Δw's are 0 in epoch 2, the system was fully trained after the first epoch.

It seems intuitively obvious that a procedure that could continue to learn to improve its weights even after the classifications are all correct would be better than a learning rule in which weight updates cease as soon as all training patterns are classified correctly. However, the foregoing example shows that the change from binary to bipolar representation improves the results rather spectacularly.

We next show that the perceptron with $\alpha = 1$ and $\theta = .1$ can solve the problem the Hebb rule could not.

Other simple examples

Example 2.13 Perceptron training is more powerful than Hebb rule training

The mapping of interest maps the first three components of the input vector onto a target value that is 1 if there are no zero inputs and that is −1 if there is one zero input. (If there are two or three zeros in the input, we do not specify the target value.) This is a portion of the parity problem for three inputs. The fourth component of the input vector is the input to the bias weight and is therefore always 1. The weight change vector is left blank if no error has occurred for a particular pattern. The learning rate is $\alpha = 1$, and the threshold $\theta = .1$. We show the following selected epochs:

INPUT	NET	OUT	TARGET	WEIGHT CHANGE	WEIGHTS
x_1 x_2 x_3 1					(w_1 w_2 w_3 b)
					(0 0 0 0)

Epoch 1:

(1 1 1 1)	0	0	1	(1 1 1 1)	(1 1 1 1)
(1 1 0 1)	3	1	−1	(−1 −1 0 −1)	(0 0 1 0)
(1 0 1 1)	1	1	−1	(−1 0 −1 −1)	(−1 0 0 −1)
(0 1 1 1)	−1	−1	−1	( )	(−1 0 0 −1)

Epoch 2:

(1 1 1 1)	−2	−1	1	(1 1 1 1)	(0 1 1 0)
(1 1 0 1)	1	1	−1	(−1 −1 0 −1)	(−1 0 1 −1)
(1 0 1 1)	−1	−1	−1	( )	(−1 0 1 −1)
(0 1 1 1)	0	0	−1	(0 −1 −1 −1)	(−1 −1 0 −2)

Epoch 3:

(1 1 1 1)	−4	−1	1	(1 1 1 1)	(0 0 1 −1)
(1 1 0 1)	−1	−1	−1	( )	(0 0 1 −1)
(1 0 1 1)	0	0	−1	(−1 0 −1 −1)	(−1 0 0 −2)
(0 1 1 1)	−2	−1	−1	( )	(−1 0 0 −2)

Epoch 4:

(1 1 1 1)	−3	−1	1	(1 1 1 1)	(0 1 1 −1)
(1 1 0 1)	0	0	−1	(−1 −1 0 −1)	(−1 0 1 −2)
(1 0 1 1)	−2	−1	−1	( )	(−1 0 1 −2)
(0 1 1 1)	−1	−1	−1	( )	(−1 0 1 −2)

Epoch 5:

(1 1 1 1)	−2	−1	1	(1 1 1 1)	(0 1 2 −1)
(1 1 0 1)	0	0	−1	(−1 −1 0 −1)	(−1 0 2 −2)
(1 0 1 1)	−1	−1	−1	( )	(−1 0 2 −2)
(0 1 1 1)	0	0	−1	(0 −1 −1 −1)	(−1 −1 1 −3)

Epoch 10:

(1 1 1 1)	−3	−1	1	(1 1 1 1)	(1 1 2 −3)
(1 1 0 1)	−1	−1	−1	( )	(1 1 2 −3)
(1 0 1 1)	0	0	−1	(−1 0 −1 −1)	(0 1 1 −4)
(0 1 1 1)	−2	−1	−1	( )	(0 1 1 −4)

Epoch 15:

```
(1 1 1 1) -1  -1      1   (1    1  1   1) (1    3 3  -4)
(1 1 0 1)  0   0     -1   (-1 -1   0  -1) (0    2 3  -5)
(1 0 1 1) -2  -1     -1   (             ) (0    2 3  -5)
(0 1 1 1)  0  -1     -1   (0   -1 -1  -1) (0    1 2  -6)
```

Epoch 20:

```
(1 1 1 1) -2  -1      1   (1    1  1   1) (2    2 4  -6)
(1 1 0 1) -2  -1     -1   (             ) (2    2 4  -6)
(1 0 1 1)  0   0     -1   (-1   0 -1  -1) (1    2 3  -7)
(0 1 1 1) -2  -1     -1   (             ) (1    2 3  -7)
```

Epoch 25:

```
(1 1 1 1)  0   0      1   (1    1  1   1) (3    4 4  -7)
(1 1 0 1)  0   0     -1   (-1 -1   0  -1) (2    3 4  -8)
(1 0 1 1) -2  -1     -1   (             ) (2    3 4  -8)
(0 1 1 1) -1  -1     -1   (             ) (2    3 4  -8)
```

Epoch 26:

```
(1 1 1 1)  1   1      1   (             ) (2    3 4  -8)
(1 1 0 1) -3  -1     -1   (             ) (2    3 4  -8)
(1 0 1 1) -2  -1     -1   (             ) (2    3 4  -8)
(0 1 1 1) -1  -1     -1   (             ) (2    3 4  -8)
```

Character recognition

Application Procedure

Step 0.   Apply training algorithm to set the weights.

Step 1.   For each input vector **x** to be classified, do Steps 2–3.

Step 2.   Set activations of input units.

Step 3.   Compute response of output unit:

$$y\_in = \sum_i x_i w_i;$$

$$y = \begin{cases} 1 & \text{if } y\_in > \theta \\ 0 & \text{if } -\theta \le y\_in \le \theta \\ -1 & \text{if } y\_in < -\theta \end{cases}$$

Example 2.14 A Perceptron to classify letters from different fonts: one output class

As the first example of using the perceptron for character recognition, consider the 21 input patterns in Figure 2.20 as examples of *A* or not-*A*. In other words, we train the perceptron to classify each of these vectors as belonging, or not belonging, to

Input from
Font 1

Input from
Font 2

Input from
Font 3

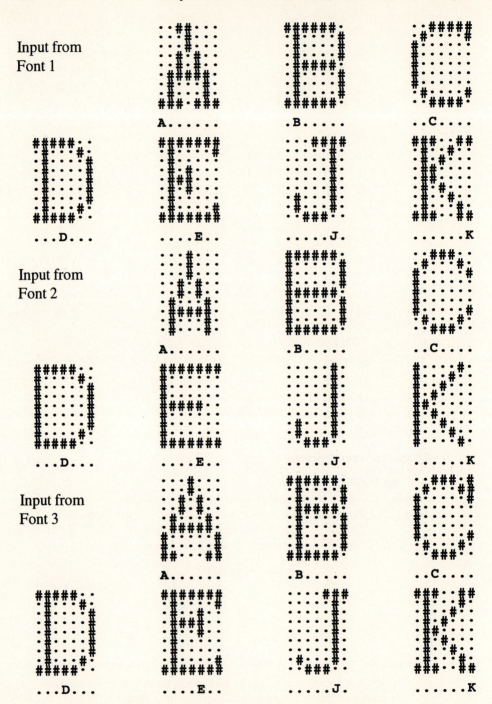

Figure 2.20   Training input and target output patterns.

the class A. In that case, the target value for each pattern is either 1 or -1; only the first component of the target vector shown is applicable. The net is as shown in Figure 2.14, and $n = 63$. There are three examples of A and 18 examples of not-A in Figure 2.20.

We could, of course, use the same vectors as examples of B or not-B and train the net in a similar manner. Note, however, that because we are using a single-layer net, the weights for the output unit signifying A do not have any interaction with the weights for the output unit signifying B. Therefore, we can solve these two problems at the same time, by allowing a column of weights for each output unit. Our net would have 63 input units and 2 output units. The first output unit would correspond to "A or not-A", the second unit to "B or not-B." Continuing this idea, we can identify 7 output units, one for each of the 7 categories into which we wish to classify our input.

Ideally, when an unknown character is presented to the net, the net's output consists of a single "yes" and six "nos." In practice, that may not happen, but the net may produce several guesses that can be resolved by other methods, such as considering the strengths of the activations of the various output units prior to setting the threshold or examining the context in which the ill-classified character occurs.

Example 2.15 A Perceptron to classify letters from different fonts: several output classes

The perceptron shown in Figure 2.14 can be extended easily to the case where the input vectors belong to one (or more) of several categories. In this type of application, there is an output unit representing each of the categories to which the input vectors may belong. The architecture of such a net is shown in Figure 2.21.

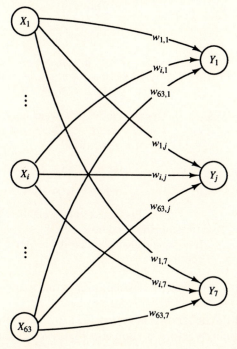

Figure 2.21 Perceptron to classify input into seven categories.

For this example, each input vector is a 63-tuple representing a letter expressed as a pattern on a 7×9 grid of pixels. The training patterns are illustrated in Figure 2.20. There are seven categories to which each input vector may belong, so there are seven components to the output vector, each representing a letter: A, B, C, D, E, K, or J. For ease of reading, we show the target output pattern indicating that the input was an "A" as $(A \cdot \cdot \cdot \cdot \cdot \cdot \cdot)$, a "B" $(\cdot B \cdot \cdot \cdot \cdot \cdot)$, etc.

The training input patterns and target responses must be converted to an appropriate form for the neural net to process. A bipolar representation has better computational characteristics than does a binary representation. The input patterns may be converted to bipolar vectors as described in Example 2.8; the target output pattern $(A \cdot \cdot \cdot \cdot \cdot \cdot \cdot)$ becomes the bipolar vector $(1, -1, -1, -1, -1, -1, -1)$ and the target pattern $(\cdot B \cdot \cdot \cdot \cdot \cdot)$ is represented by the bipolar vector $(-1, 1, -1, -1, -1, -1, -1)$.

A modified training algorithm for several output categories (threshold $= 0$, learning rate $= 1$, bipolar training pairs) is as follows:

Step 0. Initialize weights and biases
(0 or small random values).

Step 1. While stopping condition is false, do Steps 1–6.

Step 2. For each bipolar training pair **s : t**, do Steps 3–5.

Step 3. Set activation of each input unit, $i = 1, \ldots, n$:

$$x_i = s_i.$$

Step 4. Compute activation of each output unit, $j = 1, \ldots, m$:

$$y\_in_j = b_j + \sum_i x_i w_{ij}.$$

$$y_j = \begin{cases} 1 & \text{if } y\_in_j > \theta \\ 0 & \text{if } -\theta \leq y\_in_j \leq \theta \\ -1 & \text{if } y\_in_j < -\theta \end{cases}$$

Step 5. Update biases and weights, $j = 1, \ldots, m$; $i = 1, \ldots, n$:
If $t_j \neq y_j$, then

$$b_j(\text{new}) = b_j(\text{old}) + t_j;$$

$$w_{ij}(\text{new}) = w_{ij}(\text{old}) + t_j x_i.$$

Else, biases and weights remain unchanged.

Step 6. Test for stopping condition:
If no weight changes occurred in Step 2, stop; otherwise, continue.

After training, the net correctly classifies each of the training vectors.

The performance of the net shown in Figure 2.21 in classifying input vectors that are similar to the training vectors is shown in Figure 2.22. Each of the input

Input from
Font 1

Input from
Font 2

Input from
Font 3

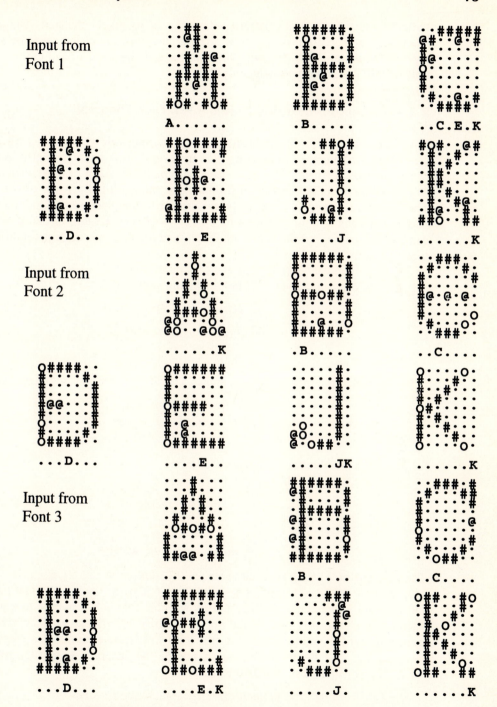

Figure 2.22   Classification of noisy input patterns using a perceptron.

patterns is a training input pattern with a few of its pixels changed. The pixels where the input pattern differs from the training pattern are indicated by @ for a pixel that is "on" now but was "off" in the training pattern, and O for a pixel that is "off" now but was originally "on."

2.3.4 Perceptron Learning Rule Convergence Theorem

The statement and proof of the perceptron learning rule convergence theorem given here are similar to those presented in several other sources [Hertz, Krogh, & Palmer, 1991; Minsky & Papert, 1988; Arbib, 1987]. Each of these provides a slightly different perspective and insights into the essential aspects of the rule. The fact that the weight vector is perpendicular to the plane separating the input patterns at each step of the learning processes [Hertz, Krogh, & Palmer, 1991] can be used to interpret the degree of difficulty of training a perceptron for different types of input.

The perceptron learning rule is as follows:

Given a finite set of P input training vectors

$$\mathbf{x}(p), \qquad p = 1, \ldots, P,$$

each with an associated target value

$$t(p), \qquad p = 1, \ldots, P,$$

which is either $+1$ or -1, and an activation function $y = f(y\_in)$, where

$$y = \begin{cases} 1 & \text{if } y\_in > \theta \\ 0 & \text{if } -\theta \le y\_in \le \theta \\ -1 & \text{if } y\_in < -\theta, \end{cases}$$

the weights are updated as follows:

If $y \ne t$, then

$$\mathbf{w}\,(\text{new}) = \mathbf{w}\,(\text{old}) + t\mathbf{x};$$

else

$$\text{no change in the weights.}$$

The perceptron learning rule convergence theorem is:

If there is a weight vector $\mathbf{w}^*$ such that $f(\mathbf{x}(p){\cdot}\mathbf{w}^*) = t(p)$ for all p, then for any starting vector $\mathbf{w}$, the perceptron learning rule will converge to a weight vector (not necessarily unique and not necessarily $\mathbf{w}^*$) that gives the correct response for all training patterns, and it will do so in a finite number of steps.

The proof of the theorem is simplified by the observation that the training set can be considered to consist of two parts:

$$F^+ = \{\mathbf{x} \text{ such that the target value is } +1\}$$

and

$$F^- = \{\mathbf{x} \text{ such that the target value is } -1\}.$$

A new training set is then defined as

$$F = F^+ \cup -F^-,$$

where

$$-F^- = \{-\mathbf{x} \text{ such that } \mathbf{x} \text{ is in } F^-\}.$$

In order to simplify the algebra slightly, we shall assume, without loss of generality, that $\theta = 0$ and $\alpha = 1$ in the proof. The existence of a solution of the original problem, namely the existence of a weight vector $\mathbf{w}^*$ for which

$$\mathbf{x}\cdot\mathbf{w}^* > 0 \qquad \text{if } \mathbf{x} \text{ is in } F^+$$

and

$$\mathbf{x}\cdot\mathbf{w}^* < 0 \qquad \text{if } \mathbf{x} \text{ is in } F^-,$$

is equivalent to the existence of a weight vector $\mathbf{w}^*$ for which

$$\mathbf{x}\cdot\mathbf{w}^* > 0 \qquad \text{if } \mathbf{x} \text{ is in } F.$$

All target values for the modified training set are $+1$. If the response of the net is incorrect for a given training input, the weights are updated according to

$$\mathbf{w}(\text{new}) = \mathbf{w}(\text{old}) + \mathbf{x}.$$

Note that the input training vectors must each have an additional component (which is always 1) included to account for the signal to the bias weight.

We now sketch the proof of this remarkable convergence theorem, because of the light that it sheds on the wide variety of forms of perceptron learning that are guaranteed to converge. As mentioned, we assume that the training set has been modified so that all targets are $+1$. Note that this will involve reversing the sign of all components (including the input component corresponding to the bias) for any input vectors for which the target was originally -1.

We now consider the sequence of input training vectors for which a weight change occurs. We must show that this sequence is finite.

Let the starting weights be denoted by $\mathbf{w}(0)$, the first new weights by $\mathbf{w}(1)$, etc. If $\mathbf{x}(0)$ is the first training vector for which an error has occurred, then

$$\mathbf{w}(1) = \mathbf{w}(0) + \mathbf{x}(0) \qquad (\text{where, by assumption, } \mathbf{x}(0)\cdot\mathbf{w}(0) \leq 0).$$

If another error occurs, we denote the vector $\mathbf{x}(1)$; $\mathbf{x}(1)$ may be the same as $\mathbf{x}(0)$ if no errors have occurred for any other training vectors, or $\mathbf{x}(1)$ may be different from $\mathbf{x}(0)$. In either case,

$$\mathbf{w}(2) = \mathbf{w}(1) + \mathbf{x}(1) \qquad (\text{where, by assumption, } \mathbf{x}(1)\cdot\mathbf{w}(1) \leq 0).$$

At any stage, say, k, of the process, the weights are changed if and only if the current weights fail to produce the correct (positive) response for the current input vector, i.e., if $\mathbf{x}(k-1)\cdot\mathbf{w}(k-1) \leq 0$. Combining the successive weight changes gives

$$\mathbf{w}(k) = \mathbf{w}(0) + \mathbf{x}(0) + \mathbf{x}(1) + \mathbf{x}(2) + \cdots + \mathbf{x}(k-1).$$

We now show that k cannot be arbitrarily large.

Let $\mathbf{w}^*$ be a weight vector such that $\mathbf{x}\cdot\mathbf{w}^* > 0$ for all training vectors in F. Let $m = \min\{\mathbf{x}\cdot\mathbf{w}^*\}$, where the minimum is taken over all training vectors in F; this minimum exists as long as there are only finitely many training vectors. Now,

$$\mathbf{w}(k)\cdot\mathbf{w}^* = [\mathbf{w}(0) + \mathbf{x}(0) + \mathbf{x}(1) + \mathbf{x}(2) + \cdots + \mathbf{x}(k-1)]\cdot\mathbf{w}^*$$

$$\geq \mathbf{w}(0)\cdot\mathbf{w}^* + km$$

since $\mathbf{x}(i)\cdot\mathbf{w}^* \geq m$ for each i, $1 \leq i \leq P$.

The Cauchy-Schwartz inequality states that for any vectors $\mathbf{a}$ and $\mathbf{b}$,

$$(\mathbf{a}\cdot\mathbf{b})^2 \leq \|\mathbf{a}\|^2 \|\mathbf{b}\|^2,$$

or

$$\|\mathbf{a}\|^2 \geq \frac{(\mathbf{a}\cdot\mathbf{b})^2}{\|\mathbf{b}\|^2} \qquad (\text{for } \|\mathbf{b}\|^2 \neq 0).$$

Therefore,

$$\|\mathbf{w}(k)\|^2 \geq \frac{(\mathbf{w}(k)\cdot\mathbf{w}^*)^2}{\|\mathbf{w}^*\|^2}$$

$$\geq \frac{(\mathbf{w}(0)\cdot\mathbf{w}^* + km)^2}{\|\mathbf{w}^*\|^2}.$$

This shows that the squared length of the weight vector grows faster than k^2, where k is the number of time the weights have changed.

However, to show that the length cannot continue to grow indefinitely, consider

$$\mathbf{w}(k) = \mathbf{w}(k-1) + \mathbf{x}(k-1),$$

together with the fact that

$$\mathbf{x}(k-1)\cdot\mathbf{w}(k-1) \leq 0.$$

By simple algebra,

$$\|\mathbf{w}(k)\|^2 = \|\mathbf{w}(k-1)\|^2 + 2\mathbf{x}(k-1)\cdot\mathbf{w}(k-1) + \|\mathbf{x}(k-1)\|^2$$

$$\leq \|\mathbf{w}(k-1)\|^2 + \|\mathbf{x}(k-1)\|^2.$$

Now let $M = \max \{\| \mathbf{x} \|^2$ for all $\mathbf{x}$ in the training set$\}$; then

$$\|\mathbf{w}(k)\|^2 \leq \|\mathbf{w}(k - 1)\|^2 + \|\mathbf{x}(k - 1)\|^2$$

$$\leq \|\mathbf{w}(k - 2)\|^2 + \|\mathbf{x}(k - 2)\|^2 + \|\mathbf{x}(k - 1)\|^2$$

$$\vdots$$

$$\leq \|\mathbf{w}(0)\|^2 + \|\mathbf{x}(0)\|^2 + \cdots + \|\mathbf{x}(k - 1)\|^2$$

$$\leq \|\mathbf{w}(0)\|^2 + kM.$$

Thus, the squared length grows less rapidly than linearly in k.

Combining the inequalities

$$\|\mathbf{w}(k)\|^2 \geq \frac{(\mathbf{w}(0)\mathbf{w}^* + km)^2}{\|\mathbf{w}^*\|^2}$$

and

$$\|\mathbf{w}(k)\|^2 \leq \|\mathbf{w}(0)\|^2 + kM$$

shows that the number of times that the weights may change is bounded. Specifically,

$$\frac{(\mathbf{w}(0){\cdot}\mathbf{w}^* + km)^2}{\|\mathbf{w}^*\|^2} \leq \|\mathbf{w}(k)\|^2 \leq \|\mathbf{w}(0)\|^2 + kM.$$

Again, to simplify the algebra, assume (without loss of generality) that $\mathbf{w}(0) = 0$. Then the maximum possible number of times the weights may change is given by

$$\frac{(km)^2}{\|\mathbf{w}^*\|^2} \leq kM,$$

or

$$k \leq \frac{M \|\mathbf{w}^*\|^2}{m^2} \,.$$

Since the assumption that $\mathbf{w}^*$ exists can be restated, without loss of generality, as the assumption that there is a solution weight vector of unit length (and the definition of m is modified accordingly), the maximum number of weight updates is M/m^2. Note, however, that many more computations may be required, since very few input vectors may generate an error during any one epoch of training. Also, since $\mathbf{w}^*$ is unknown (and therefore, so is m), the number of weight updates cannot be predicted from the preceding inequality.

The foregoing proof shows that many variations in the perceptron learning rule are possible. Several of these variations are explicitly mentioned in Chapter 11 of Minsky and Papert (1988).

The original restriction that the coefficients of the patterns be binary is unnecessary. All that is required is that there be a finite maximum norm of the training vectors (or at least a finite upper bound to the norm). Training may take a long time (a large number of steps) if there are training vectors that are very small in norm, since this would cause small m to have a small value. The argument of the proof is unchanged if a nonzero value of θ is used (although changing the value of θ may change a problem from solvable to unsolvable or vice versa). Also, the use of a learning rate other than 1 will not change the basic argument of the proof (see Exercise 2.8). Note that there is no requirement that there can be only finitely many training vectors, as long as the norm of the training vectors is bounded (and bounded away from 0 as well). The actual target values do not matter, either; the learning law simply requires that the weights be incremented by the input vector (or a multiple of it) whenever the response of the net is incorrect (and that the training vectors can be stated in such a way that they all should give the same response of the net).

Variations on the learning step include setting the learning rate α to any nonnegative constant (Minsky starts by setting it specifically to 1), setting α to $1/\|\mathbf{x}\|$ so that the weight change is a unit vector, and setting α to $(\mathbf{x} \cdot \mathbf{w})/\|\mathbf{x}\|^2$ (which makes the weight change just enough for the pattern $\mathbf{x}$ to be classified correctly at this step).

Minsky sets the initial weights equal to an arbitrary training pattern. Others usually indicate small random values.

Note also that since the procedure will converge from an arbitrary starting set of weights, the process is error correcting, as long as the errors do not occur too often (and the process is not stopped before error correction occurs).

2.4 ADALINE

The ADALINE (ADaptive LInear NEuron) [Widrow & Hoff, 1960] typically uses bipolar (1 or -1) activations for its input signals and its target output (although it is not restricted to such values). The weights on the connections from the input units to the ADALINE are adjustable. The ADALINE also has a bias, which acts like an adjustable weight on a connection from a unit whose activation is always 1.

In general, an ADALINE can be trained using the delta rule, also known as the least mean squares (LMS) or Widrow-Hoff rule. The rule (Section 2.4.2) can also be used for single-layer nets with several output units; an ADALINE is a special case in which there is only one output unit. During training, the activation of the unit is its net input, i.e., the activation function is the identity function. The learning rule minimizes the mean squared error between the activation and the target value. This allows the net to continue learning on all training patterns, even after the correct output value is generated (if a threshold function is applied) for some patterns.

After training, if the net is being used for pattern classification in which the desired output is either a $+1$ or a -1, a threshold function is applied to the net input to obtain the activation. If the net input to the ADALINE is greater than or equal to 0, the its activation is set to 1; otherwise it is set to -1. Any problem for which the input patterns corresponding to the output value $+1$ are linearly separable from input patterns corresponding to the output value -1 can be modeled successfully by an ADALINE unit. An application algorithm is given in Section 2.4.3 to illustrate the use of the activation function after the net is trained.

In Section 2.4.4, we shall see how a heuristic learning rule can be used to train a multilayer combination of ADALINES, known as a MADALINE.

2.4.1 Architecture

An ADALINE is a single unit (neuron) that receives input from several units. It also receives input from a "unit" whose signal is always $+1$, in order for the bias weight to be trained by the same process (the delta rule) as is used to train the other weights. A single ADALINE is shown in Figure 2.23.

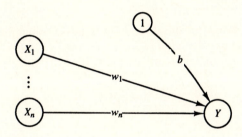

Figure 2.23 Architecture of an ADALINE.

Several ADALINES that receive signals from the same input units can be combined in a single-layer net, as described for the perceptron (Section 2.3.3). If, however, ADALINES are combined so that the output from some of them becomes input for others of them, then the net becomes multilayer, and determining the weights is more difficult. Such a multilayer net, known as a MADALINE, is considered in Section 2.4.5.

2.4.2 Algorithm

A training algorithm for an ADALINE is as follows:

Step 0.   Initialize weights.
        (Small random values are usually used.)
        Set learning rate α.
        (See comments following algorithm.)

Step 1.   While stopping condition is false, do Steps 2–6.

  Step 2.   For each bipolar training pair s:t, do Steps 3–5.

    Step 3.   Set activations of input units, $i = 1, \ldots, n$:

$$x_i = s_i.$$

    Step 4.   Compute net input to output unit:

$$y\_in = b + \sum_i x_i w_i.$$

    Step 5.   Update bias and weights, $i = 1, \ldots, n$:

$$b(\text{new}) = b(\text{old}) + \alpha(t - y\_in).$$

$$w_i(\text{new}) = w_i(\text{old}) + \alpha(t - y\_in)x_i.$$

  Step 6.   Test for stopping condition:
  If the largest weight change that occurred in Step 2 is smaller than a specified tolerance, then stop; otherwise continue.

Setting the learning rate to a suitable value requires some care. According to Hecht-Nielsen (1990), an upper bound for its value can be found from the largest eigenvalue of the correlation matrix R of the input (row) vectors $\mathbf{x}(p)$:

$$R = \frac{1}{P} \sum_{p=1}^{P} \mathbf{x}(p)^T \mathbf{x}(p),$$

namely,

$$\alpha < \text{one-half the largest eigenvalue of } R.$$

However, since R does not need to be calculated to compute the weight updates, it is common simply to take a small value for α (such as $\alpha = .1$) initially. If too large a value is chosen, the learning process will not converge; if too small a value is chosen, learning will be extremely slow [Hecht-Nielsen, 1990]. The choice of learning rate and methods of modifying it are considered further in Chapter 6. For a single neuron, a practical range for the learning rate α is $0.1 \leq n\alpha \leq 1.0$, where n is the number of input units [Widrow, Winter & Baxter, 1988].

The proof of the convergence of the ADALINE training process is essentially contained in the derivation of the delta rule, which is given in Section 2.4.4.

2.4.3 Applications

After training, an ADALINE unit can be used to classify input patterns. If the target values are bivalent (binary or bipolar), a step function can be applied as the activation function for the output unit. The following procedure shows the step function for bipolar targets, the most common case:

Step 0.   Initialize weights
          (from ADALINE training algorithm given in Section 2.4.2).
Step 1.   For each bipolar input vector **x**, do Steps 2–4.
  Step 2.   Set activations of the input units to **x**.
  Step 3.   Compute net input to output unit:

$$y\_in = b + \sum_i x_i w_i.$$

  Step 4.   Apply the activation function:

$$y = \begin{cases} 1 & \text{if } y\_in \geq 0; \\ -1 & \text{if } y\_in < 0. \end{cases}$$

Simple examples

The weights (and biases) in Examples 2.16–2.19 give the minimum total squared error for each set of training patterns. Good approximations to these values can be found using the algorithm in Section 2.4.2 with a small learning rate.

Example 2.16   **An ADALINE for the AND function: binary inputs, bipolar targets**

Even though the ADALINE was presented originally for bipolar inputs and targets, the delta rule also applies to binary input. In this example, we consider the AND function with binary input and bipolar targets. The function is defined by the following four training patterns:

x_1	x_2	t
1	1	1
1	0	−1
0	1	−1
0	0	−1

As indicated in the derivation of the delta rule (Section 2.4.4), an ADALINE is designed to find weights that minimize the total error

$$E = \sum_{p=1}^{4} (x_1(p)w_1 + x_2(p)w_2 + w_0 - t(p))^2,$$

where

$$x_1(p)w_1 + x_2(p)w_2 + w_0$$

is the net input to the output unit for pattern p and $t(p)$ is the associated target for pattern p.

Weights that minimize this error are

$$w_1 = 1$$

and

$$w_2 = 1,$$

with the bias

$$w_0 = -\frac{3}{2}.$$

Thus, the separating line is

$$x_1 + x_2 - \frac{3}{2} = 0.$$

The total squared error for the four training patterns with these weights is 1.

A minor modification to Example 2.11 (setting $\theta = 0$) shows that for the perceptron, the boundary line is

$$x_2 = -\frac{2}{3}x_1 + \frac{4}{3}.$$

(The two boundary lines coincide when $\theta = 0$.) The total squared error for the minimizing weights found by the perceptron is 10/9.

Example 2.17 An ADALINE for the AND function: bipolar inputs and targets

The weights that minimize the total error for the bipolar form of the AND function are

$$w_1 = \frac{1}{2}$$

and

$$w_2 = \frac{1}{2},$$

with the bias

$$w_0 = -\frac{1}{2}.$$

Thus, the separating line is

$$\frac{1}{2}x_1 + \frac{1}{2}x_2 - \frac{1}{2} = 0,$$

which is course the same line as

$$x_1 + x_2 - 1 = 0,$$

as found by the perceptron in Example 2.12.

Example 2.18 An ADALINE for the AND NOT function: bipolar inputs and targets

The logic function x_1 AND NOT x_2 is defined by the following bipolar input and target patterns:

x_1	x_2	t
1	1	-1
1	-1	1
-1	1	-1
-1	-1	-1

Weights that minimize the total squared error for the bipolar form of the AND NOT function are

$$w_1 = \frac{1}{2}$$

and

$$w_2 = -\frac{1}{2},$$

with the bias

$$w_0 = -\frac{1}{2}.$$

Thus, the separating line is

$$\frac{1}{2}x_1 - \frac{1}{2}x_2 - \frac{1}{2} = 0.$$

Example 2.19 An ADALINE for the OR function: bipolar inputs and targets

The logic function x_1 OR x_2 is defined by the following bipolar input and target patterns:

x_1	x_2	t
1	1	1
1	-1	1
-1	1	1
-1	-1	-1

Weights that minimize the total squared error for the bipolar form of the OR function are

$$w_1 = \frac{1}{2}$$

and

$$w_2 = \frac{1}{2},$$

with the bias

$$w_0 = \frac{1}{2}.$$

Thus, the separating line is

$$\frac{1}{2} x_1 + \frac{1}{2} x_2 + \frac{1}{2} = 0.$$

2.4.4 Derivations

Delta rule for single output unit

The delta rule changes the weights of the neural connections so as to minimize the difference between the net input to the output unit, y_in, and the target value t. The aim is to minimize the error over all training patterns. However, this is accomplished by reducing the error for each pattern, one at a time. Weight corrections can also be accumulated over a number of training patterns (so-called batch updating) if desired. In order to distinguish between the fixed (but arbitrary) index for the weight whose adjustment is being determined in the derivation that follows and the index of summation needed in the derivation, we use the index I for the weight and the index i for the summation. We shall return to the more standard lowercase indices for weights whenever this distinction is not needed. The delta rule for adjusting the Ith weight (for each pattern) is

$$\Delta w_I = \alpha(t - y\_in)x_I.$$

The nomenclature we use in the derivation is as follows:

$\mathbf{x}$ vector of activations of input units, an n-tuple.

y_in the net input to output unit Y is

$$y\_in = \sum_{i=1}^{n} x_i w_i.$$

t target output.

Derivation.   The squared error for a particular training pattern is

$$E = (t - y\_in)^2.$$

E is a function of all of the weights, w_i, $i = 1, \ldots, n$. The gradient of E is the vector consisting of the partial derivatives of E with respect to each of the weights. The gradient gives the direction of most rapid increase in E; the opposite direction gives the most rapid decrease in the error. The error can be reduced by adjusting the weight w_I in the direction of $-\dfrac{\partial E}{\partial w_I}$.

Since $y\_in = \sum\limits_{i=1}^{n} x_i w_i$,

$$\frac{\partial E}{\partial w_I} = -2(t - y\_in)\frac{\partial y\_in}{\partial w_I}$$

$$= -2(t - y\_in)x_I.$$

Thus, the local error will be reduced most rapidly (for a given learning rate) by adjusting the weights according to the delta rule,

$$\Delta w_I = \alpha(t - y\_in)x_I.$$

Delta rule for several output units

The derivation given in this subsection allows for more than one output unit. The weights are changed to reduce the difference between the net input to the output unit, y_in_J, and the target value t_J. This formulation reduces the error for each pattern. Weight corrections can also be accumulated over a number of training patterns (so-called batch updating) if desired.

The delta rule for adjusting the weight from the Ith input unit to the Jth output unit (for each pattern) is

$$\Delta w_{IJ} = \alpha(t_J - y\_in_J)x_I.$$

Derivation.   The squared error for a particular training pattern is

$$E = \sum\limits_{j=1}^{m} (t_j - y\_in_j)^2.$$

E is a function of all of the weights. The gradient of E is a vector consisting of the partial derivatives of E with respect to each of the weights. This vector gives the direction of most rapid increase in E; the opposite direction gives the direction of most rapid decrease in the error. The error can be reduced most rapidly by adjusting the weight w_{IJ} in the direction of $-\partial E/\partial w_{IJ}$.

We now find an explicit formula for $\partial E/\partial w_{IJ}$ for the arbitrary weight w_{IJ}. First, note that

$$\frac{\partial E}{\partial w_{IJ}} = \frac{\partial}{\partial w_{IJ}} \sum\limits_{j=1}^{m} (t_j - y\_in_j)^2$$

$$= \frac{\partial}{\partial w_{IJ}} (t_J - y\_in_J)^2,$$

since the weight w_{IJ} influences the error only at output unit Y_J. Furthermore, using the fact that

$$y\_in_J = \sum_{i=1}^{n} x_i w_{iJ},$$

we obtain

$$\frac{\partial E}{\partial w_{IJ}} = -2(t_J - y\_in_J) \frac{\partial y\_in_J}{\partial w_{IJ}}$$
$$= -2(t_J - y\_in_J) x_I.$$

Thus, the local error will be reduced most rapidly (for a given learning rate) by adjusting the weights according to the delta rule,

$$\Delta w_{IJ} = \alpha(t_J - y\_in_J) x_I.$$

The preceding two derivations of the delta rule can be generalized to the case where the training data are only samples from a larger data set, or probability distribution. Minimizing the error for the training set will also minimize the expected value for the error of the underlying probability distribution. (See Widrow & Lehr, 1990 or Hecht-Nielsen, 1990 for a further discussion of the matter.)

2.4.5 MADALINE

As mentioned earlier, a MADALINE consists of Many ADAptive Linear NEurons arranged in a multilayer net. The examples given for the perceptron and the derivation of the delta rule for several output units both indicate there is essentially no change in the process of training if several ADALINE units are combined in a single-layer net. In this section we will discuss a MADALINE with one hidden layer (composed of two hidden ADALINE units) and one output ADALINE unit. Generalizations to more hidden units, more output units, and more hidden layers, are straightforward.

Architecture

A simple MADALINE net is illustrated in Figure 2.24. The outputs of the two hidden ADALINES, z_1 and z_2, are determined by signals from the same input units X_1 and X_2. As with the ADALINES discussed previously, each output signal is the result of applying a threshold function to the unit's net input. Thus, y is a nonlinear function of the input vector (x_1, x_2). The use of the hidden units, Z_1 and Z_2, give the net computational capabilities not found in single layer nets, but also complicate the training process. In the next section we consider two training algorithms for a MADALINE with one hidden layer.

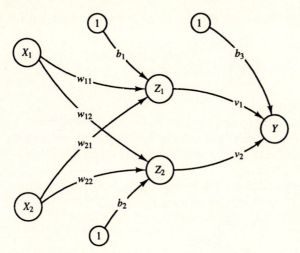

Figure 2.24   A MADALINE with two hidden ADALINES and one output ADALINE.

Algorithm

In the MRI algorithm (the original form of MADALINE training) [Widrow and Hoff, 1960], only the weights for the hidden ADALINES are adjusted; the weights for the output unit are fixed. The MRII algorithm [Widrow, Winter, & Baxter, 1987] provides a method for adjusting all weights in the net.

We consider first the MRI algorithm; the weights v_1 and v_2 and the bias b_3 that feed into the output unit Y are determined so that the response of unit Y is 1 if the signal it receives from either Z_1 or Z_2 (or both) is 1 and is -1 if both Z_1 and Z_2 send a signal of -1. In other words, the unit Y performs the logic function OR on the signals it receives from Z_1 and Z_2. The weights into Y are

$$v_1 = \frac{1}{2}$$

and

$$v_2 = \frac{1}{2},$$

with the bias

$$b_3 = \frac{1}{2}$$

(see Example 2.19). The weights on the first hidden ADALINE (w_{11} and w_{21}) and the weights on the second hidden ADALINE (w_{12} and w_{22}) are adjusted according to the algorithm.

Training Algorithm for MADALINE *(MRI).*　　The activation function for units Z_1, Z_2, and Y is

$$f(x) = \begin{cases} 1 & \text{if } x \geq 0; \\ -1 & \text{if } x < 0. \end{cases}$$

Step 0.　　Initialize weights:

Weights v_1 and v_2 and the bias b_3 are set as described; small random values are usually used for ADALINE weights.

Set the learning rate α as in the ADALINE training algorithm (a small value).

Step 1.　　While stopping condition is false, do Steps 2–8.

　　Step 2.　　For each bipolar training pair, **s**:**t**, do Steps 3–7.

　　　　Step 3.　　Set activations of input units:

$$x_i = s_i.$$

　　　　Step 4.　　Compute net input to each hidden ADALINE unit:

$$z\_in_1 = b_1 + x_1 w_{11} + x_2 w_{21},$$

$$z\_in_2 = b_2 + x_1 w_{12} + x_2 w_{22}.$$

　　　　Step 5.　　Determine output of each hidden ADALINE unit:

$$z_1 = f(z\_in_1),$$

$$z_2 = f(z\_in_2).$$

　　　　Step 6.　　Determine output of net:

$$y\_in = b_3 + z_1 v_1 + z_2 v_2;$$

$$y = f(y\_in).$$

　　　　Step 7.　　Determine error and update weights:

If $t = y$, no weight updates are performed.

Otherwise:

If $t = 1$, then update weights on Z_J, the unit whose net input is closest to 0,

$$b_J(\text{new}) = b_J(\text{old}) + \alpha(1 - z\_in_J),$$

$$w_{iJ}(\text{new}) = w_{iJ}(\text{old}) + \alpha(1 - z\_in_J)x_i;$$

If $t = -1$, then update weights on all units Z_k that have positive net input,

$$b_k(\text{new}) = b_k(\text{old}) + \alpha(-1 - z\_in_k),$$

$$w_{ik}(\text{new}) = w_{ik}(\text{old}) + \alpha(-1 - z\_in_k)x_i.$$

Step 8.   Test stopping condition.
            If weight changes have stopped (or reached an acceptable
            level), or if a specified maximum number of weight update
            iterations (Step 2) have been performed, then stop; other-
            wise continue.

Step 7 is motivated by the desire to (1) update the weights only if an error
occurred and (2) update the weights in such a way that it is more likely for the
net to produce the desired response.

If $t = 1$ and error has occurred, it means that all Z units had value -1 and
at least one Z unit needs to have a value of $+1$. Therefore, we take Z_J to be the
Z unit whose net input is closest to 0 and adjust its weights (using ADALINE training
with a target value of $+1$):

$$b_J(\text{new}) = b_J(\text{old}) + \alpha(1 - z\_in_J),$$

$$w_{iJ}(\text{new}) = w_{iJ}(\text{old}) + \alpha(1 - z\_in_J)x_i.$$

If $t = -1$ and error has occurred, it means that at least one Z unit had value
$+1$ and all Z units must have value -1. Therefore, we adjust the weights on all
of the Z units with positive net input, (using ADALINE training with a target of
-1):

$$b_k(\text{new}) = b_k(\text{old}) + \alpha(-1 - z\_in_k),$$

$$w_{ik}(\text{new}) = w_{ik}(\text{old}) + \alpha(-1 - z\_in_k)x_i.$$

MADALINES can also be formed with the weights on the output unit set to
perform some other logic function such as AND or, if there are more than two
hidden units, the "majority rule" function. The weight update rules would be
modified to reflect the logic function being used for the output unit [Widrow &
Lehr, 1990].

A more recent MADALINE training rule, called MRII [Widrow, Winter, &
Baxter, 1987], allows training for weights in all layers of the net. As in earlier
MADALINE training, the aim is to cause the least disturbance to the net at any step
of the learning process, in order to cause as little "unlearning" of patterns for
which the net had been trained previously. This is sometimes called the "don't
rock the boat" principle. Several output units may be used; the total error for
any input pattern (used in Step 7b) is the sum of the squares of the errors at each
output unit.

Training Algorithm for MADALINE (MRII).
Step 0.   Initialize weights:
            Set the learning rate α.
Step 1.   While stopping condition is false, do Steps 2–8.
            Step 2.   For each bipolar training pair, **s**:**t**, do Steps 3–7.
            Step 3–6.   Compute output of net as in the MRI algorithm.
            Step 7.   Determine error and update weights if necessary:

If $t \neq y$, do Steps 7a–b for each hidden unit whose net input is sufficiently close to 0 (say, between $-.25$ and $.25$). Start with the unit whose net input is closest to 0, then for the next closest, etc.

Step 7a.   Change the unit's ouput
(from $+1$ to -1, or vice versa).

Step 7b.   Recompute the response of the net.
If the error is reduced:
adjust the weights on this unit
(use its newly assigned output value
as target and apply the Delta Rule).

Step 8.   Test stopping condition.
If weight changes have stopped (or reached an acceptable level), or if a specified maximum number of weight update iterations (Step 2) have been performed, then stop; otherwise continue.

A further modification is the possibility of attempting to modify pairs of units at the first layer after all of the individual modifications have been attempted. Similarly adaptation could then be attempted for triplets of units.

Application

Example 2.20 Training a MADALINE for the XOR function

This example illustrates the use of the MRI algorithm to train a MADALINE to solve the XOR problem. Only the computations for the first weight updates are shown.
The training patterns are:

x_1	x_2	t
1	1	-1
1	-1	1
-1	1	1
-1	-1	-1

Step 0.
The weights into Z_1 and into Z_2 are small random values; the weights into Y are those found in Example 2.16. The learning rate, α, is .5.

Weights into Z_1			Weights into Z_2			Weights into Y		
w_{11}	w_{21}	b_1	w_{12}	w_{22}	b_2	v_1	v_2	b_3
.05	.2	.3	.1	.2	.15	.5	.5	.5

Step 1.   Begin training.

Step 2.   For the first training pair, (1, 1): -1

Step 3. $x_1 = 1$, $x_2 = 1$

Step 4. $z\_in_1 = .3 + .05 + .2 = .55$,
$z\_in_2 = .15 + .1 + .2 = .45$.

Step 5. $z_1 = 1,$
             $z_2 = 1.$
Step 6. $y\_in = .5 + .5 + .5;$
             $y\quad = 1.$
Step 7. $t - y = -1 - 1 = -2 \neq 0,$ so an error occurred.
Since $t = -1,$ and both Z units have positive net input,
update the weights on unit Z_1 as follows:

$$b_1(\text{new}) = b_1(\text{old}) + \alpha(-1 - z\_in_1)$$

$$= .3 + (.5)(-1.55)$$

$$= -.475$$

$$w_{11}(\text{new}) = w_{11}(\text{old}) + \alpha(-1 - z\_in_1)x_1$$

$$= .05 + (.5)(-1.55)$$

$$= -.725$$

$$w_{21}(\text{new}) = w_{21}(\text{old}) + \alpha(-1 - z\_in_1)x_2$$

$$= .2 + (.5)(-1.55)$$

$$= -.575$$

update the weights on unit Z_2 as follows:

$$b_2(\text{new}) = b_2(\text{old}) + \alpha(-1 - z\_in_2)$$

$$= .15 + (.5)(-1.45)$$

$$= -.575$$

$$w_{12}(\text{new}) = w_{12}(\text{old}) + \alpha(-1 - z\_in_2)x_1$$

$$= .1 + (.5)(-1.45)$$

$$= -.625$$

$$w_{22}(\text{new}) = w_{22}(\text{old}) + \alpha(-1 - z\_in_2)x_2$$

$$= .2 + (.5)(-1.45)$$

$$= -.525$$

After four epochs of training, the final weights are found to be:

$$w_{11} = -0.73 \qquad w_{12} = \quad 1.27$$
$$w_{21} = \quad 1.53 \qquad w_{22} = -1.33$$
$$b_1 = -0.99 \qquad b_2 = -1.09$$

Example 2.21 Geometric interpretation of MADALINE weights

The positive response region for the Madaline trained in the previous example is the
union of the regions where each of the hidden units have a positive response. The

decision boundary for each hidden unit can be calculated as described in Section 2.1.3.

For hidden unit Z_1, the boundary line is

$$x_2 = -\frac{w_{11}}{w_{21}} x_1 - \frac{b_1}{w_{21}}$$

$$= \frac{0.73}{1.53} x_1 + \frac{0.99}{1.53}$$

$$= 0.48 \, x_1 + 0.65$$

For hidden unit Z_2, the boundary line is

$$x_2 = -\frac{w_{12}}{w_{22}} x_1 - \frac{b_2}{w_{22}}$$

$$= \frac{1.27}{1.33} x_1 + \frac{1.09}{1.33}$$

$$= 0.96 \, x_1 - 0.82$$

These regions are shown in Figures 2.25 and 2.26. The response diagram for the MADALINE is illustrated in Figure 2.27.

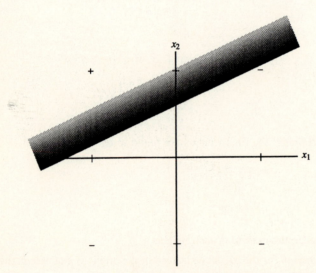

Figure 2.25 Positive response region for Z_1.

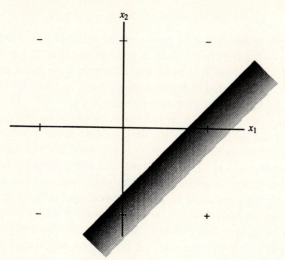

Figure 2.26 Positive response region for Z_2.

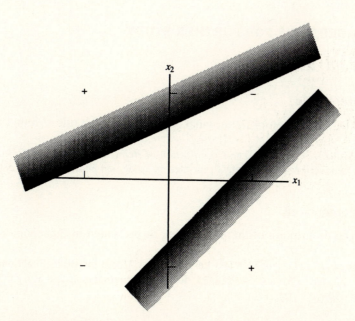

Figure 2.27 Positive response region for MADALINE for XOR function.

Discussion

The construction of sample multilayer nets may provide insights into the appropriate choice of parameters for multilayer nets in general, such as those trained using backpropagation (discussed in Chapter 6). For example, if the input patterns fall into regions that can be bounded (approximately) by a number of lines or planes, then the number of hidden units can be estimated.

It is possible to construct a net with $2p$ hidden units (in a single layer) that will learn p bipolar input training patterns (each with an associated bipolar target value) perfectly. Of course, that is not the primary (or at least not the exclusive) goal of neural nets; generalization is also important and will not be particularly good with so many hidden units. In addition, the training time and the number of interconnections will be unnecessarily large. However, $2p$ certainly gives an upper bound on the number of hidden units we might consider using.

For input that is to be assigned to different categories (the kind of input we have been considering in this chapter), we see that the regions which each neuron separates are bounded by straight lines. Closed regions (convex polygons) can be bounded by taking the intersection of several half-planes (bounded by the separating lines described earlier). Thus a net with one hidden layer (with p units) can learn a response region bounded by p straight lines. If responses in the same category occur in more than one disjoint region of the input space, an additional hidden layer to combine these regions will make training easier.

2.5 SUGGESTIONS FOR FURTHER STUDY

2.5.1 Readings

Hebb rule

The description of the original form of the Hebb rule is found in

HEBB, D. O. (1949). *The Organization of Behavior*. New York: John Wiley & Sons. Introduction and Chapter 4 reprinted in Anderson and Rosenfeld (1988), pp. 45–56.

Perceptrons

The description of the perceptron, as presented in this chapter, is based on

BLOCK, H. D. (1962). "The Perceptron: A Model for Brain Functioning, I." *Reviews of Modern Physics*, 34:123–135. Reprinted in Anderson and Rosenfeld (1988), pp. 138–150.

There are many types of perceptrons; for more complete coverage, see:

MINSKY, M. L., & S. A. PAPERT. (1988). *Perceptrons, Expanded Edition.* Cambridge, MA: MIT Press. Original Edition, 1969.

ROSENBLATT, F. (1958). "The Perceptron: A Probabilistic Model for Information Storage and Organization in the Brain." *Psychological Review,* 65:386–408. Reprinted in Anderson and Rosenfeld (1988), pp. 92–114.

ROSENBLATT, F. (1962). *Principles of Neurodynamics.* New York: Spartan.

ADALINE and MADALINE

For further discussion of ADALINE and MADALINE, see

WIDROW, B., & M. E. HOFF, JR. (1960). "Adaptive Switching Circuits." *IRE WESCON Convention Record,* part 4, pp. 96–104. Reprinted in Anderson and Rosenfeld (1988), pp. 126–134.

WIDROW, B., and S. D. STEARNS. (1985). *Adaptive Signal Processing.* Englewood Cliffs, NJ: Prentice-Hall.

WIDROW, B. & M. A. LEHR. (1990). "30 Years of Adaptive Neural Networks: Perceptron, MADALINE, and Backpropagation," *Proceeding of the IEEE,* 78(9):1415–1442.

2.5.2. Exercises

Hebb net

2.1 Apply the Hebb rule to the training patterns that define the XOR function.

2.2 There are 16 different logic functions (with two inputs and one output), of which 14 are linearly separable. Show that the Hebb rule can find the weights for all problems for which weights exist, as long as bipolar representation is used and a bias is included.

2.3 Consider character recognition using the Hebb rule. In Example 2.8, the "X" and "O" used for training differed in all but four components. Show that the net will respond correctly to an input vector formed from either the "X" or the "O" with up to 20 components missing. (Whether it responds correctly, of course, is based on your knowing which pattern you started with—you might prefer an "I don't know" response. However, since the net input to the output unit is smaller the more components are missing, the "degree of certainty" of the response can also be judged.)

Mistakes involve one or more pixels switching from the value in the original training pattern to the opposite value. Show that at approximately 10 mistakes, the net will be seriously confused. (The exact number depends on whether any of the mistakes occur in the pixels where the training patterns differ.)

Example 2.8 could be rephrased as an example of distinguishing the pattern X from not-X (rather than specifically detecting X versus O). Another pattern (for not-X) that might be added to the net is:

```
 ·  ·  #  ·  ·
 ·  #  ·  #  ·
 ·  #  #  #  ·
 #  ·  ·  ·  #
 #  ·  ·  ·  #
```

Find the new weights to store this pattern together with the two patterns used in Example 2.8. (You can work from the weights in the example.) What does the bias value tell you? How does the ability of the net to respond to noisy (missing or mistaken) data change as you add more patterns to the net? (You will probably need to try more patterns of your own choosing in order to form a good conjecture to answer this question.)

2.4 Create more letters, or different versions of X's and O's, for more training or testing of the Hebb net.

2.5 **a.** Using the Hebb rule, find the weights required to perform the following classifications: Vectors $(1, 1, 1, 1)$ and $(-1, 1, -1, -1)$ are members of the class (and therefore have target value 1); vectors $(1, 1, 1, -1)$ and $(1, -1, -1, 1)$ are not members of the class (and have target value -1).

b. Using each of the training **x** vectors as input, test the response of the net.

2.6 **a.** The Hebb rule is sometimes used by converting the binary training patterns (inputs and targets) to bipolar form to find the weight matrix. Apply this procedure to find the weights to store the following classifications:

$$s(1) = (1, 0, 1) \quad t(1) = 1$$

$$s(2) = (1, 1, 0) \quad t(2) = 0$$

b. Using the binary step function (with threshold 0) as the output unit's activation function, test the response of your network on each of the binary training patterns.

c. Using the bipolar step function (with threshold 0) as the output unit's activation function, convert the training patterns to bipolar form and test the network response again.

d. Test the response of your network on each of the following noisy versions of the bipolar form of the training patterns:

$(0, -1, 1)$ $(0, 1, -1)$ $(0, \ 0, 1)$ $(0, 0, -1)$ $(0, 1, 0)$ $(0, -1, 0)$
$(1, \ 0, 1)$ $(1, 0, -1)$ $(1, -1, 0)$ $(1, 0, \ 0)$ $(1, 1, 0)$ $(1, \ 1, 1)$

Which of the responses are correct, which are incorrect, and which are indefinite (undetermined)?

Perceptron

2.7 Graph the changes in separating lines as they occur in Example 2.12.

2.8 Explore the influence of the value of the learning rate on the speed of convergence of perceptron learning:

a. Consider different values of α in Example 2.12; explain your results.

b. Modify the proof of the perceptron learning rule convergence theorem to include an arbitrary learning rate α.

2.9 Show that the use of a bias is essential in Example 2.11. That is, show that it is impossible to find weights w_1 and w_2 for which the points $(1,1)$, $(1,0)$, $(0,1)$, and $(0,0)$ are classified correctly. First, show that $(0,0)$ will never be classified correctly, and in fact, no learning will ever occur for that point. Then, neglecting $(0,0)$, consider whether $(1,1)$, $(1,0)$, and $(0,1)$ can be classified correctly. That is, do weights w_1 and w_2 exist such that

$$(1)w_1 + (1)w_2 > \quad \theta > 0,$$

$$(1)w_1 + (0)w_2 < -\theta < 0,$$

$$(0)w_1 + (1)w_2 < -\theta < 0.$$

2.10 Show that small initial weights still allow for any position of the initial decision line for the perceptron.

2.11 Repeat Example 2.11, and show that there is no change in the training process if $\theta = 0$. Show that the separating line is

$$x_2 = -\frac{2}{3}x_1 + \frac{4}{3}.$$

2.12 Consider carefully the difference in what can be solved using the following activation functions:

$$f = \begin{cases} 1 & \text{if net} \geq \theta \\ 0 & \text{otherwise} \end{cases}$$

or

$$f = \begin{cases} 1 & \text{if net} \geq \theta \\ -1 & \text{otherwise} \end{cases}$$

or

$$f = \begin{cases} 1 & \text{if net} \geq \theta \\ 0 & \text{if } -\theta < \text{net} < \theta \\ -1 & \text{if net} \leq -\theta \end{cases}$$

2.13 Even for $\theta = 0$, the perceptron learning rule prevents the correct classification of a point on the dividing line (which is better than assigning it arbitrarily to either side of the line). If $\theta < \alpha$ (the learning rate), does the exact value of θ matter? Does it matter if $\theta > \alpha$? Does it make a difference whether we start with all initial weights equal to 0, as in Examples 2.11–2.13, or with other values (small random numbers, for instance)?

2.14 A variation of the perceptron learning rule allows active input units to increase their weights and inactive units to decrease their weights in such manner that the total weights are constant [see Block, 1962, p. 144, footnote 50]. Consider the effect this would have on the binary representation of the AND function in Example 2.11.

2.15 Using the perceptron learning rule, find the weights required to perform the following classifications: Vectors $(1, 1, 1, 1)$ and $(-1, 1, -1, -1)$ are members of the class (and therefore have target value 1); vectors $(1, 1, 1, -1)$ and $(1, -1, -1, 1)$ are not members of the class (and have target value -1). Use a learning rate of 1 and starting weights of 0. Using each of the training **x** vectors as input, test the response of the net.

ADALINE and MADALINE

2.16 Repeat Examples 2.18 and 2.19 using binary rather than bipolar vectors.

2.17 Construct a multilayer net with two hidden units that will learn a given (binary) input pattern perfectly. The first hidden unit will have its weights equal to the input pattern and its threshold equal to the number of 1's in the input pattern. The second hidden unit is designed so that it will fire if its net input is less than or equal to the number of 1's in a given pattern. Combine the output from these two hidden units so that the output unit will fire if both hidden units are on. The point of this exercise is to observe that for p input training patterns, $2p$ hidden units will allow the net to learn every training pattern perfectly.

2.18 The XOR function can be represented as

$$x_1 \text{ XOR } x_2 \Leftrightarrow (x_1 \text{ OR } x_2) \text{ AND NOT } (x_1 \text{ AND } x_2).$$

Construct a MADALINE to implement this formulation of XOR, and compare it with the MADALINE in Example 2.21.

2.19 Using the delta rule, find the weights required to perform the following classifications: Vectors $(1, 1, 1, 1)$ and $(-1, 1, -1, -1)$ are members of the class (and therefore have target value 1); vectors $(1, 1, 1, -1)$ and $(1, -1, -1, 1)$ are not members of the class (and have target value -1). Use a learning rate of .5 and starting weights of 0. Using each of the training **x** vectors as input, test the response of the net.

2.5.3 Projects

Perceptron

2.1 Write a computer program to classify letters from different fonts using perceptron learning. Use as many output units as you have different letters in your training set. Convert the letters to bipolar form. (You may wish to enter the letters as "2" if the pixel is on and "0" if it is off to facilitate testing with noisy patterns after training; your program should subtract 1 from each component of the input pattern to obtain the bipolar vector.)

    a. Repeat Example 2.15 for several values of the threshold θ. After training with each value, test the ability of the net to classify noisy versions of the training patterns. Try 5, 10, 15, 20 pixels wrong and the same levels of missing data. Do higher values of θ have any effect on how often the net is "confused"? Do you reach a value of θ for which the net cannot learn all of the training patterns?

    b. Experiment with other letters. Are some combinations harder to learn than others? Why?

ADALINE and MADALINE

2.2 Write a computer program to classify several letters using delta rule learning. Follow the directions in Project 2.1 (except for the reference to different values of θ.) Compare the ability of the trained ADALINE to classify noisy input to the results for the perceptron.

2.3 Write a computer program to train a MADALINE to perform the XOR function, using the MRI algorithm. What effect do different learning rates have on the weights?

CHAPTER 3

Pattern Association

To a significant extent, learning is the process of forming associations between related patterns. Aristotle observed that human memory connects items (ideas, sensations, etc.) that are similar, that are contrary, that occur in close proximity, or that occur in close succession [Kohonen, 1987]. The patterns we associate together may be of the same type or sensory modality (e.g., a visual image may be associated with another visual image) or of different types (e.g., the fragrance of fresh-mown grass may be associated with a visual image or a feeling). Memorization of a pattern (or a group of patterns) may be considered to be associating the pattern with itself.

An important characteristic of the associations we form is that an imput stimulus which is similar to the stimulus for the association will invoke the associated response pattern. Suppose, for example, that we have learned to read music, so that we associate with a printed note the corresponding pitch, or note on the piano keyboard, or fingering on a stringed instrument. We do not need to see precisely the same form of the musical note we originally learned in order to have the correct association; if the note is larger, or if it is handwritten, we recognize that it still means the same as before. In fact, it is not unusual after learning a few notes to be able to make a good guess at the appropriate response for a new note.

Our ability to recognize a person (either in person or from a photo) is an example of the capability of human memory to respond to new situations. It is relatively difficult to program a traditional computer to perform this task that we

101

do so easily. The variations in the person's appearance are virtually unlimited. How do we then describe to a sequential logical computer the process of deciding whether this is someone we know and, if so, which of our many acquaintances it is?

In this chapter, we shall consider some relatively simple (single-layer) neural networks that can learn a set of pattern pairs (or associations). An associative memory net may serve as a highly simplified model of human memory [see the early work by Kohonen (1972) and Anderson (1968, 1972)]; however, we shall not address the question whether they are at all realistic models. Associative memories also provide one approach to the computer engineering problem of storing and retrieving data based on content rather than storage address. Since information storage in a neural net is distributed throughout the system (in the net's weights), a pattern does not have a storage address in the same sense that it would if it were stored in a traditional computer.

Associative memory neural nets are single-layer nets in which the weights are determined in such a way that the net can store a set of pattern associations. Each association is an input-output vector pair, $s:t$. If each vector t is the same as the vector s with which it is associated, then the net is called an *autoassociative memory*. If the t's are different from the s's, the net is called a *heteroassociative memory*. In each of these cases, the net not only learns the specific pattern pairs that were used for training, but also is able to recall the desired response pattern when given an input stimulus that is similar, but not identical, to the training input. Autoassociative neural nets are presented in Sections 3.3 and 3.4. Heteroassociative neural nets are described in Sections 3.2 and 3.5.

Before training an associative memory neural net, the original patterns must be converted to an appropriate representation for computation. However, not all representations of the same pattern are equally powerful or efficient. In a simple example, the original pattern might consist of "on" and "off" signals, and the conversion could be "on" $\rightarrow +1$, "off" $\rightarrow 0$ (binary representation) or "on" $\rightarrow +1$, "off" $\rightarrow -1$ (bipolar representation). In many of our examples, we assume the conversion has already been made.

In the first section of this chapter, two common training methods for single-layer nets are presented. The Hebb rule and delta rule were introduced in Chapter 2 for pattern classification. They are described here in more general terms.

The architecture of an associative memory neural net may be feedforward or recurrent (iterative). In a feedforward net, information flows from the input units to the output units; in a recurrent net, there are connections among the units that form closed loops. Feedforward heteroassociative nets are considered in Section 3.2, and feedforward autoassociative nets are discussed in Section 3.3. Iterative autoassociative neural nets are described in Section 3.4, and iterative heteroassociative nets are considered in Section 3.5.

A key question for all associative nets is how many patterns can be stored (or pattern pairs learned) before the net starts to "forget" patterns it has learned previously. As with human memory, a number of factors influence how much the associative memory can learn. The complexity of the patterns (the number of components) and the similarity of the input patterns that are associated with significantly different response patterns both play a role. We shall consider a few of these ideas in Section 3.3.4.

3.1 TRAINING ALGORITHMS FOR PATTERN ASSOCIATION

3.1.1 Hebb Rule for Pattern Association

The Hebb rule is the simplest and most common method of determining the weights for an associative memory neural net. It can be used with patterns that are represented as either binary or bipolar vectors. We repeat the algorithm here for input and output training vectors (only a slight extension of that given in the previous chapter) and give the general procedure for finding the weights by outer products. Since we want to consider examples in which the input to the net after training is a pattern that is similar to, but not the same as, one of the training inputs, we denote our training vector pairs as $s:t$. We then denote our testing input vector as x, which may or may not be the same as one of the training input vectors.

Algorithm

Step 0.   Initialize all weights ($i = 1, \ldots, n; j = 1, \ldots, m$):

$$w_{ij} = 0.$$

Step 1.   For each input training–target output vector pair $s:t$, do Steps 2–4.

    Step 2.   Set activations for input units to current training input ($i = 1, \ldots, n$):

$$x_i = s_i$$

    Step 3.   Set activations for output units to current target output ($j = 1, \ldots, m$):

$$y_j = t_j.$$

    Step 4.   Adjust the weights ($i = 1, \ldots, n; j = 1, \ldots, m$):

$$w_{ij}(\text{new}) = w_{ij}(\text{old}) + x_i y_j.$$

The foregoing algorithm is not usually used for this simple form of Hebb learning, since weights can be found immediately, as shown in the next section. However, it illustrates some characteristics of the typical *neural* approach to learning. Algorithms for some modified forms of Hebb learning will be discussed in Chapter 7.

Outer products

The weights found by using the Hebb rule (with all weights initially 0) can also be described in terms of outer products of the input vector–output vector pairs. The outer product of two vectors

$$\mathbf{s} = (s_1, \ldots, s_i, \ldots, s_n)$$

and

$$\mathbf{t} = (t_1, \ldots, t_j, \ldots, t_m)$$

is simply the matrix product of the $n \times 1$ matrix $\mathbf{S} = \mathbf{s}^T$ and the $1 \times m$ matrix $\mathbf{T} = \mathbf{t}$:

$$\mathbf{ST} = \begin{bmatrix} s_1 \\ \vdots \\ s_i \\ \vdots \\ s_n \end{bmatrix} [t_1 \, .. \, t_j \, .. \, t_m] = \begin{bmatrix} s_1 t_1 & \cdots & s_1 t_j & \cdots & s_1 t_m \\ \vdots & \cdot & \vdots & \cdot & \vdots \\ s_i t_1 & \cdots & s_i t_j & \cdots & s_i t_m \\ \vdots & \cdot & \vdots & \cdot & \vdots \\ s_n t_1 & \cdots & s_n t_j & \cdots & s_n t_m \end{bmatrix}.$$

This is just the weight matrix to store the association $\mathbf{s}:\mathbf{t}$ found using the Hebb rule.

To store a set of associations $\mathbf{s}(p) : \mathbf{t}(p)$, $p = 1, \ldots, P$, where

$$\mathbf{s}(p) = (s_1(p), \ldots, s_i(p), \ldots, s_n(p))$$

and

$$\mathbf{t}(p) = (t_1(p), \ldots, t_j(p), \ldots, t_m(p)),$$

the weight matrix $\mathbf{W} = \{w_{ij}\}$ is given by

$$w_{ij} = \sum_{p=1}^{P} s_i(p) t_j(p).$$

This is the sum of the outer product matrices required to store each association separately.

In general, we shall use the preceding formula or the more concise vector-matrix form,

$$\mathbf{W} = \sum_{p=1}^{P} \mathbf{s}^T(p)\mathbf{t}(p),$$

to set the weights for a net that uses Hebb learning. This weight matrix is described by a number of authors [see, e.g., Kohonen, 1972, and Anderson, 1972].

Perfect recall versus cross talk

The suitability of the Hebb rule for a particular problem depends on the correlation among the input training vectors. If the input vectors are uncorrelated (orthogonal), the Hebb rule will produce the correct weights, and the response of the net when tested with one of the training vectors will be perfect recall of the input vector's associated target (scaled by the square of the norm of the input vector in question). If the input vectors are not orthogonal, the response will include a portion of each of their target values. This is commonly called *cross talk*. As shown in some of the examples of Chapter 2, in some cases the cross talk is mild enough that the correct response will still be produced for the stored vectors.

To see why cross talk occurs, recall that two vectors $s(k)$ and $s(p)$, $k \neq p$, are orthogonal if their dot product is 0. This can be written in a number of ways, including (if, as we assume, $s(k)$ and $s(p)$ are row vectors)

$$s(k)s^T(p) = 0,$$

or

$$\sum_{i=1}^{n} s_i(k)s_i(p) = 0.$$

Now consider the weight matrix $\mathbf{W}$, defined as before to store a set of input-target vector pairs. The response of the net (with the identity function for the activation function rather than the threshold function) is $\mathbf{y} = \mathbf{xW}$. If the (testing) input signal is the kth training input vector, i.e., if

$$\mathbf{x} = s(k),$$

then the response of the net is

$$s(k)\mathbf{W} = \sum_{p=1}^{P} s(k)s^T(p)t(p)$$

$$= s(k)s^T(k)t(k) + \sum_{p \neq k} s(k)s^T(p)t(p).$$

If $s(k)$ is orthogonal to $s(p)$ for $p \neq k$, then there will be no contribution to the response from any of the terms in the summation; the response will then be the target vector $t(k)$, scaled by the square of the norm of the input vector, i.e., $s(k)s^T(k)$.

However, if $s(k)$ is not orthogonal to the other s-vectors, there will be contributions to the response that involve the target values for each of the vectors to which $s(k)$ is not orthogonal.

Summary

If a threshold function is applied to the response of a net, as described here, and the cross talk is not too severe, the weights found by the Hebb rule may still be satisfactory. (See Examples 2.7, 2.8, and 3.1.)

Several authors normalize the weights found by the Hebb rule by a factor of $1/n$, where n is the number of units in the system [Hertz, Krogh, & Palmer, 1991; McClelland & Rumelhart, 1988]. As the latter observe, the use of normalization can preserve the interpretation of the weight w_{ij} as representing the correlation between the activation of unit x_i and unit y_j, even when the activations have means that are different from zero.

There are three aspects of a particular association problem that influence whether the Hebb rule will be able to find weights to produce the correct output for all training input patterns. The first is simply whether such weights exist. If the input vectors are linearly independent, they do exist. This is the extension of linear separability to the case of vector outputs. Second, there is the question of correlation between (pairs of) the input training vectors. If the input vectors are uncorrelated (orthogonal), then the weight matrix found by the Hebb rule will give perfect results for each of the training input vectors. Finally, because the weights of the Hebb rule represent the extent to which input units and output units should be ''on'' or ''off'' at the same time, patterns in which the input activations are correlated strongly, unit by unit, with the target values will be able to be learned by the Hebb rule. For this reason, the Hebb rule is also known as correlation training or encoding.

3.1.2 Delta Rule for Pattern Association

The delta rule is an iterative learning process introduced by Widrow and Hoff (1960) for the ADALINE neuron (see Chapter 2). The rule can be used for input patterns that are linearly independent but not orthogonal. Mappings in which the input vectors are linearly independent can be solved using single-layer nets as described in this chapter. However, the delta rule is needed to avoid the difficulties of cross talk which may be encountered if a simpler learning rule, such as the Hebb rule, is used. Furthermore, the delta rule will produce the least squares solution when input patterns are not linearly independent [Rumelhart, McClelland, & the PDP Research Group, 1986].

In its original form, as introduced in Chapter 2, the delta rule assumed that the activation function for the output unit was the identity function. A simple extension allows for the use of any differentiable activation function; we shall call this the extended delta rule, since some authors use the term ''generalized delta rule'' synonymously with ''backpropagation'' for multilayer nets. The nomenclature we use is as follows:

α learning rate.
$\mathbf{x}$ training input vector.
$\mathbf{t}$ target output for input vector $\mathbf{x}$.

Original delta rule

The original delta rule, for several output units, was derived in Section 2.4.4. It is repeated here for comparison with the extended delta rule described in the next section. The original rule assumes that the activation function for the output units is the identity function; or, equivalently, it minimizes the square of the difference between the net input to the output units and the target values. Thus, using **y** for the computed output for the input vector **x**, we have

$$y_j = \sum_i x_i w_{ij},$$

and the weight updates are

$$w_{ij}(\text{new}) = w_{ij}(\text{old}) + \alpha(t_j - y_j)x_i \quad (i = 1, \ldots, n; j = 1, \ldots, m).$$

This is often expressed in terms of the weight change

$$\left\{ \Delta w_{ij} = \alpha(t_j - y_j)x_i, \right\}$$

which explains why this training rule is called the delta rule.

Extended delta rule

This minor modification of the delta rule derived in Chapter 2 allows for an arbitrary, differentiable activation function to be applied to the output units. The update for the weight from the Ith input unit to the Jth output unit is

$$\Delta w_{IJ} = \alpha(t_J - y_J)x_I f'(y\_in_J).$$

Derivation.   The derivation of the extended delta rule given here follows the discussion in Section 2.4.4 closely. However, we now wish to change the weights to reduce the difference between the computed output and the target value, rather than between the net input to the output unit(s) and the target(s).

The squared error for a particular training pattern is

$$E = \sum_{j=1}^{m} (t_j - y_j)^2.$$

E is a function of all of the weights. The gradient of E is a vector consisting of the partial derivatives of E with respect to each of the weights. This vector gives the direction of most rapid increase in E; the opposite direction gives the direction of most rapid decrease in E. The error can be reduced most rapidly by adjusting the weight w_{IJ} in the direction of $-\partial E/\partial w_{IJ}$.

We now find an explicit formula for $\partial E/\partial w_{IJ}$ for the arbitrary weight w_{IJ}. First note that

$$\frac{\partial E}{\partial w_{IJ}} = \frac{\partial}{\partial w_{IJ}} \sum_{j=1}^{m} (t_j - y_j)^2$$

$$= \frac{\partial}{\partial w_{IJ}} (t_J - y_J)^2,$$

since the weight w_{IJ} only influences the error at output unit Y_J.

Furthermore, using the facts that

$$y\_in_J = \sum_{i=1}^{n} x_i w_{iJ} \quad \text{and}$$

$$y_J = f(y\_in_J),$$

we have

$$\frac{\partial E}{\partial w_{IJ}} = -2(t_J - y\_in_J) \frac{\partial y\_in_J}{\partial w_{IJ}}$$

$$= -2(t_J - y\_in_J) x_I f'(y\_in_J).$$

Thus the local error will be reduced most rapidly (for a given learning rate α) by adjusting the weights according to the delta rule

$$\Delta w_{IJ} = \alpha(t_J - y_J) x_I f'(y\_in_J).$$

3.2 HETEROASSOCIATIVE MEMORY NEURAL NETWORK

Associative memory neural networks are nets in which the weights are determined in such a way that the net can store a set of P pattern associations. Each association is a pair of vectors $(\mathbf{s}(p), \mathbf{t}(p))$, with $p = 1, 2, \ldots, P$. Each vector $\mathbf{s}(p)$ is an n-tuple (has n components), and each $\mathbf{t}(p)$ is an m-tuple. The weights may be found using the Hebb rule (Sections 3.1.1) or the delta rule (Section 3.1.2). In the examples in this section, the Hebb rule is used. The net will find an appropriate output vector that corresponds to an input vector $\mathbf{x}$ that may be either one of the stored patterns $\mathbf{s}(p)$ or a new pattern (such as one of the training patterns corrupted by noise).

3.2.1 Architecture

The architecture of a heteroassociative memory neural network is shown in Figure 3.1.

3.2.2 Application

Procedure

Step 0.   Initialize weights using either the Hebb rule (Section 3.1.1) or the delta rule (Section 3.1.2).

Step 1.   For each input vector, do Steps 2–4.

   Step 2.   Set activations for input layer units equal to the current input vector

$$x_i.$$

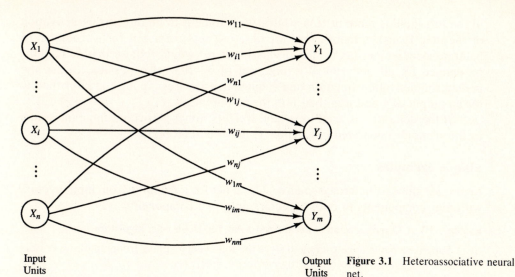

Input
Units

Output
Units

Figure 3.1 Heteroassociative neural
net.

Step 3.   Compute net input to the output units:

$$y\_in_j = \sum_i x_i w_{ij}.$$

Step 4.   Determine the activation of the output units:

$$y_j = \begin{cases} 1 & \text{if } y\_in_j > 0 \\ 0 & \text{if } y\_in_j = 0 \\ -1 & \text{if } y\_in_j < 0\,, \end{cases}$$

(for bipolar targets).

The output vector **y** gives the pattern associated with the input vector **x**. This heteroassociative memory is not iterative.

Other activation functions can also be used. If the target responses of the net are binary, a suitable activation function is given by

$$f(x) = \begin{cases} 1 & \text{if } x > 0; \\ 0 & \text{if } x \le 0. \end{cases}$$

A general form of the preceding activation function that includes a threshold θ_i and that is used in the bidirectional associative memory (BAM), an iterative net discussed in Section 3.5, is

$$y_j = \begin{cases} 1 & \text{if } y\_in_j > \theta_j \\ y_j & \text{if } y\_in_j = \theta_j\,. \\ -1 & \text{if } y\_in_j < \theta_j \end{cases}$$

The choice of the desired response for a neuron if its net input is exactly equal to the threshold is more or less arbitrary; defining it to be the current activation of the unit Y_j makes more sense for iterative nets; to use it for a feedforward heteroassociative net (as we are discussing here) would require that activations be defined for all units initially (e.g., set to 0). It is also possible to use the perceptron activation function and require that the net input be greater than θ_j for an output of 1 and less than $-\theta_j$ for an output of -1.

   If the delta rule is used to set the weights, other activation functions, such as the sigmoids illustrated in Chapter 1, may be appropriate.

Simple examples

Figure 3.2 shows a heteroassociative neural net for a mapping from input vectors with four components to output vectors with two components.

Example 3.1 A Heteroassociative net trained using the Hebb rule: algorithm

Suppose a net is to be trained to store the following mapping from input row vectors $s = (s_1, s_2, s_3, s_4)$ to output row vectors $t = (t_1, t_2)$:

		s_1	s_2	s_3	s_4			t_1	t_2
1st	s	(1,	0,	0,	0)	1st	t	(1,	0)
2nd	s	(1,	1,	0,	0)	2nd	t	(1,	0)
3rd	s	(0,	0,	0,	1)	3rd	t	(0,	1)
4th	s	(0,	0,	1,	1)	4th	t	(0,	1)

These target patterns are simple enough that the problem could be considered one in pattern classification; however, the process we describe here does not require that only one of the two output units be "on." Also, the input vectors are not mutually orthogonal. However, because the target values are chosen to be related to the input vectors in a particularly simple manner, the cross talk between the first

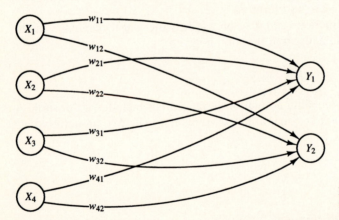

Figure 3.2 Heteroassociative neural net for simple examples.

and second input vectors does not pose any difficulties (since their target values are the same).

The training is accomplished by the Hebb rule, which is defined as

$$w_{ij}(\text{new}) = w_{ij}(\text{old}) + s_i t_j; \quad \text{i.e., } \Delta w_{ij} = s_i t_j.$$

Training

The results of applying the algorithm given in Section 3.1.1 are as follows (only the weights that change at each step of the process are shown):

Step 0.   Initialize all weights to 0.
Step 1.   For the first **s**:**t** pair $(1, 0, 0, 0):(1, 0)$:
      Step 2. $x_1 = 1; \quad x_2 = x_3 = x_4 = 0.$
      Step 3. $y_1 = 1; \quad y_2 = 0.$
      Step 4. $w_{11}(\text{new}) = w_{11}(\text{old}) + x_1 y_1 = 0 + 1 = 1.$
      (All other weights remain 0.)
Step 1.   For the second **s**:**t** pair $(1, 1, 0, 0):(1, 0)$:
      Step 2. $x_1 = 1; \quad x_2 = 1; \quad x_3 = x_4 = 0.$
      Step 3. $y_1 = 1; \quad y_2 = 0.$
      Step 4. $w_{11}(\text{new}) = w_{11}(\text{old}) + x_1 y_1 = 1 + 1 = 2;$
              $w_{21}(\text{new}) = w_{21}(\text{old}) + x_2 y_1 = 0 + 1 = 1.$
      (All other weights remain 0.)
Step 1.   For the third **s**:**t** pair $(0, 0, 0, 1):(0, 1)$:
      Step 2. $x_1 = x_2 = x_3 = 0; \quad x_4 = 1.$
      Step 3. $y_1 = 0; \quad y_2 = 1.$
      Step 4. $w_{42}(\text{new}) = w_{42}(\text{old}) + x_4 y_2 = 0 + 1 = 1.$
      (All other weights remain unchanged.)
Step 1.   For the fourth **s**:**t** pair $(0, 0, 1, 1):(0, 1)$:
      Step 2. $x_1 = x_2 = 0; \quad x_3 = 1; \quad x_4 = 1.$
      Step 3. $y_1 = 0; \quad y_2 = 1.$
      Step 4. $w_{32}(\text{new}) = w_{32}(\text{old}) + x_3 y_2 = 0 + 1 = 1;$
              $w_{42}(\text{new}) = w_{42}(\text{old}) + x_4 y_2 = 1 + 1 = 2.$
      (All other weights remain unchanged.)

The weight matrix is

$$\mathbf{W} = \begin{bmatrix} 2 & 0 \\ 1 & 0 \\ 0 & 1 \\ 0 & 2 \end{bmatrix}.$$

Example 3.2 A heteroassociative net trained using the Hebb rule: outer products

This example finds the same weights as in the previous example, but using outer products instead of the algorithm for the Hebb rule. The weight matrix to store the first pattern pair is given by the outer product of the vector

$$\mathbf{s} = (1, 0, 0, 0)$$

and

$$\mathbf{t} = (1, 0).$$

The outer product of a vector pair is simply the matrix product of the training vector written as a column vector (and treated as an $n \times 1$ matrix) and the target vector written as a row vector (and treated as a $1 \times m$ matrix):

$$\begin{bmatrix} 1 \\ 0 \\ 0 \\ 0 \end{bmatrix} \begin{bmatrix} 1 & 0 \end{bmatrix} = \begin{bmatrix} 1 & 0 \\ 0 & 0 \\ 0 & 0 \\ 0 & 0 \end{bmatrix}.$$

Similarly, to store the second pair,

$$\mathbf{s} = (1, \quad 1, \quad 0, \quad 0)$$

and

$$\mathbf{t} = (1, \quad 0),$$

the weight matrix is

$$\begin{bmatrix} 1 \\ 1 \\ 0 \\ 0 \end{bmatrix} \begin{bmatrix} 1 & 0 \end{bmatrix} = \begin{bmatrix} 1 & 0 \\ 1 & 0 \\ 0 & 0 \\ 0 & 0 \end{bmatrix}.$$

To store the third pair,

$$\mathbf{s} = (0, \quad 0, \quad 0, \quad 1)$$

and

$$\mathbf{t} = (0, \quad 1),$$

the weight matrix is

$$\begin{bmatrix} 0 \\ 0 \\ 0 \\ 1 \end{bmatrix} \begin{bmatrix} 0 & 1 \end{bmatrix} = \begin{bmatrix} 0 & 0 \\ 0 & 0 \\ 0 & 0 \\ 0 & 1 \end{bmatrix}.$$

And to store the fourth pair,

$$\mathbf{s} = (0, \quad 0, \quad 1, \quad 1)$$

and

$$\mathbf{t} = (0, \quad 1),$$

the weight matrix is

$$\begin{bmatrix} 0 \\ 0 \\ 1 \\ 1 \end{bmatrix} \begin{bmatrix} 0 & 1 \end{bmatrix} = \begin{bmatrix} 0 & 0 \\ 0 & 0 \\ 0 & 1 \\ 0 & 1 \end{bmatrix}.$$

The weight matrix to store all four pattern pairs is the sum of the weight matrices to store each pattern pair separately, namely,

$$W = \begin{bmatrix} 1 & 0 \\ 0 & 0 \\ 0 & 0 \\ 0 & 0 \end{bmatrix} + \begin{bmatrix} 1 & 0 \\ 1 & 0 \\ 0 & 0 \\ 0 & 0 \end{bmatrix} + \begin{bmatrix} 0 & 0 \\ 0 & 0 \\ 0 & 0 \\ 0 & 1 \end{bmatrix} + \begin{bmatrix} 0 & 0 \\ 0 & 0 \\ 0 & 1 \\ 0 & 1 \end{bmatrix} = \begin{bmatrix} 2 & 0 \\ 1 & 0 \\ 0 & 1 \\ 0 & 2 \end{bmatrix}.$$

Example 3.3 Testing a heteroassociative net using the training input

We now test the ability of the net to produce the correct output for each of the training inputs. The steps are as given in the application procedure at the beginning of this section, using the activation function

$$f(x) = \begin{cases} 1 & \text{if } x > 0; \\ 0 & \text{if } x \le 0. \end{cases}$$

The weights are as found in Examples 3.1 and 3.2.

Step 0. $W = \begin{bmatrix} 2 & 0 \\ 1 & 0 \\ 0 & 1 \\ 0 & 2 \end{bmatrix}.$

Step 1.   For the first input pattern, do Steps 2–4.

    Step 2. $x = (1, 0, 0, 0).$

    Step 3. $y\_in_1 = x_1 w_{11} + x_2 w_{21} + x_3 w_{31} + x_4 w_{41}$

$$= 1(2) + 0(1) + 0(0) + 0(0)$$

$$= 2;$$

$$y\_in_2 = x_1 w_{12} + x_2 w_{22} + x_3 w_{32} + x_4 w_{42}$$

$$= 1(0) + 0(0) + 0(1) + 0(2)$$

$$= 0.$$

    Step 4. $y_1 = f(y\_in_1) = f(2) = 1;$

$$y_2 = f(y\_in_2) = f(0) = 0.$$

    (This is the correct response for the first training pattern.)

Step 1.   For the second input pattern, do Steps 2–4.

    Step 2. $x = (1, 1, 0, 0).$

    Step 3. $y\_in_1 = x_1 w_{11} + x_2 w_{21} + x_3 w_{31} + x_4 w_{41}$

$$= 1(2) + 1(1) + 0(0) + 0(0)$$

$$= 3;$$

$$y\_in_2 = x_1 w_{12} + x_2 w_{22} + x_3 w_{32} + x_4 w_{42}$$

$$= 1(0) + 1(0) + 0(1) + 0(2)$$

$$= 0.$$

    Step 4. $y_1 = f(y\_in_1) = f(3) = 1;$

$$y_2 = f(y\_in_2) = f(0) = 0.$$

    (This is the correct response for the second training pattern.)

Step 1.   For the third input pattern, do Steps 2–4.
    Step 2. $\mathbf{x} = (0, 0, 0, 1)$.
    Step 3. $y\_in_1 = x_1 w_{11} + x_2 w_{21} + x_3 w_{31} + x_4 w_{41}$

$$= 0(2) + 0(1) + 0(0) + 1(0)$$

$$= 0;$$

$$y\_in_2 = x_1 w_{12} + x_2 w_{22} + x_3 w_{32} + x_4 w_{42}$$

$$= 0(0) + 0(0) + 0(1) + 1(2)$$

$$= 2.$$

    Step 4. $y_1 = f(y\_in_1) = f(0) = 0;$

$$y_2 = f(y\_in^2) = f(2) = 1.$$

(This is the correct response for the third training pattern.)

Step 1.   For the fourth input pattern, do Steps 2–4.
    Step 2. $\mathbf{x} = (0, 0, 1, 1)$.
    Step 3. $y\_in_1 = x_1 w_{11} + x_2 w_{21} + x_3 w_{31} + x_4 w_{41}$

$$= 0(2) + 0(1) + 1(0) + 1(0)$$

$$= 0;$$

$$y\_in_2 = x_1 w_{12} + x_2 w_{22} + x_3 w_{32} + x_4 w_{42}$$

$$= 0(0) + 0(0) + 1(1) + 1(2)$$

$$= 3.$$

    Step 4. $y_1 = f(y\_in_1) = f(0) = 0;$

$$y_2 = f(y\_in_2) = f(2) = 1.$$

(This is the correct response for the fourth training pattern.)
The process we have just illustrated can be represented much more succinctly using vector-matrix notation. Note first, that the net input to any particular output unit is the (dot) product of the input (row) vector with the column of the weight matrix that has the weights for the output unit in question. The (row) vector with all of the net inputs is simply the product of the input vector and the weight matrix.

We repeat the steps of the application procedure for the input vector $\mathbf{x}$, which is the first of the training input vectors $\mathbf{s}$.

Step 0. $\mathbf{W} = \begin{bmatrix} 2 & 0 \\ 1 & 0 \\ 0 & 1 \\ 0 & 2 \end{bmatrix}.$

Step 1.   For the input vector:
    Step 2. $\mathbf{x} = (1, 0, 0, 0)$.
    Step 3. $\mathbf{x\,W} = (y\_in_1, y\_in_2)$

$$(1, 0, 0, 0) \begin{bmatrix} 2 & 0 \\ 1 & 0 \\ 0 & 1 \\ 0 & 2 \end{bmatrix} = (2, 0).$$

Step 4. $f(2) = 1$; $f(0) = 0$;
$\mathbf{y} = (1, 0)$.

The entire process (Steps 2–4) can be represented by

$$\mathbf{x}\mathbf{W} = (y\_in_1, y\_in_2) \rightarrow \mathbf{y}$$

$$(1, 0, 0, 0) \begin{bmatrix} 2 & 0 \\ 1 & 0 \\ 0 & 1 \\ 0 & 2 \end{bmatrix} = (2, 0) \qquad \rightarrow (1, 0),$$

or, in slightly more compact notation,

$$(1, 0, 0, 0){\cdot}\mathbf{W} = (2, 0) \rightarrow (1, 0).$$

Note that the output activation vector is the same as the training output vector that was stored in the weight matrix for this input vector.

Similarly, applying the same algorithm, with **x** equal to each of the other three training input vectors, yields

$$(1, 1, 0, 0){\cdot}\mathbf{W} = (3, 0) \rightarrow (1, 0),$$

$$(0, 0, 0, 1){\cdot}\mathbf{W} = (0, 2) \rightarrow (0, 1),$$

$$(0, 0, 1, 1){\cdot}\mathbf{W} = (0, 3) \rightarrow (0, 1).$$

Note that the net has responded correctly to (has produced the desired vector of output activations for) each of the training patterns.

Example 3.4 Testing a heteroassociative net with input similar to the training input

The test vector $\mathbf{x} = (0, 1, 0, 0)$ differs from the training vector $\mathbf{s} = (1, 1, 0, 0)$ only in the first component. We have

$$(0, 1, 0, 0){\cdot}\mathbf{W} = (1, 0) \rightarrow (1, 0).$$

Thus, the net also associates a known output pattern with this input.

Example 3.5 Testing a heteroassociative net with input that is not similar to the training input

The test pattern $(0\ 1, 1, 0)$ differs from each of the training input patterns in at least two components. We have

$$(0\ 1, 1, 0){\cdot}\mathbf{W} = (1, 1) \rightarrow (1, 1).$$

The output is not one of the outputs with which the net was trained; in other words, the net does not recognize the pattern. In this case, we can view $\mathbf{x} = (0, 1, 1, 0)$ as differing from the training vector $\mathbf{s} = (1, 1, 0, 0)$ in the first and third components, so that the two "mistakes" in the input pattern make it impossible for the net to recognize it. This is not surprising, since the vector could equally well be viewed as formed from $\mathbf{s} = (0, 0, 1, 1)$, with "mistakes" in the second and fourth components.

In general, a bipolar representation of our patterns is computationally preferable to a binary representation. Examples 3.6 and 3.7 illustrate modifications of the previous examples to make use of the improved characteristics of bipolar vectors. In the first modification (Example 3.6), binary input and target vectors are converted to bipolar representations for the formation of the weight matrix. However, the input vectors used during testing and the response of the net are still represented in binary form. In the second modification (Example 3.7), all vectors (training input, target output, testing input, and the response of the net) are expressed in bipolar form.

Example 3.6 A heteroassociative net using hybrid (binary/bipolar) data representation

Even if one wishes to use binary input vectors, it may be advantageous to form the weight matrix from the bipolar form of training vector pairs. Specifically, to store a set of binary vector pairs $s(p):t(p)$, $p = 1, \ldots, P$, where

$$s(p) = (s_1(p), \ldots, s_i(p), \ldots, s_n(p))$$

and

$$t(p) = (t_1(p), \ldots, t_j(p), \ldots, t_m(p)),$$

using a weight matrix formed from the corresponding bipolar vectors, the weight matrix $\mathbf{W} = \{w_{ij}\}$ is given by

$$w_{ij} = \sum_p (2s_i(p) - 1)(2t_j(p) - 1).$$

Using the data from Example 3.1, we have

$$s(1) = (1, 0, 0, 0), \qquad t(1) = (1, 0);$$
$$s(2) = (1, 1, 0, 0), \qquad t(2) = (1, 0);$$
$$s(3) = (0, 0, 0, 1), \qquad t(3) = (0, 1);$$
$$s(4) = (0, 0, 1, 1), \qquad t(4) = (0, 1).$$

The weight matrix that is obtained

$$\mathbf{W} = \begin{bmatrix} 4 & -4 \\ 2 & -2 \\ -2 & 2 \\ -4 & 4 \end{bmatrix}.$$

Example 3.7 A heteroassociative net using bipolar vectors

To store a set of bipolar vector pairs $s(p):t(p)$, $p = 1, \ldots, P$, where

$$s(p) = (s_1(p), \ldots, s_i(p), \ldots, s_n(p))$$

and

$$t(p) = (t_1(p), \ldots, t_j(p), \ldots, t_m(p)),$$

the weight matrix $\mathbf{W} = \{w_{ij}\}$ is given by

$$w_{ij} = \sum_p s_i(p)t_j(p).$$

Using the data from Examples 3.1 through 3.6, we have

$$\mathbf{s}(1) = (\ \ 1, -1, -1, -1), \qquad \mathbf{t}(1) = (\ \ 1, -1);$$
$$\mathbf{s}(2) = (\ \ 1, \ \ 1, -1, -1), \qquad \mathbf{t}(2) = (\ \ 1, -1);$$
$$\mathbf{s}(3) = (-1, -1, -1, \ \ 1), \qquad \mathbf{t}(3) = (-1, \ \ 1);$$
$$\mathbf{s}(4) = (-1, -1, \ \ 1, \ \ 1), \qquad \mathbf{t}(4) = (-1, \ \ 1).$$

The same weight matrix is obtained as in Example 3.6, namely,

$$\mathbf{W} = \begin{bmatrix} 4 & -4 \\ 2 & -2 \\ -2 & 2 \\ -4 & 4 \end{bmatrix}.$$

We illustrate the process of finding the weights using outer products for this example.

The weight matrix to store the first pattern pair is given by the outer product of the vectors

$$\mathbf{s} = (1, -1, -1, -1)$$

and

$$\mathbf{t} = (1, -1).$$

The weight matrix is

$$\begin{bmatrix} 1 \\ -1 \\ -1 \\ -1 \end{bmatrix} [1 \quad -1] = \begin{bmatrix} 1 & -1 \\ -1 & 1 \\ -1 & 1 \\ -1 & 1 \end{bmatrix}.$$

Similarly, to store the second pair,

$$\mathbf{s} = (1, 1, -1, -1)$$

and

$$\mathbf{t} = (1, -1),$$

the weight matrix is

$$\begin{bmatrix} 1 \\ 1 \\ -1 \\ -1 \end{bmatrix} [1 \ -1] = \begin{bmatrix} 1 & -1 \\ 1 & -1 \\ -1 & 1 \\ -1 & 1 \end{bmatrix}.$$

To store the third pair,

$$\mathbf{s} = (-1, -1, -1, 1)$$

and

$$\mathbf{t} = (-1, \ \ 1),$$

the weight matrix is

$$\begin{bmatrix} -1 \\ -1 \\ -1 \\ 1 \end{bmatrix} \begin{bmatrix} -1 & 1 \end{bmatrix} = \begin{bmatrix} 1 & -1 \\ 1 & -1 \\ 1 & -1 \\ -1 & 1 \end{bmatrix}.$$

And to store the fourth pair,

$$\mathbf{s} = (-1, -1, 1, 1)$$

and

$$\mathbf{t} = (-1, 1),$$

the weight matrix is

$$\begin{bmatrix} -1 \\ -1 \\ 1 \\ 1 \end{bmatrix} \begin{bmatrix} -1 & 1 \end{bmatrix} = \begin{bmatrix} 1 & -1 \\ 1 & -1 \\ -1 & 1 \\ -1 & 1 \end{bmatrix}.$$

The weight matrix to store all four pattern pairs is the sum of the weight matrices to store each pattern pair separately, namely,

$$\begin{bmatrix} 1 & -1 \\ -1 & 1 \\ -1 & 1 \\ -1 & 1 \end{bmatrix} + \begin{bmatrix} 1 & -1 \\ 1 & -1 \\ -1 & 1 \\ -1 & 1 \end{bmatrix} + \begin{bmatrix} 1 & -1 \\ 1 & -1 \\ 1 & -1 \\ -1 & 1 \end{bmatrix} + \begin{bmatrix} 1 & -1 \\ 1 & -1 \\ -1 & 1 \\ -1 & 1 \end{bmatrix} = \begin{bmatrix} 4 & -4 \\ 2 & -2 \\ -2 & 2 \\ -4 & 4 \end{bmatrix}.$$

One of the computational advantages of bipolar representation of our patterns is that it gives us a very simple way of expressing two different levels of noise that may be applied to our training inputs to produce testing inputs for our net. For convenience, we shall refer informally to these levels as "missing data" and "mistakes." For instance, if each of our original patterns is a sequence of *yes* or *no* responses, "missing data" would correspond to a response of *unsure*, whereas a "mistake" would be a response of *yes* when the correct response was *no* and vice versa. With bipolar representations, *yes* would be represented by +1, *no* by −1, and *unsure* by 0.

Example 3.8 The effect of data representation: bipolar is better than binary

Example 3.5 illustrated the difficulties that a simple net (with binary input) experiences when given an input vector with "mistakes" in two components. The weight matrix formed from the bipolar representation of training patterns still cannot produce the proper response for an input vector formed from a stored vector with two "mistakes," e.g.,

$$(-1, 1, 1, -1) \cdot \mathbf{W} = (0, 0) \rightarrow (0, 0).$$

However, the net can respond correctly when given an input vector formed from a stored vector with two components "missing." For example, consider the vector

$\mathbf{x} = (0, 1, 0, -1)$, which is formed from the training vector $\mathbf{s} = (1, 1, -1, -1)$, with the first and third components "missing" rather than "wrong." We have

$$(0, 1, 0, -1) \cdot \mathbf{W} = (6, -6) \rightarrow (1, -1),$$

the correct response for the stored vector $\mathbf{s} = (1, 1, -1, -1)$. These "missing" components are really just a particular form of noise that produces an input vector which is not as dissimilar to a training vector as is the input vector produced with the more extreme "noise" denoted by the term "mistake."

Character recognition

Example 3.9 A heteroassociative net for associating letters from different fonts

A heteroassociative neural net was trained using the Hebb rule (outer products) to associate three vector pairs. The $\mathbf{x}$ vectors have 63 components, the $\mathbf{y}$ vectors 15. The vectors represent two-dimensional patterns. The pattern

is converted to a vector representation that is suitable for processing as follows: The #s are replaced by 1's and the dots by -1's, reading across each row (starting with the top row). The pattern shown becomes the vector

$$(-1, 1, -1, \quad 1, -1, 1, \quad 1, 1, 1, \quad 1, -1, 1 \quad 1, -1, 1).$$

The extra spaces between the vector components, which separate the different rows of the original pattern for ease of reading, are not necessary for the network.

Figure 3.3 shows the vector pairs in their original two-dimensional form.

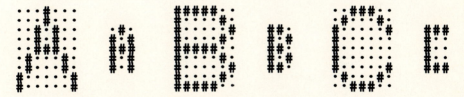

Figure 3.3 Training patterns for character recognition using heteroassociative net.

After training, the net was used with input patterns that were noisy versions of the training input patterns. The results are shown in Figures 3.4 and 3.5. The noise took the form of turning pixels "on" that should have been "off" and vice versa. These are denoted as follows:

@   Pixel is now "on," but this is a mistake (noise).

O   Pixel is now "off," but this is a mistake (noise).

Figure 3.5 shows that the neural net can recognize the small letters that are

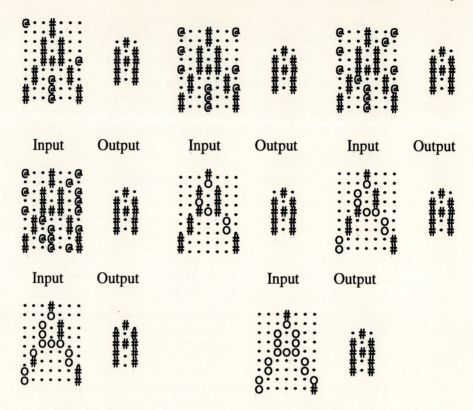

Figure 3.4 Response of heteroassociative net to several noisy versions of pattern A.

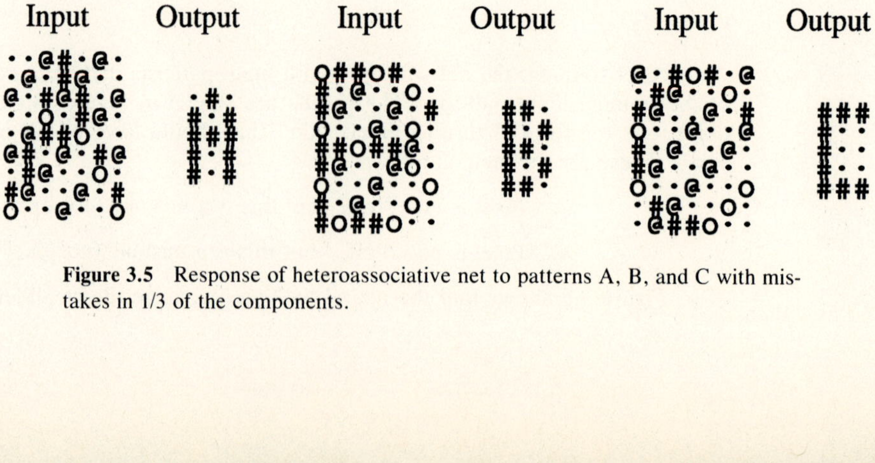

Figure 3.5 Response of heteroassociative net to patterns A, B, and C with mistakes in 1/3 of the components.

stored in it, even when given input patterns representing the large training patterns with 30% noise.

3.3 AUTOASSOCIATIVE NET

The feedforward autoassociative net considered in this section is a special case of the heteroassociative net described in Section 3.2. For an autoassociative net, the training input and target output vectors are identical. The process of training is often called *storing* the vectors, which may be binary or bipolar. A stored vector can be retrieved from distorted or partial (noisy) input if the input is sufficiently similar to it. The performance of the net is judged by its ability to reproduce a stored pattern from noisy input; performance is, in general, better for bipolar vectors than for binary vectors. In Section 3.4, several different versions of iterative autoassociative nets are discussed.

It is often the case that, for autoassociative nets, the weights on the diagonal (those which would connect an input pattern component to the corresponding component in the output pattern) are set to zero. This will be illustrated in Example 3.14. Setting these weights to zero may improve the net's ability to generalize (especially when more than one vector is stored in it) [Szu, 1989] or may increase the biological plausibility of the net [Anderson, 1972]. Setting them to zero is necessary for extension to the iterative case [Hopfield, 1982] or if the delta rule is used (to prevent the training from producing the identity matrix for the weights) [McClelland & Rumelhart, 1988].

3.3.1 Architecture

Figure 3.6 shows the architecture of an autoassociative neural net.

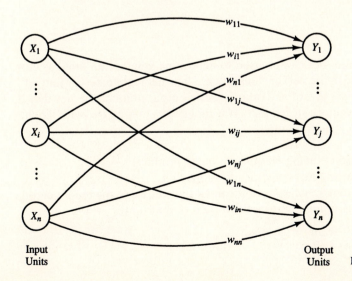

Input
Units

Output
Units

Figure 3.6 Autoassociative neural net.

3.3.2 Algorithm

For mutually orthogonal vectors, the Hebb rule can be used for setting the weights in an autoassociative net because the input and output vectors are perfectly correlated, component by component (i.e., they are the same). The algorithm is as given in Section 3.1.1; note that there are the same number of output units as input units.

Step 0. Initialize all weights, $i = 1, \ldots, n; j = 1, \ldots, m$:

$$w_{ij} = 0;$$

Step 1. For each vector to be stored, do Steps 2–4:

    Step 2. Set activation for each input unit, $i = 1, \ldots, n$:

$$x_i = s_i.$$

    Step 3. Set activation for each output unit, $j = 1, \ldots, m$:

$$y_j = s_j;$$

    Step 4. Adjust the weights, $i = 1, \ldots, n; j = 1, \ldots, m$:

$$w_{ij}(\text{new}) = w_{ij}(\text{old}) + x_i y_j.$$

As discussed earlier, in practice the weights are usually set from the formula

$$\mathbf{W} = \sum_{p=1}^{P} \mathbf{s}^{\mathrm{T}}(p)\mathbf{s}(p),$$

rather than from the algorithmic form of Hebb learning.

3.3.3 Application

An autoassociative neural net can be used to determine whether an input vector is "known" (i.e., stored in the net) or "unknown." The net recognizes a "known" vector by producing a pattern of activation on the output units of the net that is the same as one of the vectors stored in it. The application procedure (with bipolar inputs and activations) is as follows:

Step 0. Set the weights (using Hebb rule, outer product).

Step 1. For each testing input vector, do Steps 2–4.

    Step 2. Set activations of the input units equal to the input vector.

    Step 3. Compute net input to each output unit, $j = 1, \ldots, n$:

$$y\_in_j = \sum_i x_i w_{ij}.$$

Step 4.   Apply activation function ($j = 1, \ldots, n$):

$$y_j = f(y\_in_j) = \begin{cases} 1 & \text{if } y\_in_j > 0; \\ -1 & \text{if } y\_in_j \leq 0. \end{cases}$$

Simple examples

Example 3.10   An autoassociative net to store one vector: recognizing the stored vector

We illustrate the process of storing one pattern in an autoassociative net and then recalling, or recognizing, that stored pattern.

Step 0.   The vector $\mathbf{s} = (1, 1, 1, -1)$ is stored with the weight matrix:

$$\mathbf{W} = \begin{bmatrix} 1 & 1 & 1 & -1 \\ 1 & 1 & 1 & -1 \\ 1 & 1 & 1 & -1 \\ -1 & -1 & -1 & 1 \end{bmatrix}.$$

Step 1.   For the testing input vector:
    Step 2. $\mathbf{x}$ = $(1, 1, 1, -1)$.
    Step 3. $\mathbf{y}\_in$ = $(4, 4, 4, -4)$.
    Step 4. $\mathbf{y}$ = $f(4, 4, 4, -4) = (1, 1, 1, -1)$.

Since the response vector $\mathbf{y}$ is the same as the stored vector, we can say the input vector is recognized as a "known" vector.

The preceding process of using the net can be written more succinctly as

$$(1, 1, 1, -1) \cdot \mathbf{W} = (4, 4, 4, -4) \rightarrow (1, 1, 1, -1).$$

Now, if recognizing the vector that was stored were all that this weight matrix enabled the net to do, it would be no better than using the identity matrix for the weights. However, an autoassociative neural net can recognize as "known" vectors that are similar to the stored vector, but that differ slightly from it. As before, the differences take one of two forms: "mistakes" in the data or "missing" data. The only "mistakes" we consider are changes from $+1$ to -1 or vice versa. We use the term "missing" data to refer to a component that has the value 0, rather than either $+1$ or -1.

Example 3.11 Testing an autoassociative net: one mistake in the input vector

Using the succinct notation just introduced, consider the performance of the net for each of the input vectors $\mathbf{x}$ that follow. Each vector $\mathbf{x}$ is formed from the original stored vector $\mathbf{s}$ with a mistake in one component.

$$(-1, \quad 1, \quad 1, -1) \cdot \mathbf{W} = (2, 2, 2, -2) \rightarrow (1, 1, 1, -1)$$

$$(\ 1, -1, \quad 1, -1) \cdot \mathbf{W} = (2, 2, 2, -2) \rightarrow (1, 1, 1, -1)$$

$$(\ 1, \quad 1, -1, -1) \cdot \mathbf{W} = (2, 2, 2, -2) \rightarrow (1, 1, 1, -1)$$

$$(\ 1, \quad 1, \quad 1, \quad 1) \cdot \mathbf{W} = (2, 2, 2, -2) \rightarrow (1, 1, 1, -1).$$

Note that in each case the input vector is recognized as "known" after a single update of the activation vector in Step 4 of the algorithm. The reader can verify that the net also recognizes the vectors formed when one component is "missing." Those vectors are $(0, 1, 1, -1)$, $(1, 0, 1, -1)$, $(1, 1, 0, -1)$, and $(1, 1, 1, 0)$.

In general, a net is more tolerant of "missing" data than it is of "mistakes" in the data, as the examples that follow demonstrate. This is not surprising, since the vectors with "missing" data are closer (both intuitively and in a mathematical sense) to the training patterns than are the vectors with "mistakes."

Example 3.12 Testing an autoassociative net: two "missing" entries in the input vector

The vectors formed from $(1, 1, 1, -1)$ with two "missing" data are $(0, 0, 1, -1)$, $(0, 1, 0, -1)$, $(0, 1, 1, 0)$, $(1, 0, 0, -1)$, $(1, 0, 1, 0)$, and $(1, 1, 0, 0)$. As before, consider the performance of the net for each of these input vectors:

$$(0, 0, 1, -1) \cdot \mathbf{W} = (2, 2, 2, -2) \to (1, 1, 1, -1)$$

$$(0, 1, 0, -1) \cdot \mathbf{W} = (2, 2, 2, -2) \to (1, 1, 1, -1)$$

$$(0, 1, 1, 0) \cdot \mathbf{W} = (2, 2, 2, -2) \to (1, 1, 1, -1)$$

$$(1, 0, 0, -1) \cdot \mathbf{W} = (2, 2, 2, -2) \to (1, 1, 1, -1)$$

$$(1, 0, 1, 0) \cdot \mathbf{W} = (2, 2, 2, -2) \to (1, 1, 1, -1)$$

$$(1, 1, 0, 0) \cdot \mathbf{W} = (2, 2, 2, -2) \to (1, 1, 1, -1).$$

The response of the net indicates that it recognizes each of these input vectors as the training vector $(1, 1, 1, -1)$, which is what one would expect, or at least hope for.

Example 3.13 Testing an autoassociative net: two mistakes in the input vector

The vector $(-1, -1, 1, -1)$ can be viewed as being formed from the stored vector $(1, 1, 1, -1)$ with two mistakes (in the first and second components). We have:

$$(-1, -1, 1, -1) \cdot \mathbf{W} = (0, 0, 0, 0).$$

The net does not recognize this input vector.

Example 3.14 An autoassociative net with no self-connections: zeroing-out the diagonal

It is fairly common for an autoassociative network to have its diagonal terms set to zero, e.g.,

$$\mathbf{W}_0 = \begin{bmatrix} 0 & 1 & 1 & -1 \\ 1 & 0 & 1 & -1 \\ 1 & 1 & 0 & -1 \\ -1 & -1 & -1 & 0 \end{bmatrix}.$$

Consider again the input vector $(-1, -1, 1, -1)$ formed from the stored vector $(1, 1, 1, -1)$ with two mistakes (in the first and second components). We have:

$$(-1, -1, 1, -1) \cdot \mathbf{W}_0 = (-1, 1, -1, 1).$$

The net still does not recognize this input vector.

It is interesting to note that if the weight matrix $\mathbf{W}_0$ (with 0's on the diagonal) is used in the case of "missing" components in the input data (see Example 3.12), the output unit or units with the net input of largest magnitude coincide with the input unit or units whose input component or components were zero. We have:

$$(0, 0, 1, -1)\cdot\mathbf{W}_0 = (2, 2, 1, -1) \to (1, 1, 1, -1)$$

$$(0, 1, 0, -1)\cdot\mathbf{W}_0 = (2, 1, 2, -1) \to (1, 1, 1, -1)$$

$$(0, 1, 1, \quad 0)\cdot\mathbf{W}_0 = (2, 1, 1, -2) \to (1, 1, 1, -1)$$

$$(1, 0, 0, -1)\cdot\mathbf{W}_0 = (1, 2, 2, -1) \to (1, 1, 1, -1)$$

$$(1, 0, 1, \quad 0)\cdot\mathbf{W}_0 = (1, 2, 1, -2) \to (1, 1, 1, -1)$$

$$(1, 1, 0, \quad 0)\cdot\mathbf{W}_0 = (1, 1, 2, -2) \to (1, 1, 1, -1).$$

The net recognizes each of these input vectors.

3.3.4 Storage Capacity

An important consideration for associative memory neural networks is the number of patterns or pattern pairs that can be stored before the net begins to forget. In this section we consider some simple examples and theorems for noniterative autoassociative nets.

Examples

Example 3.15 Storing two vectors in an autoassociative net

More than one vector can be stored in an autoassociative net by adding the weight matrices for each vector together. For example, if $\mathbf{W}_1$ is the weight matrix used to store the vector $(1, 1, -1, -1)$ and $\mathbf{W}_2$ is the weight matrix used to store the vector $(-1, 1, 1, -1)$, then the weight matrix used to store both $(1, 1, -1, -1)$ and $(-1, 1, 1, -1)$ is the sum of $\mathbf{W}_1$ and $\mathbf{W}_2$. Because it is desired that the net respond with one of the stored vectors when it is presented with an input vector that is similar (but not identical) to a stored vector, it is customary to set the diagonal terms in the weight matrices to zero. If this is not done, the diagonal terms (which would each be equal to the number of vectors stored in the net) would dominate, and the net would tend to reproduce the input vector rather than a stored vector. The addition of $\mathbf{W}_1$ and $\mathbf{W}_2$ proceeds as follows:

$$
\mathbf{W}_1 \qquad\qquad \mathbf{W}_2 \qquad\qquad \mathbf{W}_1 + \mathbf{W}_2
$$

$$
\begin{bmatrix} 0 & 1 & -1 & -1 \\ 1 & 0 & -1 & -1 \\ -1 & -1 & 0 & 1 \\ -1 & -1 & 1 & 0 \end{bmatrix}
+
\begin{bmatrix} 0 & -1 & -1 & 1 \\ -1 & 0 & 1 & -1 \\ -1 & 1 & 0 & -1 \\ 1 & -1 & -1 & 0 \end{bmatrix}
=
\begin{bmatrix} 0 & 0 & -2 & 0 \\ 0 & 0 & 0 & -2 \\ -2 & 0 & 0 & 0 \\ 0 & -2 & 0 & 0 \end{bmatrix}.
$$

The reader should verify that the net with weight matrix $\mathbf{W}_1 + \mathbf{W}_2$ can recognize both of the vectors $(1, 1, -1, -1)$ and $(-1, 1, 1, -1)$. The number of vectors that can be stored in a net is called the *capacity* of the net.

Example 3.16 Attempting to store two nonorthogonal vectors in an autoassociative net

Not every pair of bipolar vectors can be stored in an autoassociative net with four nodes; attempting to store the vectors $(1, -1, -1, 1)$ and $(1, 1, -1, 1)$ by adding their weight matrices gives a net that cannot distinguish between the two vectors it was trained to recognize:

$$\begin{bmatrix} 0 & -1 & -1 & 1 \\ -1 & 0 & 1 & -1 \\ -1 & 1 & 0 & -1 \\ 1 & -1 & -1 & 0 \end{bmatrix} + \begin{bmatrix} 0 & 1 & -1 & 1 \\ 1 & 0 & -1 & 1 \\ -1 & -1 & 0 & -1 \\ 1 & 1 & -1 & 0 \end{bmatrix} = \begin{bmatrix} 0 & 0 & -2 & 2 \\ 0 & 0 & 0 & 0 \\ -2 & 0 & 0 & -2 \\ 2 & 0 & -2 & 0 \end{bmatrix}.$$

The difference between Example 3.15 and this example is that there the vectors are orthogonal, while here they are not. Recall that two vectors $\mathbf{x}$ and $\mathbf{y}$ are orthogonal if

$$\mathbf{x}\,\mathbf{y}^T = \sum_i x_i y_i = 0.$$

Informally, this example illustrates the difficulty that results from trying to store vectors that are too similar.

An autoassociative net with four nodes can store three orthogonal vectors (i.e., each vector is orthogonal to each of the other two vectors). However, the weight matrix for four mutually orthogonal vectors is always singular (so four vectors cannot be stored in an autoassociative net with four nodes, even if the vectors are orthogonal). These properties are illustrated in Examples 3.17 and 3.18.

Example 3.17 Storing three mutually orthogonal vectors in an autoassociative net

Let $\mathbf{W}_1 + \mathbf{W}_2$ be the weight matrix to store the orthogonal vectors $(1, 1, -1, -1)$ and $(-1, 1, 1, -1)$ and $\mathbf{W}_3$ be the weight matrix that stores $(-1, 1, -1, 1)$. Then the weight matrix to store all three orthogonal vectors is $\mathbf{W}_1 + \mathbf{W}_2 + \mathbf{W}_3$. We have

$$\underset{\mathbf{W}_1 + \mathbf{W}_2}{\begin{bmatrix} 0 & 0 & -2 & 0 \\ 0 & 0 & 0 & -2 \\ -2 & 0 & 0 & 0 \\ 0 & -2 & 0 & 0 \end{bmatrix}} + \underset{\mathbf{W}_3}{\begin{bmatrix} 0 & -1 & 1 & -1 \\ -1 & 0 & -1 & 1 \\ 1 & -1 & 0 & -1 \\ -1 & 1 & -1 & 0 \end{bmatrix}} = \underset{\mathbf{W}_1 + \mathbf{W}_2 + \mathbf{W}_3}{\begin{bmatrix} 0 & -1 & -1 & -1 \\ -1 & 0 & -1 & -1 \\ -1 & -1 & 0 & -1 \\ -1 & -1 & -1 & 0 \end{bmatrix}},$$

which correctly classifies each of the three vectors on which it was trained.

Example 3.18 Attempting to store four vectors in an autoassociative net

Attempting to store a fourth vector, $(1, 1, 1, 1)$, with weight matrix $\mathbf{W}_4$, orthogonal to each of the foregoing three, demonstrates the difficulties encountered in over training a net, namely, previous learning is erased. Adding the weight matrix for the new vector to the matrix for the first three vectors gives

$$\mathbf{W}_1 + \mathbf{W}_2 + \mathbf{W}_3 \qquad\qquad \mathbf{W}_4 \qquad\qquad\quad \mathbf{W}^*$$

$$\begin{bmatrix} 0 & -1 & -1 & -1 \\ -1 & 0 & -1 & -1 \\ -1 & -1 & 0 & -1 \\ -1 & -1 & -1 & 0 \end{bmatrix} + \begin{bmatrix} 0 & 1 & 1 & 1 \\ 1 & 0 & 1 & 1 \\ 1 & 1 & 0 & 1 \\ 1 & 1 & 1 & 0 \end{bmatrix} = \begin{bmatrix} 0 & 0 & 0 & 0 \\ 0 & 0 & 0 & 0 \\ 0 & 0 & 0 & 0 \\ 0 & 0 & 0 & 0 \end{bmatrix},$$

which cannot recognize any vector.

Theorems

The capacity of an autoassociative net depends on the number of components the stored vectors have and the relationships among the stored vectors; more vectors can be stored if they are mutually orthogonal.

Expanding on ideas suggested by Szu (1989), we prove that $n - 1$ mutually orthogonal bipolar vectors, each with n components, can always be stored using the sum of the outer product weight matrices (with diagonal terms set to zero), but that that attempting to store n mutually orthogonal vectors will result in a weight matrix that cannot reproduce any of the stored vectors. Recall again that two vectors $\mathbf{x}$ and $\mathbf{y}$ are orthogonal if $\sum_i x_i y_i = 0$.

Notation.   The kth vector to be stored is denoted by the row vector

$$\mathbf{a}(k) = (a_1(k), a_2(k), \ldots, a_n(k)).$$

The weight matrix to store $\mathbf{a}(k)$ is given by

$$\mathbf{W}(k) = \begin{bmatrix} 0 & a_1(k)a_2(k) & \cdot & \cdot & \cdot & a_1(k)a_n(k) \\ a_2(k)a_1(k) & 0 & & \cdot & \cdot & a_2(k)a_n(k) \\ \cdot & \cdot & \cdot & \cdot & \cdot & \cdot \\ \cdot & \cdot & & \cdot & \cdot & \cdot \\ \cdot & \cdot & & \cdot & \cdot & \cdot \\ a_n(k)a_1(k) & a_n(k)a_2(k) & \cdot & \cdot & \cdot & 0 \end{bmatrix}.$$

The weight matrix to store $\mathbf{a}(1), \mathbf{a}(2), \ldots, \mathbf{a}(m)$ has the general element

$$w_{ij} = \begin{cases} 0 & \text{if } i = j; \\ \displaystyle\sum_{p=1}^{m} a_i(p)a_j(p) & \text{otherwise.} \end{cases}$$

The vector $\mathbf{a}(k)$ can be recalled when it is input to a net with weight matrix $\mathbf{W}$ if $\mathbf{a}(k)$ is an eigenvector of matrix $\mathbf{W}$. To test whether $\mathbf{a}(k)$ is an eigenvector, and to determine the corresponding eigenvalue, consider the formula

$$(a_1(k), a_2(k), \ldots, a_n(k)) \, \mathbf{W}$$

$$= \left(\sum_{i=1}^{n} a_i(k)w_{i1}, \sum_{i=1}^{n} a_i(k)w_{i2}, \ldots, \sum_{i=1}^{n} a_i(k)w_{in} \right).$$

The jth component of $\mathbf{a}(k)\,\mathbf{W}$ is

$$\sum_{i=1}^{n} a_i(k)w_{ij} = \sum_{i \neq j} a_i(k) \sum_{p=1}^{m} a_i(p)a_j(p) = \sum_{p=1}^{m} a_j(p)\sum_{i \neq j} a_i(k)a_i(p).$$

Because the stored vectors are orthogonal,

$$\sum_{i=1}^{n} a_i(k)a_i(p) = 0 \qquad\qquad\qquad \text{for } k \neq p$$

and

$$\sum_{i \neq j} a_i(k)a_i(p) = \begin{cases} -a_j(k)a_j(p) & \text{for } k \neq p; \\ n - 1 & \text{for } k = p. \end{cases}$$

Since the vectors are bipolar, it is always the case that $[a_i(k)]^2 = 1$.

Combining these results, we get, for the jth component of $\mathbf{a}(k)\,\mathbf{W}$,

$$\sum_{p=1}^{m} a_j(p) \sum_{i \neq j} a_i(k)a_i(p) = \sum_{p \neq k} a_j(p) \sum_{i \neq j} a_i(k)a_i(p) + a_j(k) \sum_{i \neq j} a_i(k)a_i(p)$$

$$= \sum_{p \neq k} a_j(p) \left[-a_j(k)a_j(p) \right] + a_j(k)(n - 1)$$

$$= \sum_{p \neq k} -a_j(k) + a_j(k)(n - 1)$$

$$= -(m - 1)a_j(k) + a_j(k)(n - 1)$$

$$= (n - m)a_j(k).$$

Thus, $\mathbf{a}(k)\mathbf{W} = (n - m)\mathbf{a}(k)$, which shows that $\mathbf{a}(k)$ is an eigenvector of the weight matrix $\mathbf{W}$, with eigenvalue $(n - m)$, where n is the number of components of the vectors being stored and m is the number of stored (orthogonal) vectors. This establishes the following result.

Theorem 3.1. For $m < n$, the weight matrix is nonsingular. The eigenvalue $(n - m)$ has geometric multiplicity m, with eigenvectors $\mathbf{a}(1), \mathbf{a}(2), \ldots, \mathbf{a}(m)$. For $m = n$, zero is an eigenvalue of multiplicity n, and there are no nontrivial eigenvectors.

The following result can also be shown.

Theorem 3.2. A set of 2^k mutually orthogonal bipolar vectors can be constructed for $n = 2^k\,m$ (for m odd), and no larger set can be formed.

The proof is based on the following observations:

1. Let $[\mathbf{v}, \mathbf{v}]$ denote the concatenation of the vector $\mathbf{v}$ with itself (producing a vector with $2n$ components if $\mathbf{v}$ is an n-tuple).

2. If **a** and **b** are any two mutually orthogonal bipolar vectors (n-tuples), then [**a**, **a**], [**a**, −**a**], [**b**, **b**], and [**b**, −**b**] are mutually orthogonal $2n$-tuples.
3. Any number n can be expressed as 2^k m, where m is odd and $k \geq 0$.
4. It is clear that it is not possible to construct a pair of orthogonal bipolar n-tuples for n odd ($k = 0$), since the dot product of two bipolar n-tuples has the same parity as n.

The construction of the desired set of mutually orthogonal vectors proceeds as follows:

1. Form vector $\mathbf{v}_m(1) = (1, 1, 1, \ldots, 1)$, an m-tuple.
2. Form

$$\mathbf{v}_{2m}(1) = [\mathbf{v}_m(1), \quad \mathbf{v}_m(1)]$$

and

$$\mathbf{v}_{2m}(2) = [\mathbf{v}_m(1), \ -\mathbf{v}_m(1)];$$

$\mathbf{v}_{2m}(1)$ and $\mathbf{v}_{2m}(2)$ are othogonal $2m$-tuples.
3. Form the four orthogonal $4m$-tuples

$$\mathbf{v}_{4m}(1) = [\mathbf{v}_{2m}(1), \quad \mathbf{v}_{2m}(1)],$$

$$\mathbf{v}_{4m}(2) = [\mathbf{v}_{2m}(1), \ -\mathbf{v}_{2m}(1)],$$

$$\mathbf{v}_{4m}(3) = [\mathbf{v}_{2m}(2), \quad \mathbf{v}_{2m}(2)],$$

and

$$\mathbf{v}_{4m}(4) = [\mathbf{v}_{2m}(2), \ -\mathbf{v}_{2m}(2)].$$

4. Continue until $\mathbf{v}_n(1), \ldots, \mathbf{v}_n(2^k)$ have been formed; this is the required set of 2^k orthogonal vectors with $n = 2^k$ m components.

The method of proving that the set of orthogonal vectors constructed by the preceding procedure is maximal is illustrated here for $n = 6$. Consider the orthogonal vectors $\mathbf{v}(1) = (1, 1, 1, 1, 1, 1)$ and $\mathbf{v}(2) = (1, 1, 1, -1, -1, -1)$ constructed with the technique. Assume that there is a third vector, (a, b, c, d, e, f), which is orthogonal to both $\mathbf{v}(1)$ and $\mathbf{v}(2)$. This requires that $a + b + c + d + e + f = 0$ and $a + b + c - d - e - f = 0$. Combining these equations gives $a + b + c = 0$, which is impossible for $a, b, c \in \{1, -1\}$.

3.4 ITERATIVE AUTOASSOCIATIVE NET

We see from the next example that in some cases the net does not respond immediately to an input signal with a stored target pattern, but the response may be enough like a stored pattern (at least in the sense of having more nodes committed to values of $+1$ or -1 and fewer nodes with the "unsure" response of 0) to suggest using this first response as input to the net again.

Example 3.19 **Testing a recurrent autoassociative net: stored vector with second, third and fourth components set to zero**

The weight matrix to store the vector $(1, 1, 1, -1)$ is

$$\mathbf{W} = \begin{bmatrix} 0 & 1 & 1 & -1 \\ 1 & 0 & 1 & -1 \\ 1 & 1 & 0 & -1 \\ -1 & -1 & -1 & 0 \end{bmatrix}.$$

The vector $(1, 0, 0, 0)$ is an example of a vector formed from the stored vector with three "missing" components (three zero entries). The performance of the net for this vector is given next.

Input vector $(1, 0, 0, 0)$:

$$(1, 0, 0, 0) \cdot \mathbf{W} = (0, 1, 1, -1) \rightarrow \text{iterate}$$

$$(0, 1, 1, -1) \cdot \mathbf{W} = (3, 2, 2, -2) \rightarrow (1, 1, 1, -1).$$

Thus, for the input vector $(1, 0, 0, 0)$, the net produces the "known" vector $(1, 1, 1, -1)$ as its response in two iterations.

We can also take this iterative feedback scheme a step further and simply let the input and output units be the same, to obtain a recurrent autoassociative neural net. In Sections 3.4.1–3.4.3, we consider three that differ primarily in their activation function. Then, in Section 3.4.4, we examine a net developed by Nobel prize–winning physicist John Hopfield (1982, 1988). Hopfield's work (and his prestige) enhanced greatly the respectability of neural nets as a field of study in the 1980s. The differences between his net and the others in this section, although fairly slight, have a significant impact on the performance of the net. For iterative nets, one key question is whether the activations will converge. The weights are fixed (by the Hebb rule for example), but the activations of the units change.

3.4.1 Recurrent Linear Autoassociator

One of the simplest iterative autoassociator neural networks is known as the *linear autoassociator* [McClelland & Rumelhart, 1988; Anderson et al., 1977]. This net has n neurons, each connected to all of the other neurons. The weight matrix is symmetric, with the connection strength w_{ij} proportional to the sum over all training patterns of the product of the activations of the two units x_i and x_j. In other words, the weights can be found by the Hebb rule. McClelland and Rumelhart do not restrict the weight matrix to have zeros on the diagonal. Anderson et al. show that setting the diagonal elements in the weight matrix to zero, which they believe represents a biologically more plausible model, does not change the performance of the net significantly.

The performance of the net can be analyzed [Anderson et al., 1977] using ideas from linear algebra. An $n \times n$ nonsingular symmetric matrix (such as the weight matrix) has n mutually orthogonal eigenvectors. A recurrent linear auto-associator neural net is trained using a set of K orthogonal unit vectors $\mathbf{f}_1, \ldots, \mathbf{f}_K$, where the number of times each vector is presented, say, $\beta_1, \ldots, \beta_K$, is not necessarily the same. A formula for the components of the weight matrix could be derived as a simple generalization of the formula given before for the Hebb rule, allowing for the fact that some of the stored vectors were repeated. It is easy to see that each of these stored vectors is an eigenvector of the weight matrix. Furthermore, the number of times the vector was presented is the corresponding eigenvalue.

The response of the net, when presented with input (row) vector $\mathbf{x}$, is $\mathbf{xW}$, where $\mathbf{W}$ is the weight matrix. We know from linear algebra that the largest value of $\|\mathbf{xW}\|$ occurs when $\mathbf{x}$ is the eigenvector corresponding to the largest eigenvalue, the next largest value of $\|\mathbf{xW}\|$ occurs when $\mathbf{x}$ is the eigenvector associated with the next largest eigenvalue, etc. The recurrent linear autoassociator is intended to produce as its response (after perhaps several iterations) the stored vector (eigenvector) to which the input vector is most similar.

Any input pattern can be expressed as a linear combination of eigenvectors. The response of the net when an input vector is presented can be expressed as the corresponding linear combination of the eigenvalues (the net's response to the eigenvectors). The eigenvector to which the input vector is most similar is the eigenvector with the largest component in this linear expansion. As the net is allowed to iterate, contributions to the response of the net from eigenvectors with large eigenvalues (and with large coefficients in the input vector's eigenvector expansion) will grow relative to contributions from other eigenvectors with smaller eigenvalues (or smaller coefficients).

However, even though the net will increase its response corresponding to components of the input pattern on which it was trained most extensively (i.e., the eigenvectors associated with the largest eigenvalues), the overall response of the system may grow without bound. This difficulty leads to the modification of the next section.

3.4.2 Brain-State-in-a-Box Net

The response of the linear associator (Section 3.4.1) can be prevented from growing without bound by modifying the activation function (the identity function for the linear associator) to take on values within a cube (i.e., each component is restricted to be between -1 and 1) [Anderson, et al., 1977]. The units in the brain-state-in-a-box (BSB) net (as in the linear associator) update their activations simultaneously.

The architecture of the BSB net, as for all the nets in this section, consists of n units, each connected to every other unit. However, in this net there is a trained weight on the self-connection (i.e., the diagonal terms in the weight matrix are not set to zero). There is also a self-connection with weight 1. The algorithm given here is based the original description of the process in Anderson et al. (1977); it is similar to that given in Hecht-Nielsen (1990). Others [McClelland & Rumelhart, 1988] present a version that does not include the learning phase.

Algorithm

Step 0.   Initialize weights (small random values).
                 Initialize learning rates, α and β.
Step 1.   For each training input vector, do Steps 2–6.

> *Step 2.*   Set initial activations of net equal to the external input vector **x**:
>
> $$y_i = x_i.$$
>
> *Step 3.*   While activations continue to change, do Steps 4 and 5:
>
> > *Step 4.*   Compute net inputs:
> >
> > $$y\_in_i = y_i + \alpha \sum_j y_j w_{ji}.$$
> >
> > (Each net input is a combination of the unit's previous activation and the weighted signal received from all units.)
> >
> > *Step 5.*   Each unit determines its activation (output signal):
> >
> > $$y_i = \begin{cases} 1 & \text{if } y\_in_i > 1 \\ y_i & \text{if } -1 \le y\_in_i \le 1 \\ -1 & \text{if } y\_in_i < -1. \end{cases}$$
> >
> > (A stable state for the activation vector will be a vertex of the cube.)
>
> *Step 6.*   Update weights:
>
> $$w_{ij}(\text{new}) = w_{ij}(\text{old}) + \beta y_i y_j.$$

3.4.3 Autoassociator With Threshold Function

A threshold function can also be used as the activation function for an iterative autoassociator net. The application procedure for bipolar ($+1$ or -1) vectors and activations with symmetric weights and no self-connections, i.e.,

$$w_{ij} = w_{ji},$$

$$w_{ii} = 0,$$

is as follows:

Step 0.   Initialize weights to store patterns.
              (Use Hebbian learning.)
Step 1.   For each testing input vector, do Steps 2–5.
    Step 2.   Set activations $\mathbf{x}$.
    Step 3.   While the stopping condition is false, repeat Steps 4 and 5.
        Step 4.   Update activations of all units
                         (the threshold, θ_i, is usually taken to be zero):

$$x_i = \begin{cases} 1 & \text{if } \sum_j x_j w_{ij} > \theta_i \\ x_i & \text{if } \sum_j x_j w_{ij} = \theta_i \\ -1 & \text{if } \sum_j x_j w_{ij} < \theta_i. \end{cases}$$

        Step 5.   Test stopping condition: The net is allowed to
                         iterate until the currect vector $\mathbf{x}$ matches a
                         stored vector, or $\mathbf{x}$ matches a previous vector
                         $\mathbf{x}$, or the maximum number of iterations al-
                         lowed is reached.

The results for the input vectors described in Section 3.3 for the autoassociative net are the same if the net is allowed to iterate. Example 3.20 shows a situation in which the autoassociative net fails to recognize the input vector on the first presentation, but recognizes it when allowed to iterate.

Example 3.20 A recurrent autoassociative net recognizes *all* vectors formed from the stored vector with three "missing components"

The weight matrix to store the vector $(1, 1, 1, -1)$ is

$$\mathbf{W} = \begin{bmatrix} 0 & 1 & 1 & -1 \\ 1 & 0 & 1 & -1 \\ 1 & 1 & 0 & -1 \\ -1 & -1 & -1 & 0 \end{bmatrix}.$$

The vectors formed from the stored vector with three "missing" components (three zero entries) are $(1, 0, 0, 0)$, $(0, 1, 0, 0)$, $(0, 0, 1, 0)$, and $(0, 0, 0, -1)$. The performance of the net on each of these is as follows:

First input vector, $(1, 0, 0, 0)$

Step 4: $(1, 0, 0, 0) \cdot \mathbf{W} = (0, 1, 1, -1)$.
Step 5: $(0, 1, 1, -1)$ is neither the stored vector nor an activation vector produced
              previously (since this is the first iteration), so we allow the activations
              to be updated again.
Step 4: $(0, 1, 1, -1) \cdot \mathbf{W} = (3, 2, 2, -2) \rightarrow (1, 1, 1, -1)$.
Step 5: $(1, 1, 1, -1)$ is the stored vector, so we stop.

Thus, for the input vector $(1, 0, 0, 0)$, the net produces the "known" vector $(1, 1, 1, -1)$ as its response after two iterations.

Second input vector, $(0, 1, 0, 0)$

Step 4: $(0, 1, 0, 0) \cdot W = (1, 0, 1, -1)$.
Step 5: $(1, 0, 1, -1)$ is not the stored vector or a previous activation vector, so we iterate.
Step 4: $(1, 0, 1, -1) \cdot W = (2, 3, 2, -2) \rightarrow (1, 1, 1, -1)$.
Step 5: $(1, 1, 1, -1)$ is the stored vector, so we stop.

As with the first testing input, the net recognizes the input vector $(0, 1, 0, 0)$ as the "known" vector $(1, 1, 1, -1)$.

Third input vector, $(0, 0, 1, 0)$

Step 4: $(0, 0, 1, 0) \cdot W = (1, 1, 0, -1)$.
Step 5: $(1, 1, 0, -1)$ is neither the stored vector nor a previous activation vector, so we iterate.
Step 4: $(1, 1, 0, -1) \cdot W = (2, 2, 3, -2) \rightarrow (1, 1, 1, -1)$.
Step 5: $(1, 1, 1, -1)$ is the stored vector, so we stop.

Again, the input vector, $(0, 0, 1, 0)$, produces the "known" vector $(1, 1, 1, -1)$.

Fourth input vector, $(0, 0, 0, 1)$

Step 4: $(0, 0, 0, -1) \cdot W = (-1, -1, -1, 0)$
Step 5:   Iterate.
Step 4: $(-1, -1, 1, 0) \cdot W = (-2, -2, -2, 3) \rightarrow (-1, -1, -1, 1)$.
Step 5:   Iterate.
Step 4: $(-1, -1, -1, 1) \cdot W = (-3, -3, -3, 3) \rightarrow (-1, -1, -1, 1)$.
Step 5:   The activation vector is the same as on the previous iteration, so we stop.

For the input vector $(0, 0, 0, -1)$, the response is -1 times the "known" vector $(1, 1, 1, -1)$. Both the stored vector and its negative are stable points of the net.

Example 3.21 Testing a recurrent autoassociative net: mistakes in the first and second components of the stored vector

One example of a vector that can be formed from the stored vector $(1, 1, 1, -1)$ with mistakes in two components (the first and second) is $(-1, -1, 1, -1)$. The performance of the net (with the weight matrix given in Example 3.20) is as follows.
    For input vector $(-1, -1, 1, -1)$:

Step 4: $(-1, -1, 1, -1) \cdot W = (1, 1, -1, 1)$.
Step 5:   Iterate.
Step 4: $(1, 1, -1, 1) \cdot W = (-1, -1, 1, -1)$.
Step 5:   Since this is the input vector repeated, stop.

(Further iterations would simply alternate the two activation vectors produced already.)

The behavior of the net in this case is called a *fixed-point cycle of length two.* It has been proved [Szu, 1989] that such a cycle occurs whenever the input vector is orthogonal to all of the stored vectors in the net (where the vectors have been stored using the sum of outer products with the diagonal terms set to zero). The vector $(-1, -1, 1, -1)$ is orthogonal to the stored vector $(1, 1, 1, -1)$. In general, for a bipolar vector with $2k$ components, mistakes in k components will produce a vector that is orthogonal to the original vector. We shall consider this example further in Section 3.4.4.

3.4.4 Discrete Hopfield Net

An iterative autoassociative net similar to the nets described in this chapter has been developed by Hopfield (1982, 1984). The net is a fully interconnected neural net, in the sense that each unit is connected to every other unit. The net has symmetric weights with no self-connections, i.e.,

$$w_{ij} = w_{ji}$$

and

$$w_{ii} = 0.$$

The two small differences between this net and the iterative autoassociative net presented in Section 3.4.3 can have a significant effect in terms of whether or not the nets converge for a particular input pattern. The differences are that in the Hopfield net presented here,

1. only one unit updates its activation at a time (based on the signal it receives from each other unit) and
2. each unit continues to receive an external signal in addition to the signal from the other units in the net.

The asynchronous updating of the units allows a function, known as an *energy or Lyapunov function,* to be found for the net. The existence of such a function enables us to prove that the net will converge to a stable set of activations, rather than oscillating, as the net in Example 3.21 did [Hopfield, 1982, 1984]. Lyapunov functions, developed by the Russian mathematician and mechanical engineer Alexander Mikhailovich Lyapunov (1857–1918), are important in the study of the stability of differential equations. See *Differential Equations with Applications and Historical Notes* [Simmons, 1972] for further discussion.

The original formulation of the discrete Hopfield net showed the usefulness of the net as content-addressable memory. Later extensions [Hopfield & Tank, 1985] for continuous-valued activations can be used either for pattern association or constained optimization. Since their use in optimization problems illustrates the "value added" from the additional computation required for the continuous activation function, we shall save our discussion of the continuous Hopfield net until Chapter 7, where we discuss the use of other nets for constrained optimization problems.

Architecture

An expanded form of a common representation of the Hopfield net is shown in Figure 3.7.

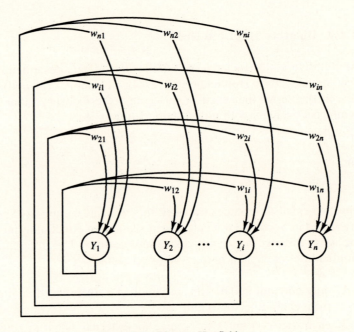

Figure 3.7 Discrete Hopfield net.

Algorithm

There are several versions of the discrete Hopfield net. Hopfield's first description [1982] used binary input vectors.

To store a set of binary patterns $s(p)$, $p = 1, \ldots, P$, where

$$s(p) = (s_1(p), \ldots, s_i(p), \ldots, s_n(p)),$$

the weight matrix $\mathbf{W} = \{w_{ij}\}$ is given by

$$w_{ij} = \sum_p [2s_i(p) - 1][2s_j(p) - 1] \qquad \text{for } i \neq j$$

and

$$w_{ii} = 0.$$

Other descriptions [Hopfield, 1984] allow for bipolar inputs. The weight matrix is found as follows:

To store a set of bipolar patterns $s(p)$, $p = 1, \ldots, P$, where

$$s(p) = (s_1(p), \ldots, s_i(p), \ldots, s_n(p)),$$

the weight matrix $\mathbf{W} = \{w_{ij}\}$ is given by

$$w_{ij} = \sum_p s_i(p)s_j(p) \qquad \text{for } i \neq j$$

and

$$w_{ii} = 0.$$

The application algorithm is stated for binary patterns; the activation function can be modified easily to accommodate bipolar patterns.

Application Algorithm for the Discrete Hopfield Net

Step 0.   Initialize weights to store patterns.
    (Use Hebb rule.)
While activations of the net are not converged, do Steps 1–7.
Step 1.   For each input vector $\mathbf{x}$, do Steps 2–6.
    Step 2.   Set initial activations of net equal to the external input vector $\mathbf{x}$:

$$y_i = x_i, \, (i = 1, \ldots n)$$

    Step 3.   Do Steps 4–6 for each unit Y_i.
    (Units should be updated in random order.)
        Step 4.   Compute net input:

$$y\_in_i = x_i + \sum_j y_j w_{ji}.$$

        Step 5.   Determine activation (output signal):

$$y_i = \begin{cases} 1 & \text{if } y\_in_i > \theta_i \\ y_i & \text{if } y\_in_i = \theta_i \\ 0 & \text{if } y\_in_i < \theta_i. \end{cases}$$

        Step 6.   Broadcast the value of y_i to all other units.
            (This updates the activation vector.)
Step 7.   Test for convergence.

The threshold, θ_i, is usually taken to be zero. The order of update of the units is random, but each unit must be updated at the same average rate. There are a number of variations on the discrete Hopfield net presented in this algorithm. Originally, Hopfield used binary activations, with no external input after the first time step [Hopfield, 1982]. Later, the external input was allowed to continue during processing [Hopfield, 1984]. Although typically, Hopfield used binary activations, the model was formulated for any two distinct activation values. Descriptions by other authors use different combinations of the features of the original model; for example, Hecht-Nielsen uses bipolar activations, but no external input [Hecht-Nielsen, 1990].

The analysis of the Lyapunov function (energy function) for the Hopfield net will show that the important features of the net that guarantee convergence are the asynchronous update of the weights and the zero weights on the diagonal. It is not important whether an external signal is maintained during processing or whether the inputs and activations are binary or bipolar.

Before considering the proof that the net will converge, we consider an example of the application of the net.

Application

A binary Hopfield net can be used to determine whether an input vector is a "known" vector (i.e., one that was stored in the net) or an "unknown" vector. The net recognizes a "known" vector by producing a pattern of activation on the units of the net that is the same as the vector stored in the net. If the input vector is an "unknown" vector, the activation vectors produced as the net iterates (repeats Step 3 in the preceding algorithm) will converge to an activation vector that is not one of the stored patterns; such a pattern is called a *spurious stable state*.

Example 3.22   Testing a discrete Hopfield net: mistakes in the first and second components of the stored vector

Consider again Example 3.21, in which the vector $(1, 1, 1, 0)$ (or its bipolar equivalent $(1, 1, 1, -1)$) was stored in a net. The binary input vector corresponding to the input vector used in that example (with mistakes in the first and second components) is $(0, 0, 1, 0)$. Although the Hopfield net uses binary vectors, the weight matrix is bipolar, the same as was used in Example 3.16. The units update their activations in a random order. For this example the update order is Y_1, Y_4, Y_3, Y_2.

Step 0.   Initialize weights to store patterns:

$$\mathbf{W} = \begin{bmatrix} 0 & 1 & 1 & -1 \\ 1 & 0 & 1 & -1 \\ 1 & 1 & 0 & -1 \\ -1 & -1 & -1 & 0 \end{bmatrix}.$$

Step 1.   The input vector is $\mathbf{x} = (0, 0, 1, 0)$. For this vector,
   Step 2. $\mathbf{y} = (0, 0, 1, 0)$.
   Step 3.   Choose unit Y_1 to update its activation:
       Step 4. $y\_in_1 = x_1 + \sum_j y_j w_{j1} = 0 + 1$.

       Step 5. $y\_in_1 > 0 \rightarrow y_1 = 1$.
       Step 6. $\mathbf{y} = (1, 0, 1, 0)$.
   Step 3.   Choose unit Y_4 to update its activation:
       Step 4. $y\_in_4 = x_4 + \sum_j y_j w_{j4} = 0 + (-2)$.

Step 5. $y\_in_4 < 0 \rightarrow y_4 = 0.$
Step 6. $\mathbf{y} = (1, 0, 1, 0).$
Step 3.   Choose unit Y_3 to update its activation:
Step 4. $y\_in_3 = x_3 + \sum_j y_j w_{j3} = 1 + 1.$

Step 5. $y\_in_3 > 0 \rightarrow y_3 = 1.$
Step 6. $\mathbf{y} = (1, 0, 1, 0).$
Step 3.   Choose unit Y_2 to update its activation:
Step 4. $y\_in_2 = x_2 + \sum_j y_j w_{j2} = 0 + 2.$

Step 5. $y\_in_2 > 0 = y_2 = 1.$
Step 6. $\mathbf{y} = (1, 1, 1, 0).$
Step 7.   Test for convergence.

Since some activations have changed during this update cycle, at least one more pass through all of the input vectors should be made. The reader can confirm that further iterations do not change the activation of any unit. The net has converged to the stored vector.

Analysis

Energy Function.   Hopfield [1984] proved that the discrete net bearing his name will converge to a stable limit point (pattern of activation of the units) by considering an energy (or Lyapunov) function for the system. An energy function is a function that is bounded below and is a nonincreasing function of the state of the system. For a neural net, the state of the system is the vector of activations of the units. Thus, if an energy function can be found for an iterative neural net, the net will converge to a stable set of activations. An energy function for the discrete Hopfield net is given by

$$E = -.5 \sum_{i \neq j} \sum_j y_i y_j w_{ij} - \sum_i x_i y_i + \sum_i \theta_i y_i.$$

If the activation of the net changes by an amount Δy_i, the energy changes by an amount

$$\Delta E = -\left[\sum_j y_j w_{ij} + x_i - \theta_i \right] \Delta y_i.$$

(This relationship depends on the fact that only one unit can update its activation at a time.)

We now consider the two cases in which a change Δy_i will occur in the activation of neuron Y_i.

If y_i is positive, it will change to zero if

$$x_i + \sum_j y_j w_{ji} < \theta_i.$$

This gives a negative change for y_i. In this case, $\Delta E < 0.$

If y_i is zero, it will change to positive if

$$x_i + \sum_j y_j w_{ji} > \theta_i.$$

This gives a positive change for y_i. In this case, $\Delta E < 0$.

Thus Δy_i is positive only if $[\sum_j y_j w_{ji} + x_i - \theta_i]$ is positive, and Δy_i is negative only if this same quantity is negative. Therefore, the energy cannot increase. Hence, since the energy is bounded, the net must reach a stable equilibrium such that the energy does not change with further iteration.

This proof shows that it is not necessary that the activations be binary. It is also not important whether the external signals are maintained during the iterations. The important aspects of the algorithm are that the energy change depend only on the change in activation of one unit and that the weight matrix be symmetric with zeros on the diagonal.

Storage Capacity.   Hopfield found experimentally that the number of binary patterns that can be stored and recalled in a net with reasonable accuracy, is given approximately by

$$P \approx 0.15n,$$

where n is the number of neurons in the net.

Abu-Mostafa and St Jacques (1985) have performed a detailed theoretical analysis of the information capacity of a Hopfield net. For a similar net using bipolar patterns, McEliece, Posner, Rodemich, and Venkatesh (1987) found that

$$P \approx \frac{n}{2 \log_2 n}.$$

3.5 BIDIRECTIONAL ASSOCIATIVE MEMORY (BAM)

We now consider several versions of the heteroassociative recurrent neural network, or bidirectional associative memory (BAM), developed by Kosko (1988, 1992a).

A bidirectional associative memory [Kosko, 1988] stores a set of pattern associations by summing bipolar correlation matrices (an n by m outer product matrix for each pattern to be stored). The architecture of the net consists of two layers of neurons, connected by directional weighted connection paths. The net iterates, sending signals back and forth between the two layers until all neurons reach equilibrium (i.e., until each neuron's activation remains constant for several steps). Bidirectional associative memory neural nets can respond to input to either layer. Because the weights are bidirectional and the algorithm alternates between updating the activations for each layer, we shall refer to the layers as the X-layer and the Y-layer (rather than the input and output layers).

Three varieties of BAM—binary, bipolar, and continuous—are considered

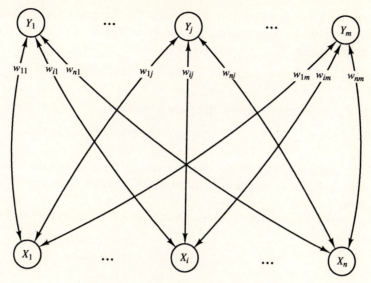

Figure 3.8 Bidirectional associative memory.

here. Several other variations exist. The architecture for each is the same and is illustrated in Figure 3.8.

3.5.1 Architecture

The single-layer nonlinear feedback BAM network (with heteroassociative content-addressable memory) has n units in its X-layer and m units in its Y-layer. The connections between the layers are bidirectional; i.e., if the weight matrix for signals sent from the X-layer to the Y-layer is $\mathbf{W}$, the weight matrix for signals sent from the Y-layer to the X-layer is $\mathbf{W}^\mathsf{T}$.

3.5.2 Algorithm

Discrete BAM

The two bivalent (binary or bipolar) forms of BAM are closely related. In each, the weights are found from the sum of the outer products of the bipolar form of the training vector pairs. Also, the activation function is a step function, with the possibility of a nonzero threshold. It has been shown that bipolar vectors improve the performance of the net [Kosko, 1988; Haines & Hecht-Nielsen, 1988].

  Setting the Weights. The weight matrix to store a set of input and target vectors $\mathbf{s}(p):\mathbf{t}(p)$, $p = 1, \ldots, P$, where

$$\mathbf{s}(p) = (s_1(p), \ldots, s_i(p), \ldots, s_n(p))$$

and

$$\mathbf{t}(p) = (t_1(p), \ldots, t_j(p), \ldots, t_m(p)),$$

can be determined by the Hebb rule. The formulas for the entries depend on whether the training vectors are binary or bipolar. For binary input vectors, the weight matrix $\mathbf{W} = \{w_{ij}\}$ is given by

$$w_{ij} = \sum_p (2s_i(p) - 1)(2t_j(p) - 1).$$

For bipolar input vectors, the weight matrix $\mathbf{W} = \{w_{ij}\}$ is given by

$$w_{ij} = \sum_p s_i(p)t_j(p).$$

Activation Functions.   The activation function for the discrete BAM is the appropriate step function, depending on whether binary or bipolar vectors are used.

For binary input vectors, the activation function for the Y-layer is

$$y_j = \begin{cases} 1 & \text{if } y\_in_j > 0 \\ y_j & \text{if } y\_in_j = 0 \\ 0 & \text{if } y\_in_j < 0, \end{cases}$$

and the activation function for the X-layer is

$$x_i = \begin{cases} 1 & \text{if } x\_in_i > 0 \\ x_i & \text{if } x\_in_i = 0 \\ 0 & \text{if } x\_in_i < 0. \end{cases}$$

For bipolar input vectors, the activation function for the Y-layer is

$$y_j = \begin{cases} 1 & \text{if } y\_in_j > \theta_j \\ y_j & \text{if } y\_in_j = \theta_j \\ -1 & \text{if } y\_in_j < \theta_j, \end{cases}$$

and the activation function for the X-layer is

$$x_i = \begin{cases} 1 & \text{if } x\_in_i > \theta_i \\ x_i & \text{if } x\_in_i = \theta_i \\ -1 & \text{if } x\_in_i < \theta_i. \end{cases}$$

Note that if the net input is exactly equal to the threshold value, the activation function "decides" to leave the activation of that unit at its previous value. For that reason, the activations of all units are initialized to zero in the algorithm that follows. The algorithm is written for the first signal to be sent from the X-layer to the Y-layer. However, if the input signal for the X-layer is the zero vector, the input signal to the Y-layer will be unchanged by the activation function, and the process will be the same as if the first piece of information had been sent from the Y-layer to the X-layer. Signals are sent only from one layer to the other at any step of the process, not simultaneously in both directions.

Algorithm.

Step 0.   Initialize the weights to store a set of P vectors;
           initialize all activations to 0.

Step 1.   For each testing input, do Steps 2–6.

  Step 2a.   Present input pattern $\mathbf{x}$ to the X-layer
               (i.e., set activations of X-layer to current input pattern).

  Step 2b.   Present input pattern $\mathbf{y}$ to the Y-layer.
               (Either of the input patterns may be the zero vector.)

  Step 3.   While activations are not converged, do Steps 4–6.

    Step 4.   Update activations of units in Y-layer.
                   Compute net inputs:

$$y\_in_j = \sum_i w_{ij} x_i.$$

                   Compute activations:

$$y_j = f(y\_in_j).$$

                   Send signal to X-layer.

    Step 5.   Update activations of units in X-layer.
                   Compute net inputs:

$$x\_in_i = \sum_j w_{ij} y_j.$$

                   Compute activations:

$$x_i = f(x\_in_i).$$

                   Send signal to Y-layer.

    Step 6.   Test for convergence:
                   If the activation vectors $\mathbf{x}$ and $\mathbf{y}$ have reached
                   equilibrium, then stop; otherwise, continue.

Continuous BAM

A continuous bidirectional associative memory [Kosko, 1988] transforms input smoothly and continuously into output in the range [0, 1] using the logistic sigmoid function as the activation function for all units.

For binary input vectors $(\mathbf{s}(p), \mathbf{t}(p))$, $p = 1, 2, \ldots, P$, the weights are determined by the aforementioned formula

$$w_{ij} = \sum_p (2s_i(p) - 1)(2t_j(p) - 1).$$

The activation function is the logistic sigmoid

$$f(x) = \frac{1}{1 + \exp(-y\_in_j)},$$

where a bias is included in calculating the net input to any unit

$$y\_in_j = b_j + \sum_i x_i w_{ij},$$

and corresponding formulas apply for the units in the X-layer.

A number of other forms of BAMs have been developed. In some, the activations change based on a differential equation known as Cohen-Grossberg activation dynamics [Cohen & Grossberg, 1983]. Note, however, that Kosko uses the term "activation" to refer to the activity level of a neuron before the output function (such as the logistic sigmoid function) is applied. (See Kosko, 1992a, for further discussion of bidirectional associative memory nets.)

3.5.3 Application

Example 3.23 A BAM net to associate letters with simple bipolar codes

Consider the possibility of using a (discrete) BAM network (with bipolar vectors) to map two simple letters (given by 5×3 patterns) to the following bipolar codes:

```
. # .          . # #
# . #          # . .
# # #          # . .
# . #          # . .
# . #          . # #

( -1, 1 )        ( 1, 1 )
```

The weight matrices are:

(to store $A \to -11$)		($C \to 11$)		(W, to store both)	
1	-1	-1	-1	0	-2
-1	1	1	1	0	2
1	-1	1	1	2	0
-1	1	1	1	0	2
1	-1	-1	-1	0	-2
-1	1	-1	-1	-2	0
-1	1	1	1	0	2
-1	1	-1	-1	-2	0
-1	1	-1	-1	-2	0
-1	1	1	1	0	2
1	-1	-1	-1	0	-2
-1	1	-1	-1	-2	0
-1	1	-1	-1	-2	0
1	-1	1	1	2	0
-1	1	1	1	0	2

To illustrate the use of a BAM, we first demonstrate that the net gives the correct Y vector when presented with the **x** vector for either the pattern A or the pattern C:

INPUT PATTERN *A*

$(-1\ 1\ -1\ 1\ -1\ 1\ 1\ 1\ 1\ 1\ -1\ 1\ 1\ -1\ 1)\mathbf{W} = (-14,\ 16) \rightarrow (-1,\ 1).$

INPUT PATTERN *C*

$(-1\ 1\ 1\ 1\ -1\ -1\ 1\ -1\ -1\ 1\ -1\ -1\ -1\ 1\ 1)\mathbf{W} = (14,\ 18) \rightarrow (1,\ 1).$

To see the bidirectional nature of the net, observe that the *Y* vectors can also be used as input. For signals sent from the *Y*-layer to the *X*-layer, the weight vector is the transpose of the vector **W**, i.e.,

$$\mathbf{W}^{\mathbf{T}} = \begin{bmatrix} 0 & 0 & 2 & 0 & 0 & -2 & 0 & -2 & -2 & 0 & 0 & -2 & -2 & 2 & 0 \\ -2 & 2 & 0 & 2 & -2 & 0 & 2 & 0 & 0 & 2 & -2 & 0 & 0 & 0 & 2 \end{bmatrix}.$$

For the input vector associated with pattern *A*, namely, $(-1,\ 1)$, we have

$(-1,\ 1)\mathbf{W}^{\mathbf{T}} =$

$(-1,\ 1)\begin{bmatrix} 0 & 0 & 2 & 0 & 0 & -2 & 0 & -2 & -2 & 0 & 0 & -2 & -2 & 2 & 0 \\ -2 & 2 & 0 & 2 & -2 & 0 & 2 & 0 & 0 & 2 & -2 & 0 & 0 & 0 & 2 \end{bmatrix}$

$= (-2\ \ 2\ \ -2\ \ 2\ \ -2\ \ 2\ \ 2\ \ 2\ \ 2\ \ -2\ \ 2\ \ 2\ \ -2\ \ 2)$

$\rightarrow (-1\ \ 1\ \ -1\ \ 1\ \ -1\ \ 1\ \ 1\ \ 1\ \ 1\ \ 1\ \ -1\ \ 1\ \ 1\ \ -1\ \ 1).$

This is pattern *A*.

Similarly, if we input the vector associated with pattern *C*, namely, $(1,\ 1)$, we obtain

$(1,\ 1)\mathbf{W}^{\mathbf{T}} =$

$(1,\ 1)\begin{bmatrix} 0 & 0 & 2 & 0 & 0 & -2 & 0 & -2 & -2 & 0 & 0 & -2 & -2 & 2 & 0 \\ -2 & 2 & 0 & 2 & -2 & 0 & 2 & 0 & 0 & 2 & -2 & 0 & 0 & 0 & 2 \end{bmatrix}$

$= (-2\ \ 2\ \ 2\ \ 2\ \ -2\ \ -2\ \ 2\ \ -2\ \ -2\ \ 2\ \ -2\ \ -2\ \ -2\ \ 2\ \ 2)$

$\rightarrow (-1\ \ 1\ \ 1\ \ 1\ \ -1\ \ -1\ \ 1\ \ -1\ \ -1\ \ 1\ \ -1\ \ -1\ \ -1\ \ 1\ \ 1),$

which is pattern *C*.

The net can also be used with noisy input for the **x** vector, the **y** vector, or both, as is shown in the next example.

Example 3.24 Testing a BAM net with noisy input

In this example, the net is given a **y** vector as input that is a noisy version of one of the training **y** vectors and no information about the corresponding **x** vector (i.e., the **x** vector is identically 0). The input vector is $(0,\ 1)$; the response of the net is

$(0,\ 1)\mathbf{W}^{\mathbf{T}} =$

$(0,\ 1)\begin{bmatrix} 0 & 0 & 2 & 0 & 0 & -2 & 0 & -2 & -2 & 0 & 0 & -2 & -2 & 2 & 0 \\ -1 & 2 & 0 & 2 & -2 & 0 & 2 & 0 & 0 & 2 & -2 & 0 & 0 & 0 & 2 \end{bmatrix}$

$= (-2\ \ 2\ \ 0\ \ 2\ \ -2\ \ 0\ \ 2\ \ 0\ \ 0\ \ 2\ \ -2\ \ 0\ \ 0\ \ 0\ \ 2)$

$\rightarrow (-1\ \ 1\ \ 0\ \ 1\ \ -1\ \ 0\ \ 1\ \ 0\ \ 0\ \ 1\ \ -1\ \ 0\ \ 0\ \ 0\ \ 1).$

Note that the units receiving 0 net input have their activations left at that value, since the initial **x** vector is 0. This **x** vector is then sent back to the Y-layer, using the weight matrix **W**:

$$(-1 \ \ 1 \ \ 0 \ \ 1 \ \ -1 \ \ 0 \ \ 1 \ \ 0 \ \ 0 \ \ 1 \ \ -1 \ \ 0 \ \ 0 \ \ 0 \ \ 1) \begin{bmatrix} 0 & -2 \\ 0 & 2 \\ 2 & 0 \\ 0 & 2 \\ 0 & -2 \\ -2 & 0 \\ 0 & 2 \\ -2 & 0 \\ -2 & 0 \\ 0 & 2 \\ 0 & -2 \\ -2 & 0 \\ -2 & 0 \\ 2 & 0 \\ 0 & 2 \end{bmatrix}$$

$$\rightarrow (0 \ 1).$$

This result is not too surprising, since the net had no information to give it a preference for either A or C. The net has converged (since, obviously, no further changes in the activations will take place) to a spurious stable state, i.e., the solution is not one of the stored pattern pairs.

If, on the other hand, the net was given both the input vector **y**, as before, and some information about the vector **x**, for example,

$$\mathbf{y} = (0 \ \ 1), \mathbf{x} = (0 \ \ 0 \ \ -1 \ \ 0 \ \ 0 \ \ 1 \ \ 0 \ \ 1 \ \ 1 \ \ 0 \ \ 0 \ \ 1 \ \ 1 \ \ -1 \ \ 0),$$

the net would be able to reach a stable set of activations corresponding to one of the stored pattern pairs.

Note that the **x** vector is a noisy version of

$$A = (-1 \ \ 1 \ \ -1 \ \ 1 \ \ -1 \ \ 1 \ \ 1 \ \ 1 \ \ 1 \ \ 1 \ \ -1 \ \ 1 \ \ 1 \ \ -1 \ \ 1),$$

where the nonzero components are those that distinguish A from

$$C = (-1 \ \ 1 \ \ 1 \ \ 1 \ \ -1 \ \ -1 \ \ 1 \ \ -1 \ \ -1 \ \ 1 \ \ -1 \ \ -1 \ \ -1 \ \ 1 \ \ 1).$$

Now, since the algorithm specifies that if the net input to a unit is zero, the activation of that unit remains unchanged, we get

$$(0, \ 1)\mathbf{W^T} =$$

$$(0, \ 1) \begin{bmatrix} 0 & 0 & 2 & 0 & 0 & -2 & 0 & -2 & -2 & 0 & 0 & -2 & -2 & 2 & 0 \\ -2 & 2 & 0 & 2 & -2 & 0 & 2 & 0 & 0 & 2 & -2 & 0 & 0 & 0 & 2 \end{bmatrix}$$

$$= (-2 \ \ 2 \ \ 0 \ \ 2 \ \ -2 \ \ 0 \ \ 2 \ \ 0 \ \ 0 \ \ 2 \ \ -2 \ \ 0 \ \ 0 \ \ 0 \ \ 2)$$

$$\rightarrow (-1 \ \ 1 \ \ -1 \ \ 1 \ \ -1 \ \ 1 \ \ 1 \ \ 1 \ \ 1 \ \ 1 \ \ -1 \ \ 1 \ \ 1 \ \ -1 \ \ 1),$$

which is pattern A.

Since this example is fairly extreme, i.e., every component that distinguishes A from C was given an input value for A, let us try something with less information given concerning **x**.

For example, let $\mathbf{y} = (0\ 1)$ and $\mathbf{x} = (0\ 0\ -1\ 0\ 0\ 1\ 0\ 1\ 0\ 0\ 0\ 0\ 0\ 0)$. Then

$(0, 1)\mathbf{W^T} =$

$$(0, 1)\begin{bmatrix} 0 & 0 & 2 & 0 & 0 & -2 & 0 & -2 & -2 & 0 & 0 & -2 & -2 & 2 & 0 \\ -2 & 2 & 0 & 2 & -2 & 0 & 2 & 0 & 0 & 2 & -2 & 0 & 0 & 0 & 2 \end{bmatrix}$$

$$= (-2\ \ 2\ \ 0\ \ 2\ \ -2\ \ 0\ \ 2\ \ 0\ \ 0\ \ 2\ \ -2\ \ 0\ \ 0\ \ 0\ \ 2)$$

$$\rightarrow (-1\ \ 1\ \ -1\ \ 1\ \ -1\ \ 1\ \ 1\ \ 1\ \ 0\ \ 1\ \ -1\ \ 0\ \ 0\ \ 0\ \ 1),$$

which is not quite pattern A.

So we try iterating, sending the $\mathbf{x}$ vector back to the Y-layer using the weight matrix W:

$$(-1\ \ 1\ \ -1\ \ 1\ \ -1\ \ 1\ \ 1\ \ 1\ \ 0\ \ 1\ \ -1\ \ 0\ \ 0\ \ 0\ \ 1)\begin{bmatrix} 0 & -2 \\ 0 & 2 \\ 2 & 0 \\ 0 & 2 \\ 0 & -2 \\ -2 & 0 \\ 0 & 2 \\ -2 & 0 \\ -2 & 0 \\ 0 & 2 \\ 0 & -2 \\ -2 & 0 \\ -2 & 0 \\ 2 & 0 \\ 0 & 2 \end{bmatrix}$$

$$\rightarrow (-6, 10) \rightarrow (-1, 1).$$

If this pattern is fed back to the X-layer one more time, the pattern A will be produced.

Hamming distance

The number of different bits in two binary or bipolar vectors $\mathbf{x}_1$ and $\mathbf{x}_2$ is called the *Hamming distance* between the vectors and is denoted by $H[\mathbf{x}_1, \mathbf{x}_2]$. The average Hamming distance between the vectors is $\frac{1}{n}H[\mathbf{x}_1, \mathbf{x}_2]$, where n is the number of components in each vector. The $\mathbf{x}$ vectors in Examples 3.23 and 3.24, namely,

differ in the 3rd, 6th, 8th, 9th, 12th, 13th, and 14th positions. This gives an average Hamming distance between these vectors of 7/15. The average Hamming distance between the corresponding $\mathbf{y}$ vectors is 1/2.

Kosko (1988) has observed that "correlation encoding" (as is used in the BAM neural net) is improved to the extent that the average Hamming distance between pairs of input patterns is comparable to the average Hamming distance between the corresponding pairs of output patterns. If that is the case, input patterns that are separated by a small Hamming distance are mapped to output vectors that are also so separated, while input vectors that are separated by a large Hamming distance go to correspondingly distant (dissimilar) output patterns. This is analogous to the behavior of a continuous function.

Erasing a stored association

The complement of a bipolar vector $\mathbf{x}$ is denoted $\mathbf{x}^c$; it is the vector formed by changing all of the 1's in vector $\mathbf{x}$ to -1's and vice versa. Encoding (storing the pattern pair) $\mathbf{s}^c : \mathbf{t}^c$ stores the same information as encoding $\mathbf{s} : \mathbf{t}$; encoding $\mathbf{s}^c : \mathbf{t}$ or $\mathbf{s} : \mathbf{t}^c$ will erase the encoding of $\mathbf{s} : \mathbf{t}$ [Kosko, 1988].

3.5.4 Analysis

Several strategies may be used for updating the activations. The algorithm described in Section 3.5.2 uses a synchronous updating procedure, namely, that all units in a layer update their activations simultaneously. Updating may also be simple asynchronous (only one unit updates its activation at each stage of the iteration) or subset asynchronous (a group of units updates all of its members' activations at each stage).

Energy function

The convergence of a BAM net can be proved using an energy or Lyapunov function, in a manner similar to that described for the Hopfield net. A Lyapunov function must be decreasing and bounded. For a BAM net, an appropriate function is the average of the signal energy for a forward and backward pass:

$$L = -0.5 \, (\mathbf{x}W\mathbf{y}^T + \mathbf{y}W^T\mathbf{x}^T).$$

However, since $\mathbf{x}W\mathbf{y}^T$ and $\mathbf{y}W^T\mathbf{x}^T$ are scalars, and the transpose of a scalar is a scalar, the preceding expression can be simplified to

$$L = -\mathbf{x}W\mathbf{y}^T$$

$$= -\sum_{j=1}^{m} \sum_{i=1}^{n} x_i w_{ij} y_j.$$

For binary or bipolar step functions, the Lyapunov function is clearly bounded below by

$$-\sum_{j=1}^{m} \sum_{i=1}^{n} |w_{ij}|.$$

Kosko [1992a] presents a proof that the Lyapunov function decreases as the net iterates, for either synchronous or subset asynchronous updates.

Storage capacity

Although the upper bound on the memory capacity of the BAM is min(n, m), where n is the number of X-layer units and m is the number of Y-layer units, Haines and Hecht-Nielsen [1988] have shown that this can be extended to min (2^n, 2^m) if an appropriate nonzero threshold value is chosen for each unit. Their choice was based on a combination of heuristics and an exhaustive search.

BAM and Hopfield nets

The discrete Hopfield net (Section 3.4.4) and the BAM net are closely related. The Hopfield net can be viewed as an autoassociative BAM with the X-layer and Y-layer treated as a single layer (because the training vectors for the two layers are identical) and the diagonal of the symmetric weight matrix set to zero.

On the other hand, the BAM can be viewed as a special case of a Hopfield net which contains all of the X- and Y-layer neurons, but with no interconnections between two X-layer neurons or between two Y-layer neurons. This requires all X-layer neurons to update their activations before any of the Y-layer neurons update theirs; then all Y field neurons update before the next round of X-layer updates. The updates of the neurons within the X-layer or within the Y-layer can be done at the same time because a change in the activation of an X-layer neuron does not affect the net input to any other X-layer unit and similarly for the Y-layer units.

3.6 SUGGESTIONS FOR FURTHER STUDY

3.6.1 Readings

The original presentation of the Hebb rule is given in *The Organization of Behavior* [Hebb, 1949)]. The Introduction and Chapter 4 are reprinted in Anderson and Rosenfeld [1988], pp. 45–56. The articles by Anderson and Kohonen that are included in the Anderson and Rosenfeld collections, as well as Kohonen's book, *Self-organization and Associative Memory,* (1989a) provide good discussions of the associative memory nets presented in this chapter.

For further discussion of the Hopfield net, the original articles included in the Anderson and Rosenfeld collection give additional background and development. The article by Tank and Hopfield (1987) in *Scientific American* is also recommended.

The discussion of BAM nets in *Neural Networks and Fuzzy Systems* [Kosko, 1992a] provides a unified treatment of these nets and their relation to Hopfield nets and to the work of Grossberg (including adaptive resonance theory nets, which we discuss in Chapter 5).

3.6.2 Exercises

Hebb rule

3.1 Show that yet another way of viewing the formation of the weight matrix for Hebb learning is to form an $n \times p$ matrix $\mathbf{S}$ whose columns are the input patterns

$$\mathbf{s}(p) = (s_1(p), \ldots, s_i(p), \ldots, s_n(p)), \text{ i.e.,}$$

$$\mathbf{S} = \begin{bmatrix} s_1(1) & \cdot & \cdot & \cdot & s_1(P) \\ \cdot & \cdot & \cdot & \cdot & \cdot \\ s_i(1) & \cdot & \cdot & \cdot & s_i(P) \\ \cdot & \cdot & \cdot & \cdot & \cdot \\ s_n(1) & \cdot & \cdot & \cdot & s_n(P) \end{bmatrix}$$

and a $p \times m$ matrix $\mathbf{T}$ whose rows are the output patterns

$$\mathbf{t}(p) = (t_1(p), \ldots, t_j(p), \ldots, t_m(p)), \text{ i.e.,}$$

$$\mathbf{T} = \begin{bmatrix} t_1(1) & \cdot & \cdot & \cdot & t_m(1) \\ \cdot & \cdot & \cdot & \cdot & \cdot \\ \cdot & \cdot & \cdot & \cdot & \cdot \\ \cdot & \cdot & \cdot & \cdot & \cdot \\ t_1(P) & \cdot & \cdot & \cdot & t_m(P) \end{bmatrix}.$$

Then the product $\mathbf{S}\,\mathbf{T}$ gives the weight matrix $\mathbf{W}$:

$$\mathbf{W} = \mathbf{S}\,\mathbf{T} = \begin{bmatrix} \sum_{p=1}^{P} s_1(p)t_1(p) & \cdot & \cdot & \cdot & \sum_{p=1}^{P} s_1(p)t_m(p) \\ & \cdot & & & \cdot \\ & \cdot & & \cdot\cdot\cdot & \cdot \\ & \cdot & & & \cdot \\ \sum_{p=1}^{P} s_n(p)t_1(p) & \cdot & \cdot & \cdot & \sum_{p=1}^{P} s_n(p)t_m(p) \end{bmatrix}$$

3.2 Show the computations for the first component of the response of a Hebb net for an input vector and how this response depends on whether the input vector is orthogonal to the other vectors stored in the net.

Heteroassociative neural net

3.3 Find the weight matrix for Example 3.1 using the results of Exercise 3.1.

3.4 Test Example 3.6 on the input training vectors. Then test on input vectors that are similar to the input training vectors.

3.5 Test Example 3.7 using the training input vectors. Then test with input vectors that are the training vector with one "mistake."

3.6 a. Use the hybrid (binary/bipolar) form of Hebb rule learning as described in Example 3.6 to find the weight matrix for the associative memory network based on the following binary input-output vector pairs:

$$s(1) = (1 \quad 0 \quad 0 \quad 0) \qquad t(1) = (1 \quad 0)$$

$$s(2) = (1 \quad 0 \quad 0 \quad 1) \qquad t(2) = (1 \quad 0)$$

$$s(3) = (0 \quad 1 \quad 0 \quad 0) \qquad t(3) = (0 \quad 1)$$

$$s(4) = (0 \quad 1 \quad 1 \quad 0) \qquad t(4) = (0 \quad 1)$$

    b. Using the unit step function (with threshold 0) as the output units' activation function, test the response of your network on each of the input patterns. Describe the results obtained.

    c. Test the response of your network on various combinations of input patterns with "mistakes" or "missing data" (as in Example 3.8). Discuss the results you observe.

3.7 Using the formation of Exercise 3.1, find the weight matrix for Example 3.7.

Autoassociative neural net

3.8 Use the Hebb rule to store the vectors $(1, 1, 1, 1)$ and $(1, 1, -1, -1)$ in an autoassociative neural net.

    a. Find the weight matrix. (Do not set diagonal terms to zero.)

    b. Test the net, using the vector $(1, 1, 1, 1)$ as input.

    c. Test the net, using $(1, 1, -1, -1)$ as input.

    d. Test the net, using $(1, 1, 1, 0)$ as input; discuss.

    e. Repeat parts a–d with the diagonal terms in the weight matrix set to zero. Discuss any differences you find in the response of the net.

3.9 Consider an autoassociative net with the bipolar step function as the activation function and weights set by the Hebb rule (outer products), with the main diagonal of the weight matrix set to zero.

    a. Find the weight matrix to store the vector

$$V_1 = (1, 1, 1, 1, 1, 1).$$

    b. Test the net, using V_1 as input.

    c. Test the net, using

$$T_1 = (1, 1, 1, 1, -1, -1).$$

    d. Find the weight matrix to store the vector

$$V_2 = (1, 1, 1, -1, -1, -1).$$

    e. Test the net, using V_2 as input.

    f. Test the net, using

$$T_2 = (1, 1, 1, -1, 0, 0).$$

    g. Find the weight matrix to store both V_1 and V_2.

    h. Test the net on V_1, V_2, T_1, T_2.

Bidirectional associative memory (BAM)

3.10 a. Use the Hebb rule as described in Section 3.5.2 to find the bipolar weight matrix to store the following (binary) input-output pattern pairs:

$$x(1) = (1 \quad 0 \quad 1) \qquad y(1) = (1 \quad 0)$$

$$x(2) = (0 \quad 1 \quad 0) \qquad y(2) = (0 \quad 1)$$

b. Using the binary step function (with threshold 0) as the activation function for both layers, test the response of your network in both directions on each of the binary training patterns. In each case, when presenting an input pattern to one layer, the initial activation of the other layer is set to zero.

c. Using the bipolar step function (with threshold 0) as the activation function for both layers, convert the training patterns to bipolar form and test the network response in both directions again. Initialize activation as in part b.

d. Test the response of your network on each of the following noisy versions of the bipolar form of the training patterns. Iterate as required for stability.

(a) (0   −1   1)   (b) (0   0   1)   (c) (1   0   0)

(d) (−1   0   −1)   (e) (−1   0   0)   (f) (0   0   −1)

(g) (1   0   −1)   (h) (1   0)   (i) (0   1)

e. In which case does the network stabilize to a correct response, in which cases to an incorrect response, and in which cases to an indefinite or indeterminate response?

3.11 a. Use the outer product version of Hebb rule learning to find the weight matrix in bipolar form for the bidirectional associative memory network based on the following binary input-output vector pairs:

$$s(1) = (1 \quad 0 \quad 0 \quad 0) \qquad t(1) = (1 \quad 0)$$

$$s(2) = (1 \quad 0 \quad 0 \quad 1) \qquad t(2) = (1 \quad 0)$$

$$s(3) = (0 \quad 1 \quad 0 \quad 0) \qquad t(3) = (0 \quad 1)$$

$$s(4) = (0 \quad 1 \quad 1 \quad 0) \qquad t(4) = (0 \quad 1)$$

b. Using the unit step function (with threshold 0) as the output units' activation function, test the response of your network on each of the input patterns. Describe the results obtained.

c. Test the response of your network on various combination of input patterns with "mistakes" or "missing data" (as in Example 3.24). Discuss the results you observe.

3.6.3 Projects

Heteroassociative neural net

3.1 Write a computer program to implement a heteroassociative neural net using the Hebb rule to set the weights (by outer products). The program should read an **x** vector from a 9 × 7 array, followed by the corresponding **y** vector from a 5 × 3 array. Start by using the patterns in Example 3.9. Expand your training set to include more letters, taking the **x** patterns from one of the fonts in Figure 2.20 (or creating your own) and using the **y** patterns in Project 3.4 (or creating your own). Explore the number of pattern pairs that can be stored in the net, and the ability of the net to respond to noisy input. You may find it convenient to represent your training patterns by arrays with an entry of "2" if the pixel is on, "0" if it is off. Your program should then subtract 1 from each entry to form the bipolar pattern vector. This approach has the advantage that missing data (for testing) can be entered as a "1" which the program will convert to zero as desired.

Autoassociative neural net

3.2 Write a computer program to implement an autoassociative neural net using the Hebb rule to set the weights (outer products). The program should read input from a 7 × 5 array into an **x** vector (a 35-tuple) and should have a training mode in which the weights are set (by outer products). The number of inputs may be specified in advance, but it will be more convenient to prompt the user interactively for more input. The program should also have a test mode in which the weights are not changed but the response of the net is determined. This response should be printed as a 7 × 5 array. It is good practice to display the input array, followed by the output array.

The input patterns to be stored are as follows:

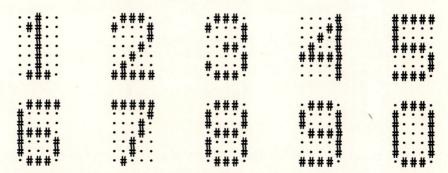

Try to answer the following questions:
1. How many patterns can you store (and recall successfully)?
   Does it matter which ones you try to store? (Consider whether any of these input vectors are orthogonal or nearly so; how is that relevant?)
   Does it matter whether you use binary or bipolar patterns?
2. How much noise can your net handle?
   Does the amount of noise that can be tolerated (i.e., for which the net will still give the original stored pattern as its response) depend on how many patterns are stored?
3. Does the performance of the net improve if it is allowed to iterate, either in the "batch mode" iteration of Section 3.4.1-3 or the "one unit at a time" form of the discrete Hopfield net of Section 3.4.4?

Discrete Hopfield net

3.3 Write a computer program to implement a discrete Hopfield net to store the letters from one of the fonts in Figure 2.20. Investigate the number of patterns that can be stored and recalled correctly, as well as the ability of the net to respond correctly to noisy input.

Bidirectional associative memory

3.4 Write a computer program to implement a bipolar BAM neural network. Allow for (at least) 15 units in the X-layer and 3 units in the Y-layer.

a. Use the program to store the association given in Example 3.23, i.e.,

$$A \rightarrow (-1\ 1),$$

$$C \rightarrow (\quad 1\ 1).$$

Try to illustrate the same cases as discussed in the text. Try some other cases of your own design. You might test the net on some noisy version of the letter C, for instance.

b. Use your program to store the following patterns (the X-layer vectors are the "letters" given in the 5×3 arrays; the associated Y-layer vectors are given below each X pattern):

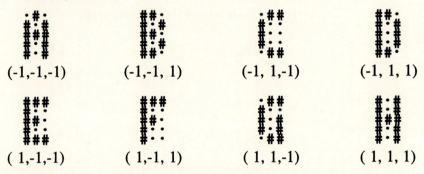

(-1,-1,-1)	(-1,-1, 1)	(-1, 1,-1)	(-1, 1, 1)
(1,-1,-1)	(1,-1, 1)	(1, 1,-1)	(1, 1, 1)

Is it possible to store all eight patterns at once? If not, how many can be stored at the same time? Try some experiments with noisy data, as in part *a*.

The following table gives the Hamming distance between the letters denoted by the foregoing patterns:

	A	B	C	D	E	F	G	H
A	0	4	7	4	6	6	5	3
B		0	7	2	4	4	7	5
C			0	7	3	5	2	8
D				0	6	6	5	5
E					0	2	5	5
F						0	8	5
G							0	6
H								0

Determine the Hamming distances between the Y-layer patterns associated with each of these letters. From the ratios of the Hamming distances, try to determine which pattern pairs will be most likely to be stored successfully.

Since the upper limit on the number of arbitrary pattern pairs that can be stored is $\min(n, m)$ (the number of components in the X- and Y-layer patterns respectively), you should consider carefully any case in which more than that number of patterns are stored. (Test whether the net can respond correctly in both directions.)

When using noisy patterns, the net is not retrained on then; they are used simply for testing the response of the net.

CHAPTER 4

Neural Networks Based on Competition

Among the simple examples of pattern classification in Chapter 3, we encountered a situation in which we had more information about the possible correct response of the net than we were able to incorporate. Specifically, when we applied a net that was trained to classify the input signal into one of the output categories, A, B, C, D, E, J, or K, the net sometimes responded that the signal was both a C and a K, or both an E and a K, or both a J and a K. In circumstances such as this, in which we know that only one of several neurons should respond, we can include additional structure in the network so that the net is forced to make a decision as to which one unit will respond. The mechanism by which this is achieved is called *competition*.

The most extreme form of competition among a group of neurons is called *Winner Take All*. As the name suggests, only one neuron in the competing group will have a nonzero output signal when the competition is completed. A specific competitive net that performs Winner-Take-All competition is the MAXNET, described in Section 4.1.1. A more general form of competition, the Mexican Hat, or On-Center-Off-Surround contrast enhancement, is described in Section 4.1.2. All of the other nets we discuss in this chapter and the next use Winner-Take-All competition as part of their operation. In computer simulations of these nets, if full neural implementation of the algorithms is not of primary importance, it is easy to replace the iterative competition phase of the process with a simple search for the neuron with the largest input (or other desired criterion) to choose as the winner.

156

With the exception of the fixed-weight competitive nets (Section 4.1), all of the other nets in Chapters 4 and 5 combine competition with some form of learning to adjust the weights of the net (i.e., the weights that are not part of any inter-connections in the competitive layer). The form of learning depends on the purpose for which the net is being trained. The learning vector quantization (LVQ) net-work, discussed in Section 4.3, and the counterpropagation network, examined in Section 4.4, are trained to perform mappings. Target values are available for the input training patterns; the learning is supervised.

Neural network learning is not restricted to supervised learning, wherein training pairs are provided, as with the pattern classification and pattern asso-ciation problems introduced in Chapters 2 and 3 and the more general input-output mappings (Sections 4.3 and 4.4, Chapter 6, and Section 7.2). A second major type of learning for neural networks is unsupervised learning, in which the net seeks to find patterns or regularity in the input data. In Section 4.2, we consider the self-organizing map, developed by Kohonen, which groups the input data into clusters, a common use for unsupervised learning. Adaptive resonance theory nets, the subject of Chapter 5, are also clustering nets. (The term "pattern clas-sification" is applied to these nets, too, but we shall reserve our use of it for situations in which the learning is supervised and target classifications are pro-vided during training.) Other examples of unsupervised learning, which do not include competition, are described in Section 7.2.

In a clustering net, there are as many input units as an input vector has components. Since each output unit represents a cluster, the number of output units will limit the number of clusters that can be formed.

The weight vector for an output unit in a clustering net (as well as in LVQ nets) serves as a representative, or exemplar, or code-book vector for the input patterns which the net has placed on that cluster. During training, the net deter-mines the output unit that is the best match for the current input vector; the weight vector for the winner is then adjusted in accordance with the net's learning al-gorithm. The training process for the adaptive resonance theory nets discussed in Chapter 5 involves a somewhat expanded form of this basic idea.

Several of the nets discussed in this chapter use the same learning algorithm, known as *Kohonen learning*. In this form of learning, the units that update their weights do so by forming a new weight vector that is a linear combination of the old weight vector and the current input vector. Typically, the unit whose weight vector was closest to the input vector is allowed to learn. The weight update for output (or cluster) unit j is given as

$$\mathbf{w}_{.j}(\text{new}) = \mathbf{w}_{.j}(\text{old}) + \alpha[\mathbf{x} - \mathbf{w}_{.j}(\text{old})]$$

$$= \alpha\mathbf{x} + (1 - \alpha)\mathbf{w}_{.j}(\text{old}),$$

where $\mathbf{x}$ is the input vector, $\mathbf{w}_{.j}$ is the weight vector for unit j (which is also the jth column of the weight matrix), and α, the learning rate, decreases as learning proceeds.

Two methods of determining the closest weight vector to a pattern vector are commonly used for self-organizing nets. Both are based on the assumption that the weight vector for each cluster (output) unit serves as an exemplar for the input vectors that have been assigned to that unit during learning.

The first method of determining the winner uses the squared Euclidean distance between the input vector and the weight vector and chooses the unit whose weight vector has the smallest Euclidean distance from the input vector.

The second method uses the dot product of the input vector and the weight vector. The dot product of an input vector with a given weight vector is simply the net input to the corresponding cluster unit, as it has been calculated in the nets presented in the previous chapters (and as defined in Chapter 1). The largest dot product corresponds to the smallest angle between the input and weight vectors if they are both of unit length. The dot product can be interpreted as giving the correlation between the input and weight vectors.

For vectors of unit length, the two methods (Euclidean and dot product) are equivalent. That is, if the input vectors and the weight vectors are of unit length, the same weight vector will be chosen as closest to the input vector, regardless of whether the Euclidean distance or the dot product method is used. In general, for consistency and to avoid the difficulties of having to normalize our inputs and weights, we shall use the Euclidean distance squared.

4.1 FIXED-WEIGHT COMPETITIVE NETS

Many neural nets use the idea of competition among neurons to enhance the contrast in activations of the neurons. In the most extreme situation, often called Winner-Take-All, only the neuron with the largest activation is allowed to remain "on." Typically, the neural implementation of this competition is not specified (and in computer simulations, the same effect can be achieved by a simple, non-neural sorting process). In Section 4.1.1, a neural subnet is given which achieves the Winner-Take-All competition. In Section 4.1.2, a more general contrast-enhancing net, known as the Mexican Hat, is described. In Section 4.1.3, the Hamming net is presented. This is a simple clustering net that uses fixed exemplars and the MAXNET subnet, discussed next.

4.1.1 MAXNET

MAXNET [Lippmann, 1987] is a specific example of a neural net based on competition. It can be used as a subnet to pick the node whose input is the largest. The m nodes in this subnet are completely interconnected, with symmetric weights. There is no training algorithm for the MAXNET; the weights are fixed.

Architecture

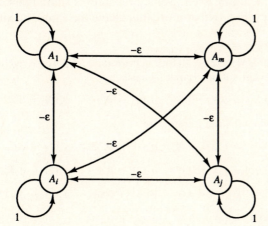

Figure 4.1 MAXNET.

Application

The activation function for the MAXNET is

$$f(x) = \begin{cases} x & \text{if } x > 0; \\ 0 & \text{otherwise.} \end{cases}$$

The application procedure is as follows:

Step 0.   Initialize activations and weights (set $0 < \epsilon < \dfrac{1}{m}$):

$a_j(0)$ input to node A_j,

$$w_{ij} = \begin{cases} 1 & \text{if } i = j; \\ -\epsilon & \text{if } i \neq j. \end{cases}$$

Step 1.   While stopping condition is false, do Steps 2–4.

   Step 2.   Update the activation of each node: For $j = 1, \ldots, m$,

$$a_j(\text{new}) = f[a_j(\text{old}) - \epsilon \sum_{k \neq j} a_k(\text{old})].$$

   Step 3.   Save activations for use in next iteration:

$$a_j(\text{old}) = a_j(\text{new}), j = 1, \ldots, m.$$

   Step 4.   Test stopping condition:
   If more than one node has a nonzero activation, continue; otherwise, stop.

Note that in Step 2, the input to the function f is simply the total input to node A_j from all nodes, including itself. Some precautions should be incorporated to handle the situation in which two or more units have the same, maximal, input.

Example 4.1 Using a MAXNET

Consider the action of a MAXNET with four neurons and inhibitory weights $\epsilon = -0.2$ when given the initial activations (input signals)

$$a_1(0) = 0.2 \quad a_2(0) = 0.4 \quad a_3(0) = 0.6 \quad a_4(0) = 0.8$$

The activations found as the net iterates are

$$a_1(1) = 0.0 \quad a_2(1) = 0.08 \quad a_3(1) = 0.32 \quad a_4(1) = 0.56$$

$$a_1(2) = 0.0 \quad a_2(2) = 0.0 \quad a_3(2) = 0.192 \quad a_4(2) = 0.48$$

$$a_1(3) = 0.0 \quad a_2(3) = 0.0 \quad a_3(3) = 0.096 \quad a_4(3) = 0.442$$

$$a_1(4) = 0.0 \quad a_2(4) = 0.0 \quad a_3(4) = 0.008 \quad a_4(4) = 0.422$$

$$a_1(5) = 0.0 \quad a_2(5) = 0.0 \quad a_3(5) = 0.0 \quad a_4(5) = 0.421$$

Although we have shown the activations as a function of the iteration, it is not necessary in general to save all of the previous values; only the activations from the previous step are actually needed, as shown in the algorithm.

4.1.2 Mexican Hat

The Mexican Hat network [Kohonen, 1989a] is a more general contrast-enhancing subnet than the MAXNET. Each neuron is connected with excitatory (positively weighted) links to a number of "cooperative neighbors," neurons that are in close proximity. Each neuron is also connected with inhibitory links (with negative weights) to a number of "competitive neighbors," neurons that are somewhat further away. There may also be a number of neurons, further away still, to which the neuron is not connected. All of these connections are within a particular layer of a neural net, so, as in the case of MAXNET in Section 4.1.1, the neurons receive an external signal in addition to these interconnection signals. The pattern of interconnections just described is repeated for each neuron in the layer. The interconnection pattern for unit X_i is illustrated in Figure 4.2. For ease of description, the neurons are pictured as arranged in a linear order, with positive connections between unit X_i and neighboring units one or two positions on either side; negative connections are shown for units three positions on either side. The size of the region of cooperation (positive connections) and the region of competition (negative connections) may vary, as may the relative magnitudes of the positive and negative weights and the topology of the regions (linear, rectangular, hexagonal, etc.).

The contrast enhancement of the signal s_i received by unit X_i is accomplished by iteration for several time steps. The activation of unit X_i at time t is given by

$$x_i(t) = f[s_i(t) + \sum_k w_k x_{i+k}(t - 1)],$$

where the terms in the summation are the weighted signals from other units (cooperative and competitive neighbors) at the previous time step. In the example illustrated in Figure 4.2, the weight v_k from unit X_i to unit X_{i+k} is positive for $k = -2, -1, 0, 1$, and 2, negative for $k = -3$, and 3, and zero for units beyond these.

Architecture

The interconnections for the Mexican Hat net involve two symmetric regions around each individual neuron. The connection weights within the closer region— weights between a typical unit X_i and units X_{i+1}, X_{i+2}, X_{i-1}, and X_{i-2}, for example—are positive (and often are taken to have the same value). These weights are shown as w_1 and w_2 in Figure 4.2. The weights between X_i and units X_{i+3} and X_{i-3} are negative (shown as w_3 in the figure). Unit X_i is not connected to units X_{i-4} and X_{i+4} in this sample architecture. In the illustration, units within a radius of 2 to the typical unit X_i are connected with positive weights; units within a radius of 3, but outside the radius of positive connections, are connected with negative weights; and units further than 3 units away are not connected.

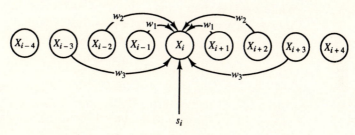

Figure 4.2   Mexican Hat interconnections for unit X_i.

Algorithm

The algorithm given here is similar to that presented by Kohonen [1989a]. The nomenclature we use is as follows:

R_1 Radius of region of interconnections; X_i is connected to units X_{i+k} and X_{i-k} for $k = 1, \ldots, R_1$.

R_2 Radius of region with positive reinforcement; $R_2 < R_1$.

$\mathbf{w}_k$ Weight on interconnections between X_i and units X_{i+k} and X_{i-k}: w_k is positive for $0 \le k \le R_2$, w_k is negative for $R_2 < k \le R_1$.

x   Vector of activations.

x_old   Vector of activations at previous time step.

t_max   Total number of iterations of contrast enhancement.

s   External signal.

As presented, the algorithm corresponds to the external signal being given only for the first iteration (Step 1) of the contrast-enhancing iterations. We have:

Step 0.   Initialize parameters t_max, R_1, R_2 as desired.
              Initialize weights:

$$w_k = C_1 \text{ for } k = 0, \ldots, R_1 \ (C_1 > 0)$$

$$w_k = C_2 \text{ for } k = R_1 + 1, \ldots, R_2 \ (C_2 < 0).$$

Initialize **x_old** to **0**.

Step 1.   Present external signal **s**:

$$\mathbf{x} = \mathbf{s}.$$

Save activations in array **x_old** (for $i = 1, \ldots, n$):

$$x\_old_i = x_i.$$

Set iteration counter: $t = 1$.

Step 2.   While t is less than t_max, do Steps 3–7.

Step 3.   Compute net input ($i = 1, \ldots, n$):

$$x_i = C_1 \sum_{k=-R_1}^{R_1} x\_old_{i+k}$$

$$+ C_2 \sum_{k=-R_2}^{-R_1-1} x\_old_{i+k} + C_2 \sum_{k=R_1+1}^{R_2} x\_old_{i+k}.$$

Step 4.   Apply activation function (ramp function from 0 to x_max, slope 1):

$$x_i = \min(x\_max, \max(0, x_i)) \ (i = 1, \ldots, n).$$

Step 5.   Save current activations in **x_old**:

$$x\_old_i = x_i \ (i = 1, \ldots, n).$$

Step 6.   Increment iteration counter:

$$t = t + 1.$$

Step 7.   Test stopping condition:
              If $t < t\_max$, continue; otherwise, stop.

In a computer implementation of the algorithm, one simple method of dealing with the units near the ends of the net, i.e., for i close to 1 or close to n, which receive input from less than the full range of units $i - R_1$ to $i + R_1$, is to dimension the array *x_old* from $1 - R_1$ to $n + R_1$ (rather than from 1 to n). Then, since only the components from 1 to n will be updated, the formulas in Step 3 will work correctly for all units.

The positive reinforcement from nearby units and negative reinforcement from units that are further away have the effect of increasing the activation of units with larger initial activations and reducing the activations of those that had a smaller external signal. This is illustrated in Example 4.2.

Application

Example 4.2 Using the Mexican Hat Algorithm

We illustrate the Mexican Hat algorithm for a simple net with seven units. The activation function for this net is

$$f(x) = \begin{cases} 0 & \text{if } x < 0 \\ x & \text{if } 0 \le x \le 2 \\ 2 & \text{if } 2 < x. \end{cases}$$

Step 0.   Initialize parameters:

$R_1 = 1;$

$R_2 = 2;$

$C_1 = 0.6;$

$C_2 = -0.4.$

Step 1. $(t = 0)$.

The external signal is (0.0, 0.5, 0.8, 1.0, 0.8, 0.5, 0.0), so

$\mathbf{x} = (0.0, 0.5, 0.8, 1.0, 0.8, 0.5, 0.0).$

Save in **x_old**:

$\mathbf{x\_old} = (0.0, 0.5, 0.8, 1.0, 0.8, 0.5, 0.0).$

Step 2. $(t = 1)$.

The update formulas used in Step 3 are listed as follows for reference:

$x_1 = 0.6\, x\_old_1 + 0.6\, x\_old_2 - 0.4\, x\_old_3$

$x_2 = 0.6\, x\_old_1 + 0.6\, x\_old_2 + 0.6\, x\_old_3 - 0.4\, x\_old_4$

$x_3 = -0.4\, x\_old_1 + 0.6\, x\_old_2 + 0.6\, x\_old_3 + 0.6\, x\_old_4 - 0.4\, x\_old_5$

$x_4 = -0.4\, x\_old_2 + 0.6\, x\_old_3 + 0.6\, x\_old_4 + 0.6\, x\_old_5 - 0.4\, x\_old_6$

$x_5 = -0.4\, x\_old_3 + 0.6\, x\_old_4 + 0.6\, x\_old_5 + 0.6\, x\_old_6 - 0.4\, x\_old_7$

$x_6 = -0.4\, x\_old_4 + 0.6\, x\_old_5 + 0.6\, x\_old_6 + 0.6\, x\_old_7$

$x_7 = -0.4\, x\_old_5 + 0.6\, x\_old_6 + 0.6\, x\_old_7.$

Step 3.   *(t = 1).*

$$x_1 = 0.6(0.0) + 0.6(0.5) - 0.4(0.8) = -0.2$$

$$x_2 = 0.6(0.0) + 0.6(0.5) + 0.6(0.8) - 0.4(1.0) = 0.38$$

$$x_3 = -0.4(0.0) + 0.6(0.5) + 0.6(0.8) + 0.6(1.0) - 0.4(0.8) = 1.06$$

$$x_4 = -0.4(0.5) + 0.6(0.8) + 0.6(1.0) + 0.6(0.8) - 0.4(0.5) = 1.16$$

$$x_5 = -0.4(0.8) + 0.6(1.0) + 0.6(0.8) + 0.6(0.5) - 0.4(0.0) = 1.06$$

$$x_6 = -0.4(1.0) + 0.6(0.8) + 0.6(0.5) + 0.6(0.0) = 0.38$$

$$x_7 = -0.4(0.8) + 0.6(0.5) + 0.6(0.0) = -0.2.$$

Step 4.

$$\mathbf{x} = (0.0, 0.38, 1.06, 1.16, 1.06, 0.38, 0.0).$$

Steps 5–7. Bookkeeping for next iteration.
Step 3.   *(t = 2).*

$$x_1 = 0.6(0.0) + 0.6(0.38) - 0.4(1.06) = -0.196$$

$$x_2 = 0.6(0.0) + 0.6(0.38) + 0.6(1.06) - 0.4(1.16) = 0.39$$

$$x_3 = -0.4(0.0) + 0.6(0.38) + 0.6(1.06) + 0.6(1.16) - 0.4(1.06) = 1.14$$

$$x_4 = -0.4(0.38) + 0.6(1.06) + 0.6(1.16) + 0.6(1.06) - 0.4(0.38) = 1.66$$

$$x_5 = -0.4(1.06) + 0.6(1.16) + 0.6(1.06) + 0.6(0.38) - 0.4(0.0) = 1.14$$

$$x_6 = -0.4(1.16) + 0.6(1.06) + 0.6(0.38) + 0.6(0.0) = 0.39$$

$$x_7 = -0.4(1.06) + 0.6(0.38) + 0.6(0.0) = -0.196$$

Step 4.

$$\mathbf{x} = (0.0, 0.39, 1.14, 1.66, 1.14, 0.39, 0.0).$$

Steps 5–7. Bookkeeping for next iteration.
The pattern of activations is shown for $t = 0$, 1, and 2 in Figure 4.3.

4.1.3 Hamming Net

A Hamming net [Lippmann, 1987; DARPA, 1988] is a maximum likelihood classifier net that can be used to determine which of several exemplar vectors is most similar to an input vector (an n-tuple). The exemplar vectors determine the weights of the net. The measure of similarity between the input vector and the stored exemplar vectors is n minus the Hamming distance between the vectors. The Hamming distance between two vectors is the number of components in which the vectors differ. For bipolar vectors $\mathbf{x}$ and $\mathbf{y}$,

$$\mathbf{x} \cdot \mathbf{y} = a - d,$$

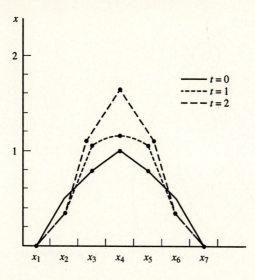

Figure 4.3 Results for Mexican Hat example.

where a is the number of components in which the vectors agree and d is the number of components in which the vectors differ, i.e., the Hamming distance. However, if n is the number of components in the vectors, then

$$d = n - a$$

and

$$\mathbf{x} \cdot \mathbf{y} = 2a - n,$$

or

$$2a = \mathbf{x} \cdot \mathbf{y} + n.$$

By setting the weights to be one-half the exemplar vector and setting the value of the bias to $n/2$, the net will find the unit with the closest exemplar simply by finding the unit with the largest net input. The Hamming net uses MAXNET as a subnet to find the unit with the largest net input.

The lower net consists of n input nodes, each connected to m output nodes (where m is the number of exemplar vectors stored in the net). The output nodes of the lower net feed into an upper net (MAXNET) that calculates the best exemplar match to the input vector. The input and exemplar vectors are bipolar.

Architecture

The sample architecture shown in Figure 4.4 assumes input vectors are 4-tuples, to be categorized as belonging to one of two classes.

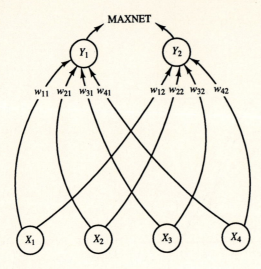

Figure 4.4 Hamming net.

Application

Given a set of m bipolar exemplar vectors, $e(1)$, $e(2)$, . . . , $e(m)$, the Hamming net can be used to find the exemplar that is closest to the bipolar input vector x. The net input y_in_j to unit Y_j, gives the number of components in which the input vector and the exemplar vector for unit Y_j $e(j)$, agree (n minus the Hamming distance between these vectors).

The nomenclature we use is as follows:

n number of input nodes, number of components of any input vector;
m number of output nodes, number of exemplar vectors;
$e(j)$ the jth exemplar vector:

$$e(j) = (e_1(j), \ldots , e_i(j), \ldots , e_n(j)).$$

The application procedure for the Hamming net is:

Step 0.   To store the m exemplar vectors, initialize the weights:

$$w_{ij} = \frac{e_i(j)}{2}, (i = 1, \ldots , n; j = 1, \ldots , m).$$

And initialize the biases:

$$b_j = \frac{n}{2}, (j = 1, \ldots , m).$$

Step 1.   For each vector x, do Steps 2–4.

Step 2.   Compute the net input to each unit Y_j:

$$y\_in_j = b_j + \sum_i x_i w_{ij}, (j = 1, \ldots, m).$$

Step 3.   Initialize activations for MAXNET:

$$y_j(0) = y\_in_j, (j = 1, \ldots, m).$$

Step 4.   MAXNET iterates to find the best match exemplar.

Example 4.3 A Hamming net to cluster four vectors

Given the exemplar vectors

$$\mathbf{e}(1) = (1, -1, -1, -1)$$

and

$$\mathbf{e}(2) = (-1, -1, -1, 1),$$

the Hamming net can be used to find the exemplar that is closest to each of the bipolar input patterns, $(1, 1, -1, -1)$, $(1, -1, -1, -1)$, $(-1, -1, -1, 1)$, and $(-1, -1, 1, 1)$.

Step 0.   Store the m exemplar vectors in the weights:

$$\mathbf{W} = \begin{bmatrix} .5 & -.5 \\ -.5 & -.5 \\ -.5 & -.5 \\ -.5 & .5 \end{bmatrix}.$$

Initialize the biases:

$$b_1 = b_2 = 2.$$

Step 1.   For the vector $\mathbf{x} = (1, 1, -1, -1)$, do Steps 2–4.

Step 2. $y\_in_1 = b_1 + \sum_i x_i w_{i1}$

$$= 2 + 1 = 3;$$

$$y\_in_2 = b_2 + \sum_i x_i w_{i2}$$

$$= 2 - 1 = 1.$$

These values represent the Hamming similarity because $(1, 1, -1, -1)$ agrees with $\mathbf{e}(1) = (1, -1, -1, -1)$ in the first, third, and fourth components and because $(1, 1, -1, -1)$ agrees with $\mathbf{e}(2) = (-1, -1, -1, 1)$ in only the third component.

Step 3. $y_1(0) = 3;$

$$y_2(0) = 1.$$

Step 4.   Since $y_1(0) > y_2(0)$, MAXNET will find that unit Y_1 has the best match exemplar for input vector $\mathbf{x} = (1, 1, -1, -1)$.

Step 1.   For the vector $\mathbf{x} = (1, -1, -1, -1)$, do Steps 2–4.

Step 2. $y\_in_1 = b_1 + \sum_i x_i w_{i1}$

$= 2 + 2 = 4;$

$y\_in_2 = b_2 + \sum_i x_i w_{i2}$

$= 2 - 0 = 2.$

Note that the input vector agrees with $\mathbf{e}(1)$ in all four components and agrees with $\mathbf{e}(2)$ in the second and third components.

Step 3. $y_1(0) = 4;$

$y_2(0) = 2.$

Step 4.   Since $y_1(0) > y_2(0)$, MAXNET will find that unit Y_1 has the best match exemplar for input vector $\mathbf{x} = (1, -1, -1, -1)$.

Step 1.   For the vector $\mathbf{x} = (-1, -1, -1, 1)$, do Steps 2–4.

Step 2. $y\_in_1 = b_1 + \sum_i x_i w_{i1}$

$= 2 + 0 = 2;$

$y\_in_2 = b_2 + \sum_i x_i w_{i2}$

$= 2 + 2 = 4.$

The input vector agrees with $\mathbf{e}(1)$ in the second and third components and agrees with $\mathbf{e}(2)$ in all four components.

Step 3. $y_1(0) = 2;$

$y_2(0) = 4.$

Step 4.   Since $y_2(0) > y_1(0)$, MAXNET will find that unit Y_2 has the best match exemplar for input vector $\mathbf{x} = (-1, -1, -1, 1)$.

Step 1.   For the vector $\mathbf{x} = (-1, -1, 1, 1)$, do Steps 2–4.

Step 2. $y\_in_1 = b_1 + \sum_i x_i w_{i1}$

$= 2 - 1 = 1;$

$y\_in_2 = b_2 + \sum_i x_i w_{i2}$

$= 2 + 1 = 3.$

The input vector agrees with $\mathbf{e}(1)$ in the second component and agrees with $\mathbf{e}(2)$ in the first, second, and fourth components.

Step 3. $y_1(0) = 1;$

$y_2(0) = 3.$

Step 4.   Since $y_2(0) > y_1(0)$, MAXNET will find that unit Y_2 has the best match exemplar for input vector $\mathbf{x} = (-1, -1, 1, 1)$.

4.2 KOHONEN SELF-ORGANIZING MAPS

The self-organizing neural networks described in this section, also called *topology-preserving maps,* assume a topological structure among the cluster units. This property is observed in the brain, but is not found in other artificial neural networks. There are *m* cluster units, arranged in a one- or two-dimensional array; the input signals are *n*-tuples [Kohonen, 1989a].

The weight vector for a cluster unit serves as an exemplar of the input patterns associated with that cluster. During the self-organization process, the cluster unit whose weight vector matches the input pattern most closely (typically, the square of the minimum Euclidean distance) is chosen as the winner. The winning unit and its neighboring units (in terms of the topology of the cluster units) update their weights. The weight vectors of neighboring units are not, in general, close to the input pattern. For example, for a linear array of cluster units, the neighborhood of radius *R* around cluster unit *J* consists of all units *j* such that $\max(1, J - R) \leq j \leq \min(J + R, m))$.

The architecture and algorithm that follow for the net can be used to cluster a set of *p* continuous-valued vectors $\mathbf{x} = (x_1, \ldots, x_i, \ldots, x_n)$ into *m* clusters. Note that the connection weights do not multiply the signal sent from the input units to the cluster units (unless the dot product measure of similarity is being used).

4.2.1 Architecture

The architecture of the Kohonen self-organizing map is shown in Figure 4.5.

Neighborhoods of the unit designated by # of radii $R = 2$, 1, and 0 in a one-dimensional topology (with 10 cluster units) are shown in Figure 4.6.

The neighborhoods of radii $R = 2$, 1 and 0 are shown in Figure 4.7 for a rectangular grid and in Figure 4.8 for a hexagonal grid (each with 49 units). In each illustration, the winning unit is indicated by the symbol "#" and the other units are denoted by "*."

Note that each unit has eight nearest neighbors in the rectangular grid, but only six in the hexagonal grid. Winning units that are close to the edge of the grid will have some neighborhoods that have fewer units than that shown in the respective figure. (Neighborhoods do not "wrap around" from one side of the grid to the other; "missing" units are simply ignored.)

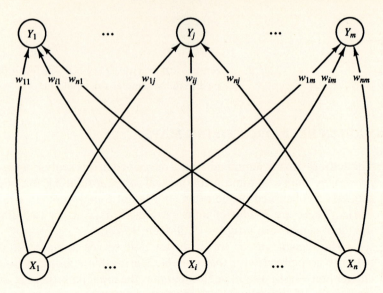

Figure 4.5 Kohonen self-organizing map.

$$*\qquad *\qquad *\qquad \{\ *\qquad (\ *\qquad [\#]\qquad *\ )\qquad *\ \}\qquad *\qquad *$$

$$\{\ \}\,R=2\qquad\qquad (\ )\,R=1\qquad\qquad [\ ]\,R=0$$

Figure 4.6 Linear array of cluster units.

4.2.2 Algorithm

Step 0.   Initialize weights w_{ij}. (Possible choices are discussed below.)
Set topological neighborhood parameters.
Set learning rate parameters.

Step 1.   While stopping condition is false, do Steps 2–8.

  Step 2.   For each input vector **x**, do Steps 3–5.

    Step 3.   For each j, compute:

$$D(j) = \sum_i (w_{ij} - x_i)^2.$$

    Step 4.   Find index J such that $D(J)$ is a minimum.

    Step 5.   For all units j within a specified neighborhood of J, and for all i:

$$w_{ij}(\text{new}) = w_{ij}(\text{old}) + \alpha[x_i - w_{ij}(\text{old})].$$

  Step 6.   Update learning rate.

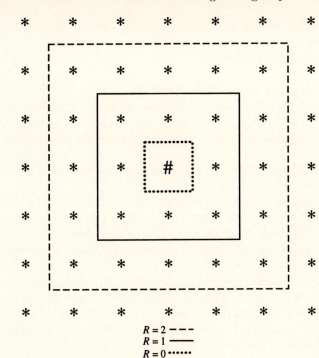

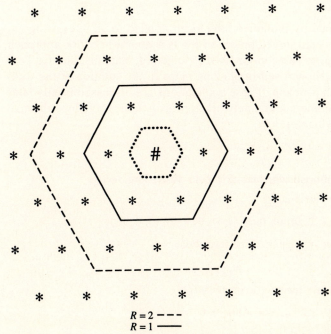

$R = 2$ - - -
$R = 1$ ———
$R = 0$ ••••••

Figure 4.7 Neighborhoods for rectangular grid.

$R = 2$ - - -
$R = 1$ ———
$R = 0$ ••••••

Figure 4.8 Neighborhoods for hexagonal grid.

Step 7.   Reduce radius of topological neighborhood at specified times.
Step 8.   Test stopping condition.

Alternative structures are possible for reducing R and α.

The learning rate α is a slowly decreasing function of time (or training epochs). Kohonen (1989a, p. 133) indicates that a linearly decreasing function is satisfactory for practical computations; a geometric decrease would produce similar results.

The radius of the neighborhood around a cluster unit also decreases as the clustering process progresses.

The formation of a map occurs in two phases: the initial formation of the correct order and the final convergence. The second phase takes much longer than the first and requires a small value for the learning rate. Many iterations through the training set may be necessary, at least in some applications [Kohonen, 1989a].

Random values may be assigned for the initial weights. If some information is available concerning the distribution of clusters that might be appropriate for a particular problem, the initial weights can be taken to reflect that prior knowledge. In Examples 4.4–4.9, the weights are initialized to random values (chosen from the same range of values as the components of the input vectors).

4.2.3 Application

Neural networks developed by Kohonen have been applied to an interesting variety of problems. One recent development of his is a neural network approach to computer-generated music [Kohonen, 1989b]. Angeniol, Vaubois, and Le Texier (1988) have applied Kohonen self-organizing maps to the solution of the well-known traveling salesman problem. These applications are discussed briefly later in this section. A more common neural network approach to the traveling salesman problem is discussed in Chapter 7.

Simple example

Example 4.4 A Kohonen self-organizing map (SOM) to cluster four vectors

Let the vectors to be clustered be

$$(1, 1, 0, 0); (0, 0, 0, 1); (1, 0, 0, 0); (0, 0, 1, 1).$$

The maximum number of clusters to be formed is

$$m = 2.$$

Suppose the learning rate (geometric decrease) is

$$\alpha(0) = .6,$$

$$\alpha(t + 1) = .5 \, \alpha(t).$$

With only two clusters available, the neighborhood of node J (Step 4) is set so that only one cluster updates its weights at each step (i.e., $R = 0$).

Step 0.   Initial weight matrix:

$$\begin{bmatrix} .2 & .8 \\ .6 & .4 \\ .5 & .7 \\ .9 & .3 \end{bmatrix}.$$

Initial radius:

$R = 0.$

Initial learning rate:

$\alpha(0) = 0.6.$

Step 1.   Begin training.

Step 2.   For the first vector, $(1, 1, 0, 0)$, do Steps 3–5.

Step 3. $D(1) = (.2 - 1)^2 + (.6 - 1)^2$
$$+ (.5 - 0)^2 + (.9 - 0)^2 = 1.86;$$

$D(2) = (.8 - 1)^2 + (.4 - 1)^2$
$$+ (.7 - 0)^2 + (.3 - 0)^2 = 0.98.$$

Step 4.   The input vector is closest to output node 2, so
$$J = 2.$$

Step 5.   The weights on the winning unit are updated:
$$w_{i2}(\text{new}) = w_{i2}(\text{old}) + .6\,[x_i - w_{i2}(\text{old})]$$
$$= .4\,w_{i2}(\text{old}) + .6\,x_i.$$

This gives the weight matrix

$$\begin{bmatrix} .2 & .92 \\ .6 & .76 \\ .5 & .28 \\ .9 & .12 \end{bmatrix}.$$

Step 2.   For the second vector, $(0, 0, 0, 1)$, do Steps 3–5.

Step 3.
$$D(1) = (.2 - 0)^2 + (.6 - 0)^2$$
$$+ (.5 - 0)^2 + (.9 - 1)^2 = 0.66;$$

$$D(2) = (.92 - 0)^2 + (.76 - 0)^2$$
$$+ (.28 - 0)^2 + (.12 - 1)^2 = 2.2768.$$

Step 4.   The input vector is closest to output node 1, so
$$J = 1.$$

Step 5.   Update the first column of the weight matrix:

$$\begin{bmatrix} .08 & .92 \\ .24 & .76 \\ .20 & .28 \\ .96 & .12 \end{bmatrix}.$$

Step 2.   For the third vector, (1, 0, 0, 0), do Steps 3–5.

Step 3.

$$D(1) = (.08 - 1)^2 + (.24 - 0)^2$$
$$+ (.2 - 0)^2 + (.96 - 0)^2 = 1.8656;$$

$$D(2) = (.92 - 1)^2 + (.76 - 0)^2$$
$$+ (.28 - 0)^2 + (.12 - 0)^2 = 0.6768.$$

Step 4.   The input vector is closest to output node 2, so

$$J = 2.$$

Step 5.   Update the second column of the weight matrix:

$$\begin{bmatrix} .08 & .968 \\ .24 & .304 \\ .20 & .112 \\ .96 & .048 \end{bmatrix}.$$

Step 2.   For the fourth vector, (0, 0, 1, 1), do Steps 3–5.

Step 3.

$$D(1) = (.08 - 0)^2 + (.24 - 0)^2$$
$$+ (.2 - 1)^2 + (.96 - 1)^2 = 0.7056;$$

$$D(2) = (.968 - 0)^2 + (.304 - 0)^2$$
$$+ (.112 - 1)^2 + (.048 - 1)^2 = 2.724.$$

Step 4.

$$J = 1.$$

Step 5.   Update the first column of the weight matrix:

$$\begin{bmatrix} .032 & .968 \\ .096 & .304 \\ .680 & .112 \\ .984 & .048 \end{bmatrix}.$$

Step 6.   Reduce the learning rate:

$$\alpha = .5 \, (0.6) = .3$$

The weight update equations are now

$$w_{ij}(\text{new}) = w_{ij}(\text{old}) + .3 \, [x_i - w_{ij}(\text{old})]$$

$$= .7 w_{ij}(\text{old}) + .3 x_i.$$

The weight matrix after the second epoch of training is

$$\begin{bmatrix} .016 & .980 \\ .047 & .360 \\ .630 & .055 \\ .999 & .024 \end{bmatrix}$$

Modifying the adjustment procedure for the learning rate so that it decreases geometrically from .6 to .01 over 100 iterations (epochs) gives the following results:

Iteration 0:   Weight matrix: $\begin{bmatrix} .2 & .8 \\ .6 & .4 \\ .5 & .7 \\ .9 & .3 \end{bmatrix}$

Iteration 1:   Weight matrix: $\begin{bmatrix} .032 & .970 \\ .096 & .300 \\ .680 & .110 \\ .980 & .048 \end{bmatrix}$

Iteration 2:   Weight matrix: $\begin{bmatrix} .0053 & .9900 \\ -.1700 & .3000 \\ .7000 & .0200 \\ 1.0000 & .0086 \end{bmatrix}$

Iteration 10:   Weight matrix: $\begin{bmatrix} 1.5e\text{-}7 & 1.0000 \\ 4.6e\text{-}7 & .3700 \\ .6300 & 5.4e\text{-}7 \\ 1.0000 & 2.3e\text{-}7 \end{bmatrix}$

Iteration 50:   Weight matrix: $\begin{bmatrix} 1.9e\text{-}19 & 1.0000 \\ 5.7e\text{-}15 & .4700 \\ .5300 & 6.6e\text{-}15 \\ 1.0000 & 2.8e\text{-}15 \end{bmatrix}$

Iteration 100: Weight matrix: $\begin{bmatrix} 6.7e\text{-}17 & 1.0000 \\ 2.0e\text{-}16 & .4900 \\ .5100 & 2.3e\text{-}16 \\ 1.0000 & 1.0e\text{-}16 \end{bmatrix}$

These weight matrices appear to be converging to the matrix

$$\begin{bmatrix} 0.0 & 1.0 \\ 0.0 & 0.5 \\ 0.5 & 0.0 \\ 1.0 & 0.0 \end{bmatrix},$$

column vectors define the center of each cluster

the first column of which is the average of the two vectors placed in cluster 1 and the second column of which is the average of the two vectors placed in cluster 2.

Character Recognition

Examples 4.5–4.7 show typical results from using a Kohonen self-organizing map to cluster input patterns representing letters in three different fonts. The input patterns for fonts 1, 2, and 3 are given in Figure 4.9. In each of the examples, 25 cluster units are available, which means that a maximum of 25 clusters may be formed. Results are shown only for the units that are actually the winning unit for some input pattern after training. The effect of the topological structure is seen in the contrast between Example 4.5 (in which there is no structure), Example 4.6 (in which there is a linear structure as described before), and Example 4.7 (in which a rectangular structure is used). In each example, the learning rate is reduced linearly from an initial value of .6 to a final value of .01.

Example 4.5 A SOM to cluster letters from different fonts: no topological structure

If no structure is assumed for the cluster units, i.e., if only the winning unit is allowed to learn the pattern presented, the 21 patterns form 5 clusters:

UNIT	PATTERNS
3	C1, C2, C3
13	B1, B3, D1, D3, E1, K1, K3
16	A1, A2, A3
18	J1, J2, J3
24	B2, D2, E2, K2

Example 4.6 A SOM to cluster letters from different fonts: linear structure

A linear structure (with $R = 1$) gives a better distribution of the patterns onto the available cluster units. The winning node J and its topological neighbors ($J + 1$ and $J - 1$) are allowed to learn on each iteration. Note that in general, the neighboring nodes that learn do not initially have weight vectors that are particularly close to the input pattern.

UNIT	PATTERNS	UNIT	PATTERNS
6	K2	20	C1, C2, C3
10	J1, J2, J3	22	D2
14	E1, E3	23	B2, E2
16	K1, K3	25	A1, A2, A3
18	B1, B3, D1, D3		

Note also that in many cases there are unused units between a pair of units that have clusters of patterns associated with them. This suggests that units which are being pulled in opposite directions during training do not learn any pattern very well. (In other words, in most cases, these input patterns form very distinct classes.)

Example 4.7 A SOM to cluster letters from different fonts: diamond structure

In this example, a simple two-dimensional topology is assumed for the cluster units, so that each cluster unit is indexed by two subscripts. If unit X_{IJ} is the winning unit, the units $X_{I+1,J}$, $X_{I-1,J}$, $X_{I,J+1}$, and $X_{I,J-1}$ also learn. This gives a diamond to-

Input from
Font 1

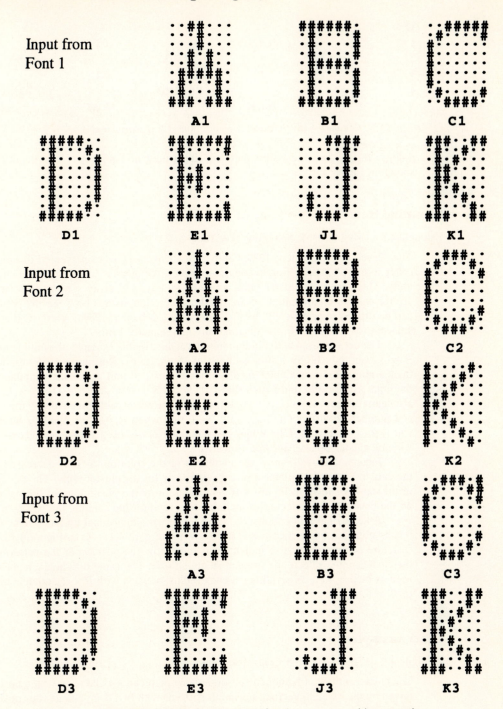

Figure 4.9 Training input patterns for character recognition examples.

i\j	1	2	3	4	5
1		J1, J2, J3		D2	
2	C1, C2, C3		D1, D3		B2, E2
3		B1		K2	
4			E1, E3, B3		A3
5		K1, K3		A1, A2	

Figure 4.10   Character recognition with rectangular grid.

pology, rather than the entire rectangle illustrated in Figure 4.7. The results are shown in Figure 4.10.

Spanning tree

Example 4.8   Using a SOM: Spanning Tree Data

The 32 vectors [Kohonen, 1989a] shown in Figure 4.11 were presented in random order to a Kohonen self-organizing map with a rectangular topology on its cluster units. There were 70 cluster units arranged in a 10×7 array. The pattern names are for ease of identification of the results. The relationships between the patterns can be displayed graphically, as in Figure 4.12 [Kohonen, 1989a]; patterns that are adjacent to each other in the diagram differ by exactly 1 bit.

The net was used with random initial weights. In this example, the initial radius, $R = 3$, was reduced by 1 after each set of 75 iterations. During these 75 iterations, the learning rate was reduced linearly from .6 to .01. If unit $X_{I,J}$ is the winning unit, the units $X_{i,j}$ for all i and j such that $I - R \leq i \leq I + R$ and $J - R \leq j \leq J + R$ also learn (unless the value of i or j falls outside the permissible range for the topology and number of cluster units chosen). Note that when $R = 3$, as many as 49 units will learn (see Figure 4.7). When the Kohonen net is used with $R = 0$, only the winning cluster node is allowed to learn.

Figures 4.13–4.16 show the evolution of the solution, as R is decreased, for the data in Figure 4.11, using a rectangular array for the cluster units. The structure of the data is shown in Figure 4.16 to indicate how the positioning of the patterns on the cluster units reflects the spanning tree relationships among the patterns.

A hexagonal grid can also be used for a two-dimensional topology. The final results obtained using such a grid are shown in Figure 4.17. As in Figure 4.16, the structure of the data is also indicated to show how the position of the patterns on the cluster units reflects the original spanning tree. The same iteration scheme was used as before, i.e., 75 iterations at each radius, starting with $R = 3$ and decreasing to $R = 0$.

Other examples

Example 4.9   Using a SOM: A Geometric Example

The cluster units in a Kohonen self-organizing map can viewed as having a position (given by their weight vector). For input patterns with two components, this position is easy to represent graphically. The topological relationships between cluster units

PATTERN	COMPONENTS				
A	1	0	0	0	0
B	2	0	0	0	0
C	3	0	0	0	0
D	4	0	0	0	0
E	5	0	0	0	0
F	3	1	0	0	0
G	3	2	0	0	0
H	3	3	0	0	0
I	3	4	0	0	0
J	3	5	0	0	0
K	3	3	1	0	0
L	3	3	2	0	0
M	3	3	3	0	0
N	3	3	4	0	0
O	3	3	5	0	0
P	3	3	6	0	0
Q	3	3	7	0	0
R	3	3	8	0	0
S	3	3	3	1	0
T	3	3	3	2	0
U	3	3	3	3	0
V	3	3	3	4	0
W	3	3	6	1	0
X	3	3	6	2	0
Y	3	3	6	3	0
Z	3	3	6	4	0
1	3	3	6	2	1
2	3	3	6	2	2
3	3	3	6	2	3
4	3	3	6	2	4
5	3	3	6	2	5
6	3	3	6	2	6

Figure 4.11   Spanning tree test data [Kohonen, 1989a].

```
A  B  C  D  E
      F
      G
      H  K  L  M  N  O  P  Q  R
      I        S        W
      J        T        X  1  2  3  4  5  6
               U        Y
               V        Z
```

Figure 4.12   Spanning tree test data structure [Kohonen, 1989a].

I,J	H,K	–	G	F	–	A,B,C,D,E
–	–	–	–	–	–	–
–	–	–	–	–	–	–
L,M	–	–	–	–	–	–
S	T,U	–	–	V	–	Y,Z
N	–	–	–	–	X	–
O	–	–	–	–	–	–
–	–	–	1	–	2	–
P	–	W	–	–	–	3
Q,R	–	–	–	–	–	4,5,6

Figure 4.13   Results after 75 iterations with $R = 3$.

H,I,J	–	G	A,B	F	C	D,E
K	–	–	–	–	–	–
–	–	–	–	–	–	–
L,M	S	–	T,U	–	–	Y,Z
–	–	–	V	–	–	X
N	–	–	–	–	–	–
O	–	–	1	–	–	–
P	W	–	2	3	–	–
–	–	–	–	–	–	–
Q,R	–	–	–	–	4	5,6

Figure 4.14   Results after 75 additional iterations with $R = 2$.

I,J	–	G	A	B	C	D,E
–	H	–	–	–	F	–
–	K	–	–	–	–	–
L	–	–	–	V	–	Z
M	–	S	T	U	–	Y
N	–	–	–	–	–	X
O	–	–	W	–	1	–
P	–	–	–	–	2	–
Q	–	6	–	4	–	H
R	–	–	5	–	–	3

Figure 4.15   Results after 75 more iterations with $R = 1$.

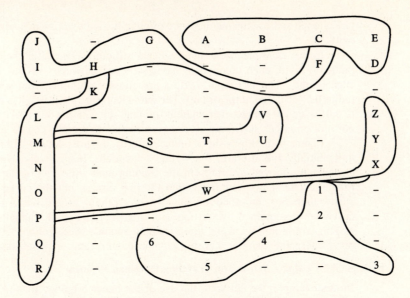

Figure 4.16   Results after another 75 iterations with $R = 0$.

Figure 4.17   Results of spanning tree example using hexagonal array.

in Kohonen self-organizing maps are often indicated by drawing lines connecting the units.

In this example, we assume a linear structure. The initial weights are chosen randomly, with each component having a value between -1 and 1. There are 50 cluster units. The 100 input vectors are chosen randomly from within a circle of radius 0.5 (centered at the origin). The initial learning rate is 0.5; it is reduced linearly to 0.01 over 100 epochs. Throughout training, the winning unit and its nearest neighbor unit on either side (units J, $J + 1$, and $J - 1$) are allowed to learn.

Figure 4.18 shows the training patterns. Figures 4.19–4.23 show the cluster units initially and after 10, 20, 30, and 100 epochs, respectively. Not only have the cluster units moved to represent the training inputs (i.e., all of the weight vectors for the cluster units now fall within the unit circle), but the curve connecting the cluster units has smoothed out somewhat as training progresses. An even smoother curve can be obtained by starting with a larger radius and gradually reducing it to 0. This would involve using more training epochs. (See Kohonen, 1989a for many other interesting examples of this geometric interpretation of self-organizing maps.)

Example 4.10 Using a SOM: The Traveling Salesman Problem

In this example, we illustrate the use of the linear topology for the cluster units in a Kohonen self-organizing map to solve a classic problem in constrained optimization, the so-called traveling salesman problem (TSP). Several nets that are designed for constrained optimization problems are discussed in Chapter 7. The aim of the TSP is to find a tour of a given set of cities that is of minimum length. A tour consists

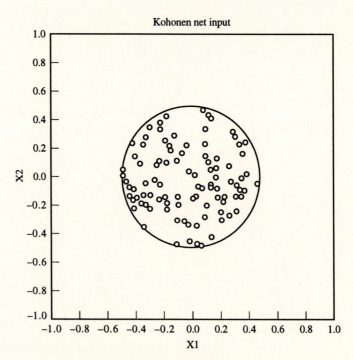

Figure 4.18 Input patterns.

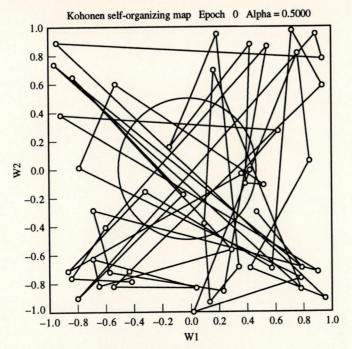

Figure 4.19   Initial cluster units.

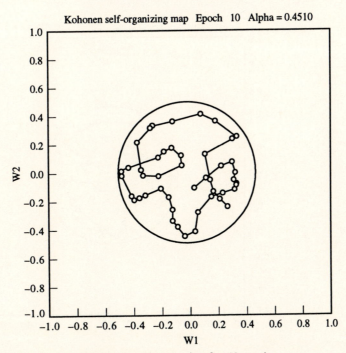

Figure 4.20   Cluster units after 10 epochs.

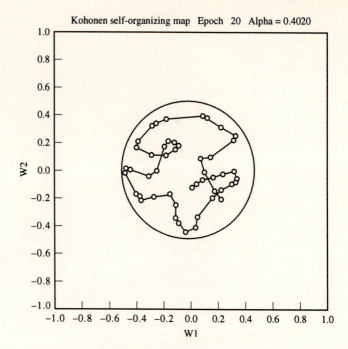

Figure 4.21   Cluster units after 20 epochs.

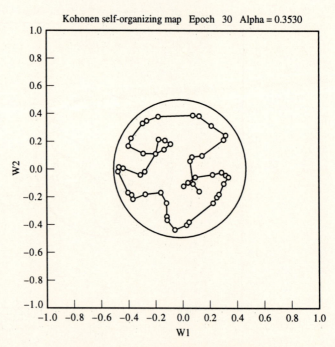

Figure 4.22   Cluster units after 30 epochs.

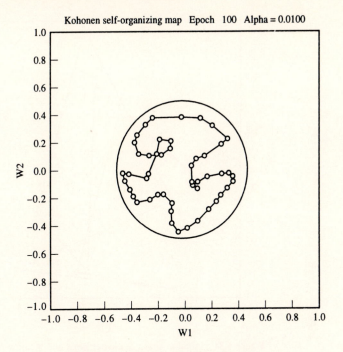

Kohonen self-organizing map   Epoch   100   Alpha = 0.0100

Figure 4.23   Cluster units after 100 epochs.

of visiting each city exactly once and returning to the starting city. Angeniol, Vaubois, and Le Texier (1988) have illustrated the use of a Kohonen net to solve the TSP. The net uses the city coordinates as input; there are as many cluster units as there are cities to be visited. The net has a linear topology, with the first and last cluster unit also connected. Figure 4.24 shows the initial random position of the cluster units; Figure 4.25 shows the results after 100 epochs of training with $R = 1$ (learning rate decreasing from 0.5 to 0.4). The final tour after 100 epochs of training with $R = 0$ is shown in Figure 4.26.

This tour is ambiguous in terms of the order in which city B and city C are visited, because one cluster unit is positioned midway between the cities (rather than being directly on one city). Another unit has been trapped between city J and cities B and C; it is not being chosen as the winner when any input is presented and is therefore "wasted." However, the results can easily be interpreted as representing one of the tours

$$A\ D\ E\ F\ G\ H\ I\ J\ B\ C$$

and

$$A\ D\ E\ F\ G\ H\ I\ J\ C\ B.$$

The coordinates of and distances between the cities are given in Chapter 7.

The same tour (with the same ambiguity) was found, using a variety of initial weights. Choosing initial weights within a small region of the input space (the center or any of the four corners), as is often done, did not change the results.

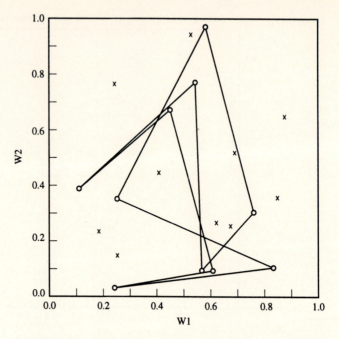

Figure 4.24   Initial position of cluster units and location of cities

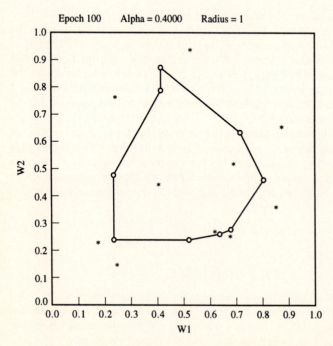

Figure 4.25   Position of cluster units and location of cities after 100 epochs with $R = 1$.

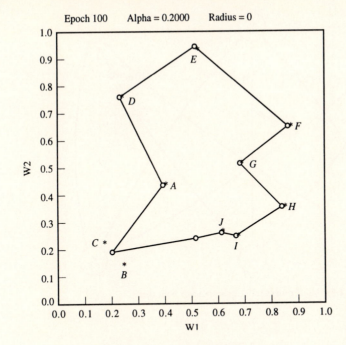

Figure 4.26 Position of cluster units and location of cities after additional 100 epochs with $R = 0$.

4.3 LEARNING VECTOR QUANTIZATION

Learning vector quantization (LVQ) [Kohonen, 1989a, 1990a] is a pattern classification method in which each output unit represents a particular class or category. (Several output units should be used for each class.) The weight vector for an output unit is often referred to as a *reference* (or *codebook*) vector for the class that the unit represents. During training, the output units are positioned (by adjusting their weights through supervised training) to approximate the decision surfaces of the theoretical Bayes classifier. It is assumed that a set of training patterns with known classifications is provided, along with an initial distribution of reference vectors (each of which represents a known classification).

After training, an LVQ net classifies an input vector by assigning it to the same class as the output unit that has its weight vector (reference vector) closest to the input vector.

4.3.1 Architecture

The architecture of an LVQ neural net, shown in Figure 4.27, is essentially the same as that of a Kohonen self-organizing map (without a topological structure being assumed for the output units). In addition, each output unit has a known class that it represents.

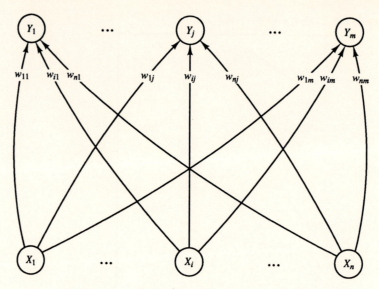

Figure 4.27 Learning vector quantization neural net.

4.3.2 Algorithm

The motivation for the algorithm for the LVQ net is to find the output unit that is closest to the input vector. Toward that end, if $\mathbf{x}$ and $\mathbf{w}_c$ belong to the same class, then we move the weights toward the new input vector; if $\mathbf{x}$ and $\mathbf{w}_c$ belong to different classes, then we move the weights away from this input vector.

The nomenclature we use is as follows:

$\mathbf{x}$	training vector $(x_1, \ldots, x_i, \ldots, x_n)$.
T	correct category or class for the training vector.
$\mathbf{w}_j$	weight vector for jth output unit $(w_{1j}, \ldots, w_{ij}, \ldots, w_{nj})$.
C_j	category or class represented by jth output unit.
$\|\mathbf{x} - \mathbf{w}_j\|$	Euclidean distance between input vector and (weight vector for) jth output unit.

Step 0. Initialize reference vectors (several strategies are discussed shortly); initialize learning rate, $\alpha(0)$.

Step 1. While stopping condition is false, do Steps 2–6.

    Step 2. For each training input vector $\mathbf{x}$, do Steps 3–4.

        Step 3. Find J so that $\|\mathbf{x} - \mathbf{w}_J\|$ is a minimum.

Step 4.   Update $\mathbf{w}_J$ as follows:
if $T = C_J$, then

$$\mathbf{w}_J(\text{new}) = \mathbf{w}_J(\text{old}) + \alpha[\mathbf{x} - \mathbf{w}_J(\text{old})];$$

if $T \neq C_J$, then

$$\mathbf{w}_J(\text{new}) = \mathbf{w}_J(\text{old}) - \alpha[\mathbf{x} - \mathbf{w}_J(\text{old})].$$

Step 5.   Reduce learning rate.

Step 6.   Test stopping condition:
The condition may specify a fixed number of iterations
(i.e., executions of Step 1) or the learning rate reaching a
sufficiently small value.

4.3.3 Application

The simplest method of initializing the weight (reference) vectors is to take the
first m training vectors and use them as weight vectors; the remaining vectors
are then used for training [Kohonen, 1989a]. This is the method of Example 4.11.
Another simple method, illustrated in Example 4.12, is to assign the initial weights
and classifications randomly.

Another possible method of initializing the weights is to use K-means clustering [Makhoul, Roucos, & Gish, 1985] or the self-organizing map [Kohonen,
1989a] to place the weights. Each weight vector is then calibrated by determining
the input patterns that are closest to it, finding the class that the largest number
of these input patterns belong to, and assigning that class to the weight vector.

Simple Example

Example 4.11 Learning vector quantization (LVQ): five vectors assigned to two classes

In this very simple example, two reference vectors will be used. The following input
vectors represent two classes, 1 and 2:

VECTOR	CLASS
(1, 1, 0, 0)	1
(0, 0, 0, 1)	2
(0, 0, 1, 1)	2
(1, 0, 0, 0)	1
(0, 1, 1, 0)	2

The first two vectors will be used to initialize the two reference vectors. Thus, the
first output unit represents class 1, the second class 2 (symbolically, $C_1 = 1$ and
$C_2 = 2$). This leaves vectors (0, 0, 1, 1), (1, 0, 0, 0), and (0, 1, 1, 0) as the training
vectors. Only one iteration (one epoch) is shown:

Step 0.   Initialize weights:

$$\mathbf{w}_1 = (1, 1, 0, 0);$$

$$\mathbf{w}_2 = (0, 0, 0, 1).$$

Initialize the learning rate: $\alpha = .1$.

Step 1.   Begin computations.

Step 2.   For input vector $\mathbf{x} = (0, 0, 1, 1)$ with $T = 2$, do Steps 3–4.

Step 3. $J = 2$, since $\mathbf{x}$ is closer to $\mathbf{w}_2$ than to $\mathbf{w}_1$.

Step 4.   Since $T = 2$ and $C_2 = 2$, update $\mathbf{w}_2$ as follows:

$$\mathbf{w}_2 = (0,0,0,1) + .1\,[(0,0,1,1) - (0,0,0,1)]$$

$$= (0,0,.1,1).$$

Step 2.   For input vector $\mathbf{x} = (1, 0, 0, 0)$ with $T = 1$, do Steps 3–4.

Step 3. $J = 1$.

Step 4.   Since $T = 1$ and $C_1 = 1$, update $\mathbf{w}_1$ as follows:

$$\mathbf{w}_1 = (1,1,0,0) + .1\,[(1,0,0,0) - (1,1,0,0)]$$

$$= (1,.9,0,0).$$

Step 2.   For input vector $\mathbf{x} = (0, 1, 1, 0)$ with $T = 2$, do Steps 3–4.

Step 3. $J = 1$.

Step 4.   Since $T = 2$, but $C_1 = 1$, update $\mathbf{w}_1$ as follows:

$$\mathbf{w}_1 = (1,.9,0,0) - .1\,[(0,1,1,0) - (1,.9,0,0)]$$

$$= (1.1,.89, -.1,0).$$

Step 5.   This completes one epoch of training.
Reduce the learning rate.

Step 6.   Test the stopping condition.

Geometric example

Example 4.12 Using LVQ: a geometric example with four cluster units

This example shows the use of LVQ to represent points in the unit square as belonging to one of four classes, indicated in Figures 4.28–4.33 by the symbols +, 0, #, and @. There are four cluster units, one for each class. The weights are initialized so that the cluster units are positioned in the four corners of the input region when training starts:

	INITIAL WEIGHTS	
Class 1 (+)	0	0
Class 2 (0)	1	0
Class 3 (@)	1	1
Class 4 (#)	0	1

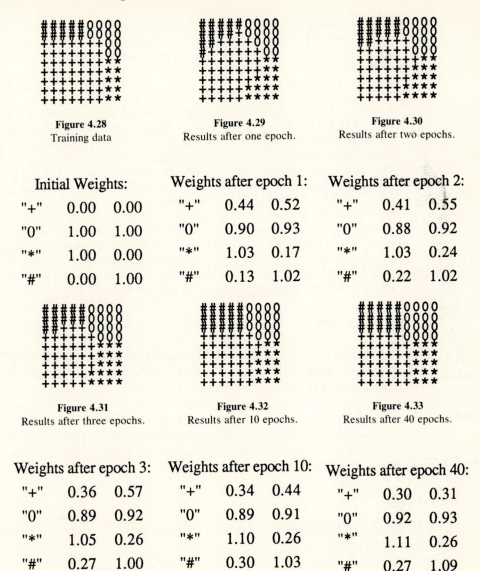

Figure 4.28
Training data

Figure 4.29
Results after one epoch.

Figure 4.30
Results after two epochs.

Initial Weights:			Weights after epoch 1:			Weights after epoch 2:		
"+"	0.00	0.00	"+"	0.44	0.52	"+"	0.41	0.55
"0"	1.00	1.00	"0"	0.90	0.93	"0"	0.88	0.92
"*"	1.00	0.00	"*"	1.03	0.17	"*"	1.03	0.24
"#"	0.00	1.00	"#"	0.13	1.02	"#"	0.22	1.02

Figure 4.31
Results after three epochs.

Figure 4.32
Results after 10 epochs.

Figure 4.33
Results after 40 epochs.

Weights after epoch 3:			Weights after epoch 10:			Weights after epoch 40:		
"+"	0.36	0.57	"+"	0.34	0.44	"+"	0.30	0.31
"0"	0.89	0.92	"0"	0.89	0.91	"0"	0.92	0.93
"*"	1.05	0.26	"*"	1.10	0.26	"*"	1.11	0.26
"#"	0.27	1.00	"#"	0.30	1.03	"#"	0.27	1.09

The training data are shown in Figure 4.28; the results of testing the net on the same input points as are used for training are shown in Figures 4.29–4.33.

Example 4.13 Using LVQ: more cluster units improves performance

Using the training data as shown in Figure 4.28 (training input for points (x, y): $x = 0.1i, i = 1, \ldots, 9; y = 0.1j, j = 1, \ldots, 9$), we now use 20 output units, with their weights and class assignments initialized randomly. Of course, this ignores available information during initialization, but we do so for purposes of demonstration. In practice, one would select a representative sample of patterns from each class to use as the initial reference vectors. Using a fixed learning rate of .1, 1,000 epochs of training were performed. The larger number of epochs required is a result of the random initialization of the weights. In order to show the regions more clearly, we test on all points (x, y) for $x = 0.05i, i = 2, \ldots, 18; y = 0.05j, j = 2, \ldots, 18$.

Figures 4.34–4.40 show the results at selected stages of training. Notice that much of the redistributing of the cluster vectors occurs during the first 100 epochs. However, three cluster vectors (one for each of Classes 2, 3, and 4) are "caught" in the region where the input vectors are from Class 1 (shown with the symbol " + "). The results after 100, 200, 400, 600, and 800 epochs show that the net shifts these vectors in various directions, before they are pushed off the right hand side of the figure. The final classification of the test points does not resolve the L-shaped region shown with the symbol "0" much more distinctly than in the previous example. This is due, at least in part, to the random initialization of the weights. Although more vectors were available for each class, since they were poorly positioned when training began, many of the clusters were forced to positions in which they failed to be the "winner" for any of the input points. Suitable initialization of the weights for the cluster units greatly improves the performance of LVQ.

4.3.4 Variations

We now consider several improved LVQ algorithms, called LVQ2, LVQ2.1 [Kohonen, 1990a], and LVQ3 [Kohonen, 1990b]. In the original LVQ algorithm, only the reference vector that is closest to the input vector is updated. The direction it is moved depends on whether the winning reference vector belongs to the same class as the input vector. In the improved algorithms, two vectors (the winner and a runner-up) learn if several conditions are satisfied. The idea is that if the input is approximately the same distance from both the winner and the runner-up, then each of them should learn.

LVQ2

In the first modification, LVQ2, the conditions under which both vectors are modified are that:

1. The winning unit and the runner-up (the next closest vector) represent different classes.
2. The input vector belongs to the same class as the runner-up.
3. The distances from the input vector to the winner and from the input vector to the runner-up are approximately equal. This condition is expressed in terms of a window, using the following notation:

$$\mathbf{x} \quad \text{current input vector;}$$
$$\mathbf{y}_c \quad \text{reference vector that is closest to } \mathbf{x};$$

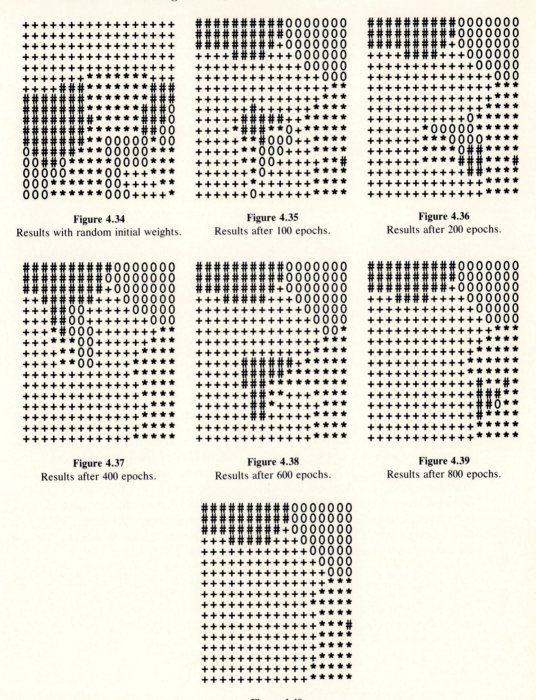

Figure 4.34
Results with random initial weights.

Figure 4.35
Results after 100 epochs.

Figure 4.36
Results after 200 epochs.

Figure 4.37
Results after 400 epochs.

Figure 4.38
Results after 600 epochs.

Figure 4.39
Results after 800 epochs.

Figure 4.40
Results after 1,000 epochs.

$\mathbf{y}_r$ reference vector that is next to closest to $\mathbf{x}$ (the runner-up);

$\mathbf{d}_c$ distance from $\mathbf{x}$ to $\mathbf{y}_c$;

d_r distance from $\mathbf{x}$ to $\mathbf{y}_r$.

To be used in updating the reference vectors, a window is defined as follows: The input vector $\mathbf{x}$ falls in the window if

$$\frac{d_c}{d_r} > 1 - \epsilon$$

and

$$\frac{d_r}{d_c} < 1 + \epsilon,$$

where the value of ϵ depends on the number of training samples; a value of .35 is typical [Kohonen, 1990a].

In LVQ2, the vectors $\mathbf{y}_c$ and $\mathbf{y}_r$ are updated if the input vector $\mathbf{x}$ falls in the window, $\mathbf{y}_c$ and $\mathbf{y}_r$ belong to different classes, and $\mathbf{x}$ belongs to the same class as $\mathbf{y}_r$. If these conditions are met, the closest reference vector and the runner up are updated:

$$\mathbf{y}_c(t + 1) = \mathbf{y}_c(t) - \alpha(t)[\mathbf{x}(t) - \mathbf{y}_c(t)];$$

$$\mathbf{y}_r(t + 1) = \mathbf{y}_r(t) + \alpha(t)[\mathbf{x}(t) - \mathbf{y}_r(t)].$$

LVQ2.1

In the modification called LVQ2.1 Kohonen (1990a) considers the two closest reference vectors, $\mathbf{y}_{c1}$ and $\mathbf{y}_{c2}$. The requirement for updating these vectors is that one of them, say, $\mathbf{y}_{c1}$, belongs to the correct class (for the current input vector $\mathbf{x}$) and the other ($\mathbf{y}_{c2}$) does not belong to the same class as $\mathbf{x}$. Unlike LVQ, LVQ2.1 does not distinguish between whether the closest vector is the one representing the correct class or the incorrect class for the given input. As with LVQ2, it is also required that $\mathbf{x}$ fall in the window in order for an update to occur. The test for the window condition to be satisfied becomes

$$\min \left[\frac{d_{c1}}{d_{c2}}, \frac{d_{c2}}{d_{c1}} \right] > 1 - \epsilon$$

and

$$\max \left[\frac{d_{c1}}{d_{c2}}, \frac{d_{c2}}{d_{c1}} \right] < 1 + \epsilon.$$

The more complicated expressions result from the fact that we do not know whether $\mathbf{x}$ is closer to $\mathbf{y}_{c1}$ or to $\mathbf{y}_{c2}$.

If these conditions are met, the reference vector that belongs to the same class as **x** is updated according to

$$\mathbf{y}_{c1}(t + 1) = \mathbf{y}_{c1}(t) + \alpha(t)[\mathbf{x}(t) - \mathbf{y}_{c1}(t)],$$

and the reference vector that does not belong to the same class as **x** is updated according to

$$\mathbf{y}_{c2}(t + 1) = \mathbf{y}_{c2}(t) - \alpha(t)[\mathbf{x}(t) - \mathbf{y}_{c2}(t)].$$

LVQ3

A further refinement, LVQ3 [Kohonen, 1990b], allows the two closest vectors to learn as long as the input vector satisfies the window condition

$$\min\left[\frac{d_{c1}}{d_{c2}}, \frac{d_{c2}}{d_{c1}}\right] > (1 - \epsilon)(1 + \epsilon),$$

where typical values of $\epsilon = 0.2$ are indicated. (Note that this window condition is also used for LVQ2 in Kohonen, 1990b.) If one of the two closest vectors, $\mathbf{y}_{c1}$, belongs to the same class as the input vector **x**, and the other vector $\mathbf{y}_{c2}$ belongs to a different class, the weight updates are as for LVQ2.1. However, LVQ3 extends the training algorithm to provide for training if **x**, $\mathbf{y}_{c1}$, and $\mathbf{y}_{c2}$ belong to the same class. In this case, the weight updates are

$$\mathbf{y}_c(t + 1) = \mathbf{y}_c(t) + \beta(t)[\mathbf{x}(t) - \mathbf{y}_c(t)]$$

for both $\mathbf{y}_{c1}$ and $\mathbf{y}_{c2}$. The learning rate $\beta(t)$ is a multiple of the learning rate $\alpha(t)$ that is used if $\mathbf{y}_{c1}$ and $\mathbf{y}_{c2}$ belong to different classes. The appropriate multiplier is typically between 0.1 and 0.5, with smaller values corresponding to a narrower window. Symbolically,

$$\beta(t) = m\,\alpha(t) \quad \text{for } 0.1 < m < 0.5.$$

This modification to the learning process ensures that the weights (codebook vectors) continue to approximate the class distributions and prevents the codebook vectors from moving away from their optimal placement if learning continues.

4.4 COUNTERPROPAGATION

Counterpropagation networks [Hecht-Nielsen, 1987a, 1987b, 1988] are multilayer networks based on a combination of input, clustering, and output layers. Counterpropagation nets can be used to compress data, to approximate functions, or to associate patterns.

A counterpropagation net approximates its training input vector pairs by adaptively constructing a look-up table. In this manner, a large number of training data points can be compressed to a more manageable number of look-up table entries. If the training data represent function values, the net will approximate a function. A heteroassociative net is simply one interpretation of a function from a set of vectors (patterns) **x** to a set of vectors **y**. The accuracy of the approximation is determined by the number of entries in the look-up table, which equals the number of units in the cluster layer of the net.

Counterpropagation nets are trained in two stages. During the first stage, the input vectors are clustered. In the original definition of counterpropagation nets, no topology was assumed for the cluster units. However, the addition of a linear topology, as discussed in Section 4.2, can improve the performance of the net. The clusters that are formed may be based on either the dot product metric or the Euclidean norm metric. During the second stage of training, the weights from the cluster units to the output units are adapted to produce the desired response. More details on training counterpropagation nets are given in Sections 4.4.1 and 4.4.2.

There are two types of counterpropagation nets: full and forward only. Full counterpropagation is the subject of Section 4.4.1, forward-only counterpropagation of Section 4.4.2.

4.4.1 Full Counterpropagation

Full counterpropagation was developed to provide an efficient method of representing a large number of vector pairs, **x**:**y** by adaptively constructing a look-up table. It produces an approximation **x***:**y*** based on input of an **x** vector (with no information about the corresponding **y** vector), or input of a **y** vector only, or input of an **x**:**y** pair, possibly with some distorted or missing elements in either or both vectors.

Full counterpropagation uses the training vector pairs **x**:**y** to form the clusters during the first phase of training. In the original definition, the competition in the cluster (or Kohonen) layer chose the unit that had the largest net input as the winner; this corresponds to using the dot product metric. Whenever vectors are to be compared using the dot product metric, they should be normalized. Although it is possible to normalize them without losing information by adding an extra component (see Exercise 4.10), to avoid this effort and provide the most direct comparison between full and forward-only counterpropagation, we shall use the Euclidean norm for both (as well as for Kohonen self-organizing maps and LVQ).

Architecture

Figure 4.41 illustrates the architecture of a full counterpropagation network; for simplicity, the weights are not shown. Figures 4.42 and 4.43, which indicate the units that are active during each of the two phases of training a full counterpropagation net, show the weights.

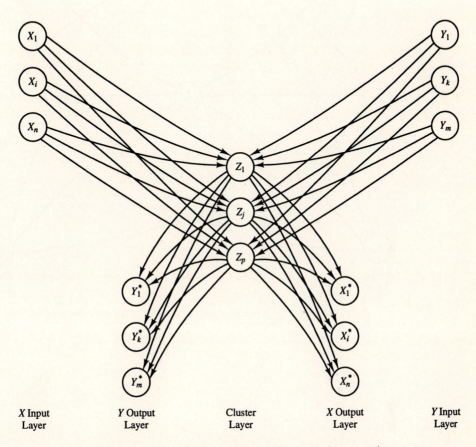

Figure 4.41   Architecture of full counterpropagation network.

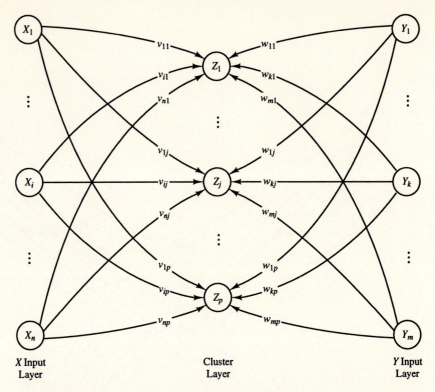

X Input
Layer

Cluster
Layer

Y Input
Layer

Figure 4.42 Active units during the first phase of counterpropagation training.

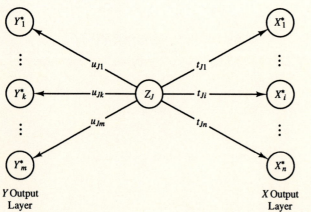

Y Output
Layer

X Output
Layer

Figure 4.43 Second phase of
counterpropagation training.

Algorithm

Training a counterpropagation network occurs in two phases. During the first phase, the units in the X input, cluster, and Y input layers are active. The units in the cluster layer compete; the interconnections are not shown.

In the basic definition of counterpropagation, no topology is assumed for the cluster layer units; only the winning unit is allowed to learn. The learning rule for weight updates on the winning cluster unit is

$$v_{iJ}(\text{new}) = (1 - \alpha)v_{iJ}(\text{old}) + \alpha x_i, \quad i = 1, \ldots, n;$$

$$w_{kJ}(\text{new}) = (1 - \beta)w_{kJ}(\text{old}) + \beta y_k, \quad k = 1, \ldots, m.$$

This is standard Kohonen learning, which consists of both the competition among the units and the weight updates for the winning unit.

During the second phase of the algorithm, only unit J remains active in the cluster layer. The weights from the winning cluster unit J to the output units are adjusted so that the vector of activations of the units in the Y output layer, $\mathbf{y}^*$, is an approximation to the input vector $\mathbf{y}$; $\mathbf{x}^*$ is an approximation to $\mathbf{x}$.

The weight updates for the units in the Y output and X output layers are

$$u_{Jk}(\text{new}) = (1 - a)u_{Jk}(\text{old}) + ay_k, \quad k = 1, \ldots, m;$$

$$t_{Ji}(\text{new}) = (1 - b)t_{Ji}(\text{old}) + bx_i, \quad i = 1, \ldots, n.$$

This is known as Grossberg learning, which, as used here, is a special case of the more general outstar learning [Hecht-Nielsen, 1990]. Outstar learning occurs for all units in a particular layer; no competition among those units is assumed. However, the forms of the weight updates for Kohonen learning and Grossberg learning are closely related.

The learning rules for the output layers can also be viewed as delta rule learning. To see this, observe that y_k is the target value for unit Y_k^* and $u_{Jk}(\text{old})$ is the computed activation of the unit (since the signal sent by unit Z_J is 1). Now, simple algebra gives

$$u_{Jk}(\text{new}) = (1 - a)u_{Jk}(\text{old}) + ay_k$$

$$= u_{Jk}(\text{old}) + a(y_k - u_{Jk}(\text{old})).$$

Thus, the weight change is simply the learning rate a times the error. Exactly the same comments apply to the weight updates for the units in the X output layer [Dayhoff, 1990].

The nomenclature we use in the counterpropagation algorithm is as follows:

$\mathbf{x}$ input training vector:

$$\mathbf{x} = (x_1, \ldots, x_i, \ldots, x_n).$$

$\mathbf{y}$ target output corresponding to input $\mathbf{x}$:

$$\mathbf{y} = (y_1, \ldots, y_k, \ldots, y_m).$$

z_j activation of cluster layer unit Z_j.

$\mathbf{x}^*$ computed approximation to vector $\mathbf{x}$.

$\mathbf{y}^*$ computed approximation to vector $\mathbf{y}$.

v_{ij} weight from X input layer, unit X_i, to cluster layer, unit Z_j.

w_{kj} weight from Y input layer, unit Y_k, to cluster layer, unit Z_j.

u_{jk} weight from cluster layer, unit Z_j, to Y output layer, unit Y_k^*.

t_{ji} weight from cluster layer, unit Z_j, to X output layer, unit X_i^*.

α, β learning rates for weights into cluster layer (Kohonen learning).

a, b learning rates for weights out from cluster layer (Grossberg learning).

The training algorithm for full counterpropagation is:

Step 0.   Initialize weights, learning rates, etc.

Step 1.   While stopping condition for phase 1 training is false, do Steps 2–7.

       Step 2.   For each training input pair $\mathbf{x}:\mathbf{y}$, do Steps 3–5.

             Step 3.   Set X input layer activations to vector $\mathbf{x}$; set Y input layer activations to vector $\mathbf{y}$.

             Step 4.   Find winning cluster unit; call its index J;

             Step 5.   Update weights for unit Z_J:

$$v_{iJ}(\text{new}) = (1 - \alpha)v_{iJ}(\text{old}) + \alpha x_i,$$
$$i = 1, \ldots, n;$$
$$w_{kJ}(\text{new}) = (1 - \beta)w_{kJ}(\text{old}) + \beta y_k,$$
$$k = 1, \ldots, m.$$

       Step 6.   Reduce learning rates α and β.

       Step 7.   Test stopping condition for phase 1 training.

Step 8.   While stopping condition for phase 2 training is false, do Steps 9–15. (*Note:* α and β are small, constant values during phase 2.)

       Step 9.   For each training input pair $\mathbf{x}:\mathbf{y}$, do Steps 10–13.

             Step 10.   Set X input layer activations to vector $\mathbf{x}$; set Y input layer activations to vector $\mathbf{y}$.

             Step 11.   Find winning cluster unit; call its index J.

             Step 12.   Update weights into unit Z_J:

$$v_{iJ}(\text{new}) = (1 - \alpha)v_{iJ}(\text{old}) + \alpha x_i,$$
$$i = 1, \ldots, n;$$
$$w_{kJ}(\text{new}) = (1 - \beta)w_{kJ}(\text{old}) + \beta y_k,$$
$$k)1, \ldots, m.$$

Step 13.   Update weights from unit Z_J to the output layers:

$$u_{Jk}(\text{new}) = (1 - a)u_{Jk}(\text{old}) + ay_k,$$

$$k = 1, \ldots, m.$$

$$t_{Ji}(\text{new}) = (1 - b)t_{Ji}(\text{old}) + bx_i,$$

$$i = 1, \ldots, n.$$

Step 14.   Reduce learning rate.

Step 15.   Test stopping condition for phase 2 training.

In Steps 4 and 11:

In case of a tie, take the unit with the smallest index.

To use the dot product metric, find the cluster unit Z_j with the largest net input:

$$z\_in_j = \sum_i x_i v_{ij} + \sum_k y_k w_{kj}$$

The weight vectors and input vectors should be normalized to use the dot product metric.

To use the Euclidean distance metric, find the cluster unit Z_j, the square of whose distance from the input vectors is smallest:

$$D_j = \sum_i (x_i - v_{ij})^2 + \sum_k (y_k - w_{kj})^2.$$

Application

After training, a counterpropagation neural net can be used to find approximations $\mathbf{x}^*$ and $\mathbf{y}^*$ to the input, output vector pair $\mathbf{x}$ and $\mathbf{y}$. Hecht-Nielsen [1990] refers to this process as accretion, as opposed to interpolation between known values of a function. The application procedure for counterpropagation is as follows:

Step 0.   Initialize weights.

Step 1.   For each input pair $\mathbf{x}:\mathbf{y}$, do Steps 2–4.

Step 2.   Set X input layer activations to vector $\mathbf{x}$; set Y input layer activations to vector $\mathbf{y}$;

Step 3.   Find the cluster unit Z_J that is closest to the input pair.

Step 4.   Compute approximations to $\mathbf{x}$ and $\mathbf{y}$:

$$x_i^* = t_{Ji},$$

$$y_k^* = u_{Jk}.$$

The net can also be used in an interpolation mode; in this case, several units are allowed to be active in the cluster layer. The activations are set so that

$\sum\limits_{j} z_j = 1$ (in order to form a convex combination of values). The interpolated approximations to **x** and **y** are then

$$x_i^* = \sum_j z_j t_{ji},$$

$$y_k^* = \sum_j z_j u_{jk}.$$

The accuracy of approximation is increased by using interpolation.

For testing with only an **x** vector for input (i.e., there is no information about the corresponding **y**), it may be preferable to find the winning unit J based on comparing only the **x** vector and the first n components of the weight vector for each cluster layer unit.

Example 4.14 A full counterpropagation net for the function $y = \dfrac{1}{x}$

In this example, we consider the performance of a full counterpropagation net to form a look-up table for the function $y = 1/x$ on the interval $[0.1, 10.0]$. Suppose we have 10 cluster units (in the Kohonen layer); there is 1 X input layer unit, 1 Y input layer unit, 1 X output layer unit, and 1 Y output layer unit. Suppose further that we have a large number of training points (perhaps 1,000), with x values between 0.1 and 10.0 and the corresponding y values given by $y = 1/x$. The training input points, which are uniformly distributed along the curve, are presented in random order.

If our initial weights (on the cluster units) are chosen appropriately, then after the first phase of training, the clusters units will be uniformly distributed along the curve. If we use a linear structure (as for a Kohonen self-organizing map) on the cluster units, this will improve the chances that the weights will represent the points on the curve in a statistically optimal manner.

Typical results give the following weights for the cluster units. These can be interpreted as the positions (in the x-y plane) that the cluster units represent. The first weight for each cluster unit is the weight from the X input unit, the second weight the weight from the Y input unit. We have:

CLUSTER UNIT	v	w
Z_1	0.11	9.0
Z_2	0.14	7.0
Z_3	0.20	5.0
Z_4	0.30	3.3
Z_5	0.6	1.6
Z_6	1.6	0.6
Z_7	3.3	0.30
Z_8	5.0	0.20
Z_9	7.0	0.14
Z_{10}	9.0	0.11

After the second phase of training, the weights to the output units will be approximately the same as the weights into the cluster units.

The weights are shown on a diagram of the net in Figure 4.44 and on a graph of the function in Figure 4.45.

We can use this net to obtain the approximate value of y for $x = 0.12$ as follows:

Step 0.   Initialize weights.
Step 1.   For the input $x = 0.12$, $y = 0.0$, do Steps 2–4.
   Step 2.   Set X input layer activations to vector $\mathbf{x}$;
                   set Y input layer activations to vector $\mathbf{y}$;
   Step 3.   Find the index J of the winning cluster unit;
                   the squares of the distances from the input to each of the cluster units are:

$$D_1 = (0.12 - 0.11)^2 + (0.00 - 9.00)^2 = 81$$

$$D_2 = (0.12 - 0.14)^2 + (0.00 - 7.00)^2 = 49$$

$$D_3 = (0.12 - 0.20)^2 + (0.00 - 5.00)^2 = 25$$

$$D_4 = (0.12 - 0.30)^2 + (0.00 - 3.30)^2 = 11$$

$$D_5 = (0.12 - 0.60)^2 + (0.00 - 1.60)^2 = 2.8$$

$$D_6 = (0.12 - 1.60)^2 + (0.00 - 0.60)^2 = 2.6$$

$$D_7 = (0.12 - 3.30)^2 + (0.00 - 0.30)^2 = 10.2$$

$$D_8 = (0.12 - 5.00)^2 + (0.00 - 0.20)^2 = 23.9$$

$$D_9 = (0.12 - 7.00)^2 + (0.00 - 0.14)^2 = 47.4$$

$$D_{10} = (0.12 - 9.00)^2 + (0.00 - 0.11)^2 = 78.9$$

   Thus, based on the total input, the closest cluster unit is $J = 6$.
   Step 4.   Compute approximations to $\mathbf{x}$ and $\mathbf{y}$:

$$\mathbf{x}^* = t_J = 1.6$$

$$\mathbf{y}^* = u_J = 0.6.$$

Clearly, this is not really the approximation we wish to find. Since we only have information about the x input, we should use the earlier mentioned modification to the application procedure. Thus, if we base our search for the winning cluster unit on distance from the x input to the corresponding weight for each cluster unit, we find the following in Steps 3 and 4:

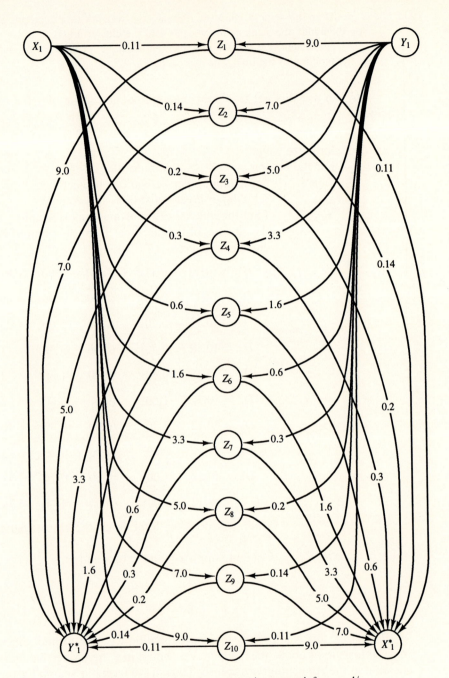

Figure 4.44 Full counterpropagation network for $y = 1/x$.

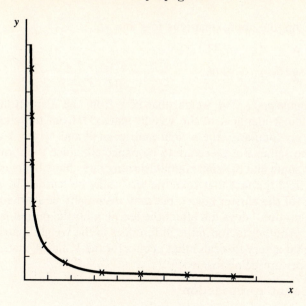

Figure 4.45 Graph of $y = 1/x$, showing position of cluster units.

Step 3.   Find the index J of the winning cluster unit; the squares of the distances from the input to each of the cluster units are:

$$D_1 = (0.12 - 0.11)^2 = 0.0001$$

$$D_2 = (0.12 - 0.14)^2 = 0.0004$$

$$D_3 = (0.12 - 0.20)^2 = 0.064$$

$$D_4 = (0.12 - 0.30)^2 = 0.032$$

$$D_5 = (0.12 - 0.60)^2 = 0.23$$

$$D_6 = (0.12 - 1.60)^2 = 2.2$$

$$D_7 = (0.12 - 3.30)^2 = 10.1$$

$$D_8 = (0.12 - 5.00)^2 = 23.8$$

$$D_9 = (0.12 - 7.00)^2 = 47.3$$

$$D_{10} = (0.12 - 9.00)^2 = 81$$

Thus, based on the input from x only, the closest cluster unit is $J = 1$.

Step 4.   Compute approximations to **x** and **y**:

$$\mathbf{x}^* = t_J = 0.11,$$

$$\mathbf{y}^* = u_J = 9.00.$$

Observations and Comments.   The weight matrix V from the X input layer to the cluster layer is almost identical to the weight matrix T from the cluster layer to the X output layer. Similarly, the weight matrices (W and U) for Y and Y^* are also essentially the same. But these are to be expected, since the form of the learning rules are the same and the same initial learning rates have been used. The slight differences reflect the fact that some patterns may be learned by one unit early in the training (of the cluster layer), but may eventually be learned by a different unit. This "migration" does not affect the learning for the output layer (T matrix and U matrix). Another factor in the differences in the weight matrices is the additional learning (at a very low rate) that occurs for the V and W matrices while the U and T matrices are being formed.

4.4.2 Forward-Only Counterpropagation

Forward-only counterpropagation nets are a simplified version of the full counterpropagation nets discussed in Section 4.4.1. Forward-only nets are intended to approximate a function $\mathbf{y} = f(\mathbf{x})$ that is not necessarily invertible; that is, forward-only counterpropagation nets may be used if the mapping from **x** to **y** is well defined, but the mapping from **y** to **x** is not.

Forward-only counterpropagation differs from full counterpropagation in using only the **x** vectors to form the clusters on the Kohonen units during the first stage of training. The original presentation of forward-only counterpropagation used the Euclidean distance between the input vector and the weight (exemplar) vector for the Kohonen unit (rather than the dot product metric used in the original development of full counterpropagation). However, either metric can be used for either form of counterpropagation.

Architecture

Although the architecture of a forward-only counterpropagation net, as illustrated in Figure 4.46, appears to be the same as the architecture of a backpropagation net, the counterpropagation net has interconnections among the units in the cluster layer, which are not shown. In general, in forward-only counterpropagation, after competition, only one unit in that layer will be active and send a signal to the output layer.

Algorithm

The training procedure for the forward-only counterpropagation net consists of several steps, as indicated in the algorithm that follows. First, an input vector is presented to the input units. The units in the cluster layer compete (winner take

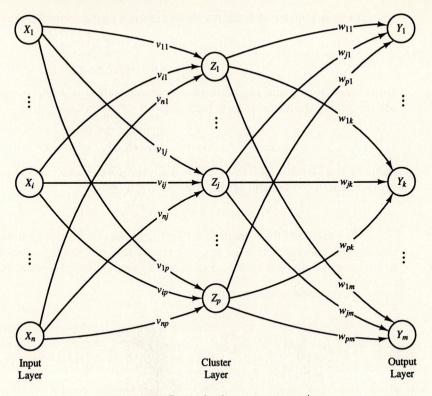

Figure 4.46   Forward-only counterpropagation.

all) for the right to learn the input vector. After the entire set of training vectors has been presented, the learning rate is reduced and the vectors are presented again; this continues through several iterations.

After the weights from the input layer to the cluster layer have been trained (the learning rate has been reduced to a small value), the weights from the cluster layer to the output layer are trained. Now, as each training input vector is presented to the input layer, the associated target vector is presented to the output layer. The winning cluster unit (call it J) sends a signal of 1 to the output layer. Each output unit k has a computed input signal w_{Jk} and target value y_k. Using the difference between these values, the weights between the winning cluster unit and the output layer are updated. The learning rule for these weights is similar to the learning rule for the weights from the input units to the cluster units:

LEARNING RULE FOR WEIGHTS FROM INPUT UNITS TO CLUSTER UNITS

$$v_{iJ}(\text{new}) = v_{iJ} + \alpha(x_i - v_{iJ})$$

$$= (1 - \alpha)v_{iJ}(\text{old}) + \alpha y_k.$$

LEARNING RULE FOR WEIGHTS FROM CLUSTER UNITS TO OUTPUT UNITS

$$w_{Jk}(\text{new}) = w_{Jk} + a(y_k - w_{Jk})$$

$$= (1 - a)w_{Jk}(\text{old}) + ax_i.$$

However, if w_{Jk} is interpreted as the computed output (i.e., $y_k = w_{Jk}$), and the activation of the cluster units is included, viz.,

$$z_j = \begin{cases} 1 & \text{if } j = J \\ 0 & \text{otherwise,} \end{cases}$$

then the learning rule for the weights from the cluster units to the output units can be written in the form of the delta rule:

$$w_{jk}(\text{new}) = w_{jk} + az_j(y_k - w_{Jk}).$$

The training of the weights from the input units to the cluster units continues at a low learning rate while the learning rate for the weights from the cluster units to the output units is gradually reduced. The nomenclature used is as follows:

α, a   learning rate parameters:

$$.5 < \alpha < .8,$$

$$0 < a < 1$$

(suggested initial values, $a = .1$, $\alpha = .6$ [Hecht-Nielsen, 1988]).

$\mathbf{x}$ activation vector for input units:

$$\mathbf{x} = (x_1, \ldots, x_i, \ldots, x_n).$$

$\|\mathbf{x} - \mathbf{v}\|$ Euclidean distance between vectors $\mathbf{x}$ and $\mathbf{v}$.

As with full counterpropagation, no topological structure was assumed for the cluster units in the original formulation of forward-only counterpropagation. However, in many cases, training can be improved by using a linear structure on the cluster units, as described in Section 4.2. The structure helps to ensure that, after training, the weights for the cluster units are distributed in a statistically optimal manner.

The training algorithm for the forward-only counterpropagation net is as follows:

Step 0.   Initialize weights, learning rates, etc.

Step 1.   While stopping condition for phase 1 is false, do Steps 2–7.

        Step 2.   For each training input $\mathbf{x}$, do Steps 3–5.

               Step 3.   Set X input layer activations to vector $\mathbf{x}$.

               Step 4.   Find winning cluster unit; call its index J.

               Step 5.   Update weights for unit Z_J:

$$v_{iJ}(\text{new}) = (1 - \alpha)v_{iJ}(\text{old}) + \alpha x_i,$$
$$i = 1, \ldots, n.$$

Step 6.   Reduce learning rate α.

Step 7.   Test stopping condition for phase 1 training.

Step 8.   While stopping condition for phase 2 training is false, do Steps 9–15.
(*Note:* α is a small, constant value during phase 2.)

Step 9.   For each training input pair **x** : **y**, do Steps 10–13.

Step 10.   Set X input layer activations to vector **x**;
set Y output layer activations to vector **y**.

Step 11.   Find the winning cluster unit; call its index J.

Step 12.   Update weights into unit Z_J (α is small):

$$v_{iJ}(\text{new}) = (1 - \alpha)v_{iJ}(\text{old}) + \alpha x_i,$$
$$i = 1, \ldots, n.$$

Step 13.   Update weights from unit Z_J to the output units:

$$w_{Jk}(\text{new}) = (1 - a)w_{Jk}(\text{old}) + a y_k,$$
$$k = 1, \ldots, m.$$

Step 14.   Reduce learning rate a.

Step 15.   Test stopping condition for phase 2 training.

In Steps 4 and 11:

In case of a tie, take the unit with the smallest index.

To use the dot product metric, find the cluster unit Z_J with the largest net input:

$$z\_in_j = \sum_i x_i v_{ij}.$$

To use the Euclidean distance metric, find the cluster unit Z_J, the square of whose distance from the input pattern,

$$D_j = \sum_i (x_i - v_{ij})^2,$$

is smallest.

Applications

The application procedure for foward-only counterpropagation is:

Step 0.   Initialize weights (by training as in previous subsection).

Step 1.   Present input vector **x**.

Step 2.   Find unit J closest to vector **x**.

Step 3.   Set activations of output units:

$$y_k = w_{Jk}.$$

A forward-only counterpropagation net can also be used in an "interpolation mode. In this case, more than one Kohonen unit has a nonzero activation with

$$\sum_j z_j = 1.$$

The activation of the output units is then given by

$$y_k = \sum_j z_j w_{jk}.$$

Again, accuracy is increased by using the interpolation mode.

Example 4.15 A forward-only counterpropagation net for the function $y = \dfrac{1}{x}$

In this example, we consider the performance of a forward-only counterpropagation net to form a look-up table for the function $y = 1/x$ on the interval [0.1, 10.0]. Suppose we have 10 cluster units (in the cluster layer); there is 1 X input layer unit and 1 Y output layer unit. Suppose further that we have a large number of training points (the x values for our function) uniformly distributed between 0.1 and 10.0 and presented in a random order.

 If we use a linear structure on the cluster units, the weights (from the input unit to the 10 cluster units) will be approximately 0.5, 1.5, 2.5, 3.5, . . . , 9.5 after the first phase of training.

 After the second phase of training, the weights to the Y output units will be approximately 5.5, 0.75, 0.4, . . . , 0.1. Thus, the approximations to the function values will be much more accurate for large values of x than for small values. Figure 4.47 illustrates the weights associated with each cluster unit.

 Comparing these results with those of Example 4.14 (for full counterpropagation), we see that even if the net is intended only for approximating the mapping from x to y, the full counterpropagation net may distribute the cluster units in a

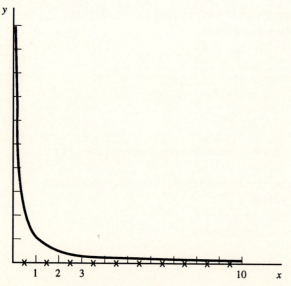

Figure 4.47 Results for $y = 1/x$ using forward-only counterpropagation.

manner that produces more accurate approximations over the entire range of input values.

4.5 SUGGESTIONS FOR FURTHER STUDY

4.5.1 Readings

Fixed-weight nets

LIPPMANN, R. P. (1987). "An Introduction to Computing with Neural Nets." *IEEE ASSP Magazine*, 4:4–22.

Kohonen self-organizing maps

ANGENIOL, B., G. VAUBOIS, and J-Y. LE TEXIER. (1988). "Self-organizing Feature Maps and the Travelling Salesman Problem." *Neural Networks*, 1(4):289–293.

KOHONEN, T. (1982). "Self-organized Formation of Topologically Correct Feature Maps." *Biological Cybernetics*, 43:59–69. Reprinted in Anderson and Rosenfeld [1988] pp. 511–521.

KOHONEN, T. (1989a). *Self-organization and Associative Memory*, 3rd ed. Berlin: Springer-Verlag.

KOHONEN, T. (1990a). "Improved Versions of Learning Vector Quantization." *International Joint Conference on Neural Networks, I*, pp. 545–550.

KOHONEN, T. (1990b). "The Self-Organizing Map." *Proceedings of the IEEE*, 78(9):1464–1480.

Counterpropagation

HECHT-NIELSEN, R. (1987a). "Counterpropagation Networks." *Applied Optics*, 26(23), 4979–4984.

HECHT-NIELSEN, R. (1987b). "Counterpropagation Networks." *IEEE International Conference on Neural Networks, II*, pp. 19–32.

4.5.2 Exercises

Introduction

4.1 Show that if the weight vectors are not the same length, it is possible that the weight vector that appears to be closest to the input vector will not be the weight vector that is chosen when the dot product metric is used. More specifically, consider two weight vectors $\mathbf{w}_1$ and $\mathbf{w}_2$ with lengths $\|\mathbf{w}_1\|$ and $\|\mathbf{w}_2\|$ and angles with a horizontal axis of θ_1 and θ_2, respectively. For an arbitrary input vector $\mathbf{s}$, what is the inequality relation (in terms of the lengths and angles of the three vectors $\mathbf{w}_1$, $\mathbf{w}_2$, and $\mathbf{s}$) that determines when the neuron represented by $\mathbf{w}_1$ would be chosen as the winner (using the dot product metric). Give an example where this might not be the desired choice.

Kohonen self-organizing maps

4.2 a. Given a Kohonen self-organizing map with weights as shown in the following diagram, use the square of the Euclidean distance to find the cluster unit C_J that is closest to the input vector (.5, .2).

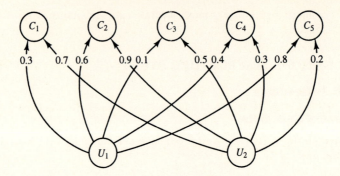

b. Using a learning rate of .2, find the new weights for unit C_J.

c. If units C_{J-1} and C_{J+1} are also allowed to learn the input pattern, find their new weights.

4.3 Repeat the preceding exercise for the input vector (.5, .5), with $\alpha = .1$.

4.4 Consider a Kohonen net with two cluster units and five input units. The weight vectors for the cluster units are

$$w_1 = (1.0, 0.8, 0.6, 0.4, 0.2)$$

and

$$w_2 = (0.2, 0.4, 0.6, 0.8, 1.0).$$

Use the square of the Euclidean distance to find the winning cluster unit for the input pattern

$$x = (0.5, 1.0, 0.5 \ 0.0, 0.0).$$

Using a learning rate of .2, find the new weights for the winning unit.

Learning vector quantization

4.5 Consider an LVQ net with two input units and four target classes: C_1, C_2, C_3, and C_4. There are 16 classification units, with weight vectors indicated by the coordinates on the following chart, read in row-column order. For example, the unit with weight vector (0.2, 0.4) is assigned to represent Class 3, and the classification units for Class 1 have initial weight vectors of (0.2, 0.2), (0.2, 0.6), (0.6, 0.8), and (0.6, 0.4).

x_2						
1.0						
0.8		C_3	C_4	C_1	C_2	
0.6		C_1	C_2	C_3	C_4	
0.4		C_3	C_4	C_1	C_2	
0.2		C_1	C_2	C_3	C_4	
0.0						
	0.0	0.2	0.4	0.6	0.8	1.0 x_1

Using the square of the Euclidean distance (and the geometry of the diagram, to avoid having to do any distance calculations), determine the changes that occur as you do the following:

a. Present an input vector of (0.25, 0.25) representing Class 1. Using a learning rate of $\alpha = 0.5$, show which classification unit moves where (i.e., determine its new weight vector).
b. Present an input vector of (0.4, 0.35) representing Class 1. What happens?
c. Instead of presenting the second vector as in part b, present the vector (0.4, 0.45). What happens?
d. Suppose the training inputs are drawn from the following regions:

$$\text{Class 1}\quad 0.0 \le x_1 < 0.5 \quad 0.0 \le x_2 < 0.5$$
$$\text{Class 2}\quad 0.5 \le x_1 \le 1.0 \quad 0.0 \le x_2 < 0.5$$
$$\text{Class 3}\quad 0.0 \le x_1 < 0.5 \quad 0.5 \le x_2 \le 1.0$$
$$\text{Class 4}\quad 0.5 \le x_1 \le 1.0 \quad 0.5 \le x_2 \le 1.0.$$

From a short-term point of view, which of the second vectors presented—(0.4, 0.35) in part b or (0.4, 0.45) in part c—has a better effect in moving the classi- fication units toward their desired positions to represent the input data?

Counterpropagation

4.6 Consider the following full counterpropagation net:

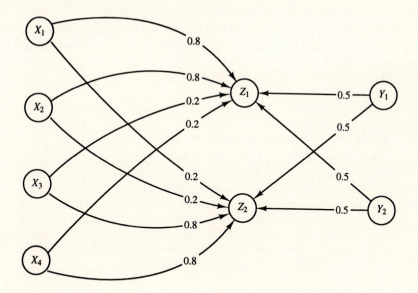

Using the input pair

$$x = (1, 0, 0, 0), \qquad y = (1, 0),$$

perform the first phase of training (one step only). Find the activation of the cluster layer units. Update the weights using a learning rate of .3.

4.7 Repeat Exercise 4.6, except use $x = (1, 0, 1, 1)$ and $y = (0, 1)$.

4.8 Modify Exercise 4.6 to use forward-only counterpropagation.

4.9 Design a counterpropagation net to solve the problem of associating a group of binary x vectors (sextuples) with the appropriate binary y vectors (with two components), defined as follows:

If two or more of the first three components of the x vector are 1's, then the first component of the y vector is to be 1 (otherwise it is to be 0). Similarly, if two or more of the last three components of the x vector are 1's, then the second component of the y vector is to be 1 (otherwise it is to be 0).

Discuss how to choose the number of units for the Kohonen layer. Describe the process of training the net. Illustrate the process for the pair

$$(1, 1, 0, 0, 1, 0) \leftrightarrow (1, 0).$$

4.10 Show that input vectors (n-tuples) can be converted to normalized ($n + 1$)-tuples by the following process:

Find N such that $N > \|v\|$ for every vector $\mathbf{v}$. For each vector, its ($n + 1$)st component is $\sqrt{N^2 - \|\mathbf{v}\|^2}$.

Show also that the augmented vector has norm N.

4.11 Show that forward-only and full counterpropagation are equivalent if the training input pairs x, y for full counterpropagation are concatenated and the concatenated vector is treated as both the training input and the target pattern for a forward-only counterpropagation net.

4.5.3 Projects

Kohonen self-organizing maps

4.1 Write a computer program to implement a Kohonen self-organizing map neural net. Use 2 input units, 50 cluster units, and a linear topology for the cluster units. Allow the winner and its nearest topological neighbor to learn. (In other words, if unit J is the winner, then units $J - 1$ and $J + 1$ also learn, unless $J - 1 < 1$ or $J + 1 > 50$.)

Use an initial learning rate of 0.5, and gradually reduce it to 0.01 (over 100 epochs). The initial weights on all cluster units are to be random numbers between -1 and 1 (for each component of the weight vector for each unit).

Generate a training data file as follows:

Choose two random numbers between -0.5 and 0.5, and call them x_1 and x_2. Put the point (x_1, x_2) in the training set if

$$x_1^2 + x_2^2 < 0.25.$$

Repeat until you have a set of 100 training points.

After every 10 epochs of training, graph the cluster units (by using their weight vector as a position in the two-dimensional Euclidean plane); draw a line connecting

C_1 to C_2, C_2 to C_3, etc., to show their topological relationships. You should start with a real mess for the initial positions of the weights, which will gradually smooth out to give a line that winds its way through the region from which the training points were chosen.

LVQ

4.2 Write a computer program to implement the LVQ net described in Exercise 4.5. Train the net with the data given in part d of that exercise. Experiment with different learning rates, different numbers of classification units, different geometries for the input data, etc.

4.3 Repeat Example 4.13, using a random input order for the training points and using the first five training points from each class to initialize the weights for the five cluster vectors for that class.

4.4 Repeat Example 4.13, using variant LVQ2 of the learning vector quantization method. Repeat again, this time using LVQ2.1.

Counterpropagation

4.5 Write a program to implement the counterpropagation algorithm with 63 input units, 26 units in the cluster layer, and 15 units in the Y-layer. Read initial weights from a file.

    a. In the training mode, use a linear reduction of learning rates 0.9, 0.8, . . . , 0.1. Input the training vector pairs from a file. (Print their values out upon first presentation.) Save the final weights to a file also.

    b. In testing mode, input a test pattern, print it out, and determine the approximate pair of patterns. Use inputs that correspond to each of the x vectors used in training (with 0's for the corresponding y's), inputs that correspond to each of the y vectors used in training (with 0's for the corresponding x's), and noisy versions of the training patterns.

    c. Try the dot product metric, with inputs normalized to unit length (the Euclidean distance). Repeat, using the Hamming distance to normalize the vectors.

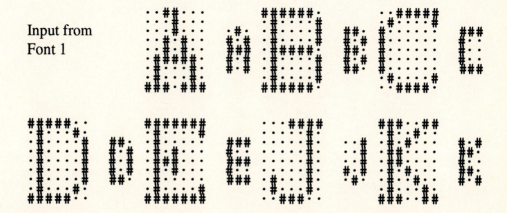

Input from Font 1

Input from
Font 2

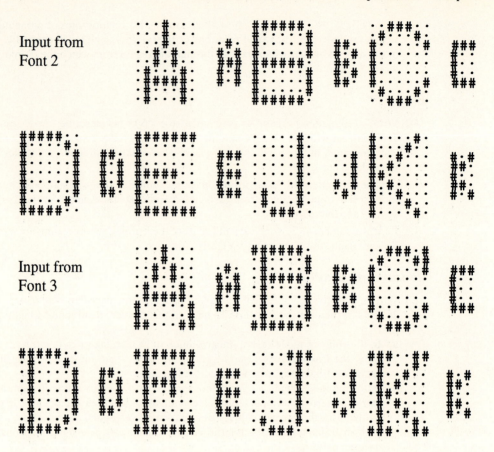

Input from
Font 3

4.6 Implement a full counterpropagation program for Example 4.14.

4.7 Implement a forward-only counterpropagation program for Example 4.15.

4.8 Let the digits 0, 1, 2, . . . , 7 be represented as

0:	1	0	0	0	0	0	0	0
1:	0	1	0	0	0	0	0	0
2:	0	0	1	0	0	0	0	0
3:	0	0	0	1	0	0	0	0
4:	0	0	0	0	1	0	0	0
5:	0	0	0	0	0	1	0	0
6:	0	0	0	0	0	0	1	0
7:	0	0	0	0	0	0	0	1

Use a counterpropagation net (full or forward only) to map these digits to their binary representations:

0:	0	0	0
1:	0	0	1
2:	0	1	0
3:	0	1	1
4:	1	0	0
5:	1	0	1
6:	1	1	0
7:	1	1	1

   a. Use the Euclidean distance metric.
   b. Repeat for the dot product metric, with inputs and targets normalized.

4.9 Use counterpropagation to solve the problem in Example 4.12, and compare the results with the preceding ones.

CHAPTER 5

Adaptive Resonance Theory

5.1 INTRODUCTION

Adaptive resonance theory (ART) was developed by Carpenter and Grossberg [1987a]. One form, ART1, is designed for clustering binary vectors; another, ART2 [Carpenter & Grossberg, 1987b], accepts continuous-valued vectors. These nets cluster inputs by using unsupervised learning. Input patterns may be presented in any order. Each time a pattern is presented, an appropriate cluster unit is chosen and that cluster's weights are adjusted to let the cluster unit learn the pattern. As is often the case in clustering nets, the weights on a cluster unit may be considered to be an exemplar (or code vector) for the patterns placed on that cluster.

5.1.1 Motivation

Adaptive resonance theory nets are designed to allow the user to control the degree of similarity of patterns placed on the same cluster. However, since input patterns may differ in their level of detail (number of components that are nonzero), the *relative similarity* of an input pattern to the weight vector for a cluster unit, rather than the *absolute difference* between the vectors, is used. (A difference in one component is more significant in patterns that have very few nonzero components than it is in patterns with many nonzero components).

As the net is trained, each training pattern may be presented several times. A pattern may be placed on one cluster unit the first time it is presented and then

218

placed on a different cluster when it is presented later (due to changes in the weights for the first cluster if it has learned other patterns in the meantime.) A stable net will not return a pattern to a previous cluster; in other words, a pattern oscillating among different cluster units at different stages of training indicates an unstable net.

Some nets achieve stability by gradually reducing the learning rate as the same set of training patterns is presented many times. However, this does not allow the net to learn readily a new pattern that is presented for the first time after a number of training epochs have already taken place. The ability of a net to respond to (learn) a new pattern equally well at any stage of learning is called *plasticity*. Adaptive resonance theory nets are designed to be both stable and plastic.

Also, attention has been paid to structuring ART nets so that neural processes can control the rather intricate operation of these nets. This requires a number of neurons in addition to the input units, cluster units, and units for the comparison of the input signal with the cluster unit's weights.

5.1.2 Basic Architecture

The basic architecture of an adaptive resonance neural net involves three groups of neurons: an input processing field (called the F_1 layer), the cluster units (the F_2 layer), and a mechanism to control the degree of similarity of patterns placed on the same cluster (a reset mechanism). The F_1 layer can be considered to consist of two parts: the input portion and the interface portion. Some processing may occur in the input portion (especially in ART2). The interface portion combines signals from the input portion and the F_2 layer, for use in comparing the similarity of the input signal to the weight vector for the cluster unit that has been selected as a candidate for learning. We shall denote the input portion of the F_1 layer as $F_1(a)$ and the interface portion as $F_1(b)$.

To control the similarity of patterns placed on the same cluster, there are two sets of connections (each with its own weights) between each unit in the interface portion of the input field and each cluster unit. The $F_1(b)$ layer is connected to the F_2 layer by bottom-up weights; the bottom-up weight on the connection from the ith F_1 unit to the jth F_2 unit is designated b_{ij}. The F_2 layer is connected to the $F_1(b)$ layer by top-down weights; the top-down weight on the connection from the jth F_2 unit to the ith F_1 unit is designated t_{ji}.

The F_2 layer is a competitive layer: The cluster unit with the largest net input becomes the candidate to learn the input pattern. The activations of all other F_2 units are set to zero. The interface units now combine information from the input and cluster units. Whether or not this cluster unit is allowed to learn the input pattern depends on how similar its top-down weight vector is to the input vector. This decision is made by the reset unit, based on signals it receives from the input (a) and interface (b) portions of the F_1 layer. If the cluster unit is not allowed to learn, it is inhibited and a new cluster unit is selected as the candidate.

The supplemental units needed to control the processing of information in the ART nets are described for ART1 and ART2 in Sections 5.2.1 and 5.3.1, respectively.

5.1.3 Basic Operation

It is difficult to describe even the basic architecture of adaptive resonance theory nets without discussing the operation of the nets. Details of the operation of ART1 and ART2 are presented later in this chapter.

A *learning trial* in ART consists of the presentation of one input pattern. Before the pattern is presented, the activations of all units in the net are set to zero. All F_2 units are inactive. (Any F_2 units that had been inhibited on a previous learning trial are again available to compete.) Once a pattern is presented, it continues to send its input signal until the learning trial is completed.

The degree of similarity required for patterns to be assigned to the same cluster unit is controlled by a user-specified parameter, known as the *vigilance parameter*. Although the details of the reset mechanism for ART1 and ART2 differ, its function is to control the state of each node in the F_2 layer. At any time, an F_2 node is in one of three states:

active ("on," activation = d; d = 1 for ART1, $0 < d < 1$ for ART2),
inactive ("off," activation = 0, but available to participate in competition), or
inhibited ("off," activation = 0, and prevented from participating in any further competition during the presentation of the current input vector).

A summary of the steps that occur in ART nets in general is as follows:

Step 0.   Initialize parameters.
Step 1.   While stopping condition is false, do Steps 2–9.
    Step 2.   For each input vector, do Steps 3–8.
        Step 3.   Process F_1 layer.
        Step 4.   While reset condition is true, do Steps 5–7.
            Step 5.   Find candidate unit to learn the current input pattern:
                F_2 unit (which is not inhibited) with largest input.
            Step 6. $F_1(b)$ units combine their inputs from $F_1(a)$ and F_2.
            Step 7.   Test reset condition (details differ for ART1 and ART2):
                If reset is true, then the current candidate unit is rejected (inhibited); return to Step 4.
                If reset is false, then the current candidate unit is accepted for learning; proceed to Step 8.

> *Step 8.*   Learning: Weights change according to differential equations.
>
> *Step 9.*   Test stopping condition.

The calculations in Step 2 constitute a "learning trial," i.e., one presentation of one pattern. Although ART does not require that all training patterns be presented in the same order, or even that all patterns be presented with the same frequency, we shall refer to the calculations of Step 1 (one presentation of each training pattern) as an epoch. Thus, the foregoing algorithmic structure, while convenient, is not the most general form for ART. The learning process may involve many weight updates and/or many epochs.

Learning

In adaptive resonance theory, the changes in the activations of units and in weights are governed by coupled differential equations. The net is a continuously changing (dynamical) system, but the process can be simplified because the activations are assumed to change much more rapidly than the weights. Once an acceptable cluster unit has been selected for learning, the bottom-up and top-down signals are maintained for an extended period, during which time the weight changes occur. This is the "resonance" that gives the net its name.

Two types of learning that differ both in their theoretical assumptions and in their performance characteristics can be used for ART nets. In the *fast learning* mode, it is assumed that weight updates during resonance occur rapidly, relative to the length of time a pattern is presented on any particular trial. Thus, in fast learning, the weights reach equilibrium on each trial. In the *slow learning* mode the weight changes occur slowly relative to the duration of a learning trial; the weights do not reach equilibrium on a particular trial. Many more presentations of the patterns are required for slow learning than for fast learning, but fewer calculations occur on each learning trial in slow learning. In order to achieve the performance characteristics of slow learning to the fullest extent, we shall assume that only one weight update, with a relatively small learning rate, occurs on each learning trial in the slow learning mode.

In fast learning, the net is considered stabilized when each pattern chooses the correct cluster unit when it is presented (without causing any unit to reset). For ART1, because the patterns are binary, the weights associated with each cluster unit also stabilize in the fast learning mode. The resulting weight vectors are appropriate for the type of input patterns used in ART1. Also, the equilibrium weights are easy to determine, and the iterative solution of the differential equations that control the weight updates is not necessary. This is the form of learning that is typically used for ART1 and is the only algorithm we consider.

In general, for ART2, the weights produced by fast learning continue to change each time a pattern is presented. The net stabilizes after only a few presentations of each training pattern. However, since the differential equations for the weight updates depend on the activation of units whose activations change during the resonance process, it is not as straightforward to find the equilibrium weights immediately for ART2 as it is for ART1. The process of solving the weight update equations by letting the net iterate as resonance occurs is illustrated in several examples in Section 5.3.3.

In the slow learning mode, weight changes do not reach equilibrium during any particular learning trial and more trials are required before the net stabilizes [Carpenter & Grossberg, 1987a, 1987b]. Although slow learning is theoretically possible for ART1, it is not typically used in this form. For ART2, however, the weights produced by slow learning may be much more appropriate than those produced by fast learning for certain types of data. Examples of the use of slow learning for ART2 are given in Section 5.3.3.

5.2 ART1

ART1 is designed to cluster binary input vectors, allowing for great variation in the number of nonzero components, and direct user control of the degree of similarity among patterns placed on the same cluster unit. The architecture of an ART1 net consists of two fields of units—the F_1 units and the F_2 (cluster) units—together with a reset unit to control the degree of similarity of patterns placed on the same cluster unit. (For convenience, we have expanded the description of the F_1 units from the original presentation of the net in Carpenter and Grossberg (1987a) to include an explicit set of input units, the $F_1(a)$ units, as well as the $F_1(b)$ units, which exchange information with the F_2 units.) The F_1 and F_2 layers are connected by two sets of weighted pathways. In addition, several supplemental units are included in the net to provide for neural control of the learning process.

The learning process is designed so that it is not required either that patterns be presented in a fixed order or that the number of patterns to be clustered be known in advance (i.e., more patterns can be added to the data set during the training process if desired). Updates for both the bottom-up and top-down weights are controlled by differential equations. However, it is assumed that the ART1 net is being operated in the fast learning mode, in which the weights reach equilibrium during each learning trial (presentation of a pattern). Since the activations of the F_1 units do not change during this resonance phase, the equilibrium weights can be determined exactly, and the iterative solution of the differential equations is not necessary.

5.2.1 Architecture

The architecture of ART1 consists of computational units and supplemental units.

Computational units

The architecture of the computational units for ART1 consists of F_1 units (input and interface units), F_2 units (cluster units), and a reset unit that implements user control over the degree of similarity of patterns placed on the same cluster. This main portion of the ART1 architecture is illustrated in Figure 5.1.

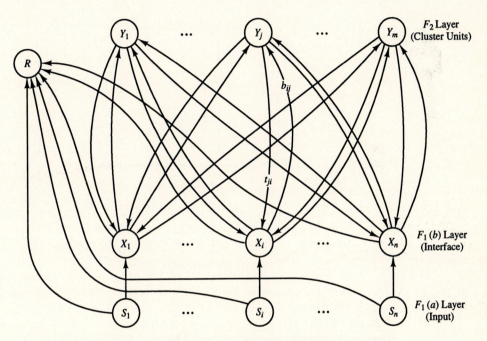

Figure 5.1   Basic structure of ART1.

Each unit in the $F_1(a)$ (input) layer is connected to the corresponding unit in the $F_1(b)$ (interface) layer. Each unit in the $F_1(a)$ and $F_1(b)$ layer is connected to the reset unit, which in turn is connected to every F_2 unit. Each unit in the $F_1(b)$ layer is connected to each unit in the F_2 (cluster) layer by two weighted pathways. The $F_1(b)$ unit X_i is connected to the F_2 unit Y_j by bottom-up weights b_{ij}. Similarly, unit Y_j is connected to unit X_i by top-down weights t_{ji}. Only one representative weight b_{ij} is indicated on the connections between the F_1 and F_2 layer; similarly, t_{ji} is a representative weight for the connections between the F_2 and F_1 layer. The F_2 layer is a competitive layer in which only the uninhibited node with the largest net input has a nonzero activation.

Supplemental units

The supplemental units shown in Figure 5.2 are important from a theoretical point of view. They provide a mechanism by which the computations performed by the algorithm in Section 5.2.2 can be accomplished using neural network principles. However, this theoretical discussion of the necessary supplemental units is not required to be able to use the computational algorithm. The reader may proceed directly to the algorithm if desired.

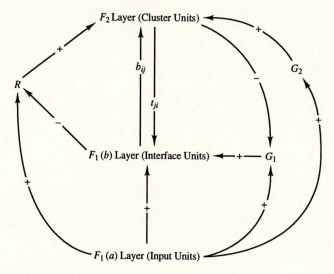

Figure 5.2   The supplemental units for ART1. Adapted from [Carpenter & Grossberg, 1987a]

    The computational algorithm for ART1 represents a more involved neural process than can be represented in terms of just the F_1 units, the F_2 units, and the reset unit in Figure 5.1. The difficulty is that these units are required to respond differently at different stages of the process, and a biological neuron does not have a method for deciding what to do when. For example, units in the F_1 interface region ($F_1(b)$ units) should be "on" when an input signal is received (from $F_1(a)$) and no F_2 units are active. However, when an F_2 unit is active, an $F_1(b)$ unit should remain "on" only if it receives (nonzero) signals from both the F_2 layer and the input units.

    The operation of the reset mechanism also poses a challenge in terms of implementation within a neural processing system. F_2 units must be inhibited (and prevented from competing) under certain conditions, but returned to availability on subsequent learning trials.

    Both of these problems can be solved by introducing two supplemental units (called *gain control units*) G_1 and G_2, in addition to the reset unit R shown in Figure 5.1. Each of these three special units receives signals from, and sends its signal to, all of the units in the layers indicated in Figure 5.2. Excitatory signals are indicated by "+" and inhibitory signals by "−". A signal is sent whenever any unit in the designated layer is "on."

Each unit in either the $F_1(b)$ (interface region) or F_2 layer of the ART1 net has three sources from which it can receive a signal. $F_1(b)$ interface units can receive signals from an $F_1(a)$ unit (an input signal), from an F_2 node (a top-down signal), or from the G_1 unit. Similarly, an F_2 unit can receive a signal from the F_1 interface units, unit R, or the G_2 unit. An $F_1(b)$ (interface) or F_2 unit must receive two excitatory signals in order to be "on." Since there are three possible sources of signals, this requirement is called the two-thirds rule.

The $F_1(b)$ (interface) nodes are required to send a signal (to the F_2 nodes and the reset unit) whenever an input vector is presented and no F_2 node is active. However, after an F_2 node has been chosen in the competition, it is necessary that only the $F_1(b)$ nodes whose top-down signal and input signal match remain active. This is accomplished through the use of the G_1 and G_2 units and the two-thirds rule. The G_1 unit is inhibited whenever any F_2 unit is on. When no F_2 unit is on, each F_1 interface unit receives a signal from the G_1 unit; in this case, all of the units that receive a positive input signal from the input vector that is presented will fire. In a similar manner, the G_2 unit controls the firing of the F_2 units; they must obey a two-thirds rule, too. The two-thirds rule also plays a role in the choice of parameters and initial weights, considered in Section 5.2.4.

The reset unit R controls the vigilance matching. As indicated in Figure 5.2 when any unit in the $F_1(a)$ input layer is on, an excitatory signal is sent to unit R. The strength of that signal depends on how many F_1 (input) units are on. However, R also receives inhibitory signals from the F_1 interface units that are on. If enough $F_1(b)$ interface units are on (as determined by the vigilance parameter set by the user), unit R is prevented from firing. If unit R does fire, it inhibits any F_2 unit that is on. This forces the F_2 layer to choose a new winning node.

5.2.2 Algorithm

In the first two parts of this section, we provide a description of the training process for ART1 (using fast learning) and a step-by-step algorithm for the training. A discussion of the choice of parameter values and initial weights follows the training algorithm. The notation we use is as follows:

n	number of components in the input vector.
m	maximum number of clusters that can be formed.
b_{ij}	bottom-up weights (from $F_1(b)$ unit X_i to F_2 unit Y_j).
t_{ji}	top-down weights (from F_2 unit Y_j to F_1 unit X_i).
ρ	vigilance parameter.
s	binary input vector (an n-tuple).
x	activation vector for $F_1(b)$ layer (binary).
$\|\mathbf{x}\|$	norm of vector **x**, defined as the sum of the components x_i.

Description

A binary input vector **s** is presented to the $F_1(a)$ layer, and the signals are sent to the corresponding X units. These $F_1(b)$ units then broadcast to the F_2 layer over connection pathways with bottom-up weights. Each F_2 unit computes its net input, and the units compete for the right to be active. The unit with the largest net input sets its activation to 1; all others have an activation of 0. We shall denote the index of the winning unit as J. This winning unit becomes the candidate to learn the input pattern. A signal is then sent down from F_2 to $F_1(b)$ (multiplied by the top-down weights). The X units (in the interface portion of the F_1 layer) remain "on" only if they receive nonzero signals from both the $F_1(a)$ and F_2 units.

    The norm of the vector **x** (the activation vector for the interface portion of F_1) gives the number of components in which the top-down weight vector for the winning F_2 unit $\mathbf{t}_J$ and the input vector **s** are both 1. (This quantitity is sometimes referred to as the *match*.) If the ratio of $\|\mathbf{x}\|$ to $\|\mathbf{s}\|$ is greater than or equal to the vigilance parameter, the weights (top down and bottom up) for the winning cluster unit are adjusted.

    However, if the ratio is less than the vigilance parameter, the candidate unit is rejected, and another candidate unit must be chosen. The current winning cluster unit becomes inhibited, so that it cannot be chosen again as a candidate on this learning trial, and the activations of the F_1 units are reset to zero. The same input vector again sends its signal to the interface units, which again send this as the bottom-up signal to the F_2 layer, and the competition is repeated (but without the participation of any inhibited units).

    The process continues until either a satisfactory match is found (a candidate is accepted) or all units are inhibited. The action to be taken if all units are inhibited must be specified by the user. It might be appropriate to reduce the value of the vigilance parameter, allowing less similar patterns to be placed on the same cluster, or to increase the number of cluster units, or simply to designate the current input pattern as an outlier that could not be clustered.

    In the theoretical formulation of the ART process, learning occurs during the resonance phase, in which the signals continue to flow throughout the net. Care is taken in setting the parameter values so that a reset cannot suddenly occur as the weights change, nor will a new winning unit be chosen after a candidate unit is accepted for learning (see Section 5.2.4). Since the activations of the F_1 units do not change as learning progresses in ART1, the equilibrium weight values can be found directly (see Section 5.2.4). The algorithm given shortly uses these equilibrium values in Step 8.

    At the end of each presentation of a pattern (normally, after the weights have been adjusted), all cluster units are returned to inactive status (zero activations, but available to participate in the next competition).

    The use of the ratio of the match to the norm of the input vector in the reset calculations described allows an ART1 net to respond to relative differences. This reflects the fact that a difference of one component in vectors with only a few nonzero components is much more significant than a difference of one component in vectors with many nonzero components.

Training algorithm

The training algorithm for an ART1 net is presented next. A discussion of the role of the parameters and an appropriate choice of initial weights follows.

Step 0.   Initialize parameters:

$$L > 1,$$

$$0 < \rho \leq 1.$$

Initialize weights:

$$0 < b_{ij}(0) < \frac{L}{L - 1 + n},$$

$$t_{ji}(0) = 1.$$

Step 1.   While stopping condition is false, do Steps 2–13.

    Step 2.   For each training input, do Steps 3–12.

        Step 3.   Set activations of all F_2 units to zero.
Set activations of $F_1(a)$ units to input vector **s**.

        Step 4.   Compute the norm of **s**:

$$\|\mathbf{s}\| = \sum_i s_i.$$

        Step 5.   Send input signal from $F_1(a)$ to the $F_1(b)$ layer:

$$x_i = s_i.$$

        Step 6.   For each F_2 node that is not inhibited:
If $y_j \neq -1$, then

$$y_j = \sum_i b_{ij} x_i.$$

        Step 7.   While reset is true, do Steps 8–11.

            Step 8.   Find J such that $y_J \geq y_j$ for all nodes j.
If $y_J = -1$, then all nodes are inhibited and this pattern cannot be clustered.

            Step 9.   Recompute activation **x** of $F_1(b)$:

$$x_i = s_i t_{Ji}.$$

            Step 10.   Compute the norm of vector **x**:

$$\|\mathbf{x}\| = \sum_i x_i.$$

Step 11.   Test for reset:

$$\text{If } \frac{\|\mathbf{x}\|}{\|\mathbf{s}\|} < \rho, \text{ then}$$

$$y_J = -1 \text{ (inhibit node } J\text{) (and continue,}$$
$$\text{executing Step 7 again).}$$

$$\text{If } \frac{\|\mathbf{x}\|}{\|\mathbf{s}\|} \geq \rho,$$

then proceed to Step 12.

Step 12.   Update the weights for note J (fast learning):

$$b_{iJ}(\text{new}) = \frac{L x_i}{L - 1 + \|\mathbf{x}\|},$$

$$t_{Ji}(\text{new}) = x_i.$$

Step 13.   Test for stopping condition.

Comments.   Step 3 removes all inhibitions from the previous learning trial (presentation of a pattern).

Setting $y = -1$ for an inhibited node (in Step 6) will prevent that node from being a winner. Since all weights and signals in the net are nonnegative a unit with a negative activation can never have the largest activation.

In Step 8, in case of a tie, take J to be the smallest such index.

In Step 9, unit X_i is "on" only if it receives both an external signal s_i and a signal sent down from F_2 to F_1, t_{Ji}.

The stopping condition in Step 13 might consist of any of the following:

No weight changes,
no units reset, or
maximum number of epochs reached.

The ART1 training process is often described in terms of setting the activation of the winning F_2 unit to 1 and all others to 0. However, it is easy to implement the algorithm without making explicit use of those activations, so in the interest of computational simplicity, they are omitted here.

The user must specify the desired action to be taken in the event that all the F_2 nodes are inhibited in Step 8. The possibilities include adding more cluster units, reducing the vigilance, and classifying the pattern as an outlier that cannot be clustered (with the given parameters).

Note that t_{ji} is either 0 or 1, and once it is set to 0 during learning, it can never be set back to 1 (which provides the stable learning mentioned before).

Parameters

User-defined parameters with restrictions indicated for permissible values [Carpenter & Grossberg, 1987a] and sample values [Lippmann, 1987] are as follows:

PARAMETER	PERMISSIBLE RANGE	SAMPLE VALUE
L	$L > 1$	2
ρ	$0 < \rho \leq 1$ (vigilance parameter)	.9
b_{ij}	$0 < b_{ij}(0) < \dfrac{L}{L-1+n}$ (bottom-up weights)	$\dfrac{1}{1+n}$
t_{ji}	$t_{ji}(0) = 1$ (top-down weights)	1

The derivation of the restrictions on the initial bottom-up and top-down weights is given in Section 5.2.4. Larger values for the initial bottom-up weights may encourage the net to form more clusters. The sample values of $b_{ij}(0)$ (one-half of the maximum allowed value for $L = 2$) are used in the examples of Section 5.2.3.

The equilibrium weights, which are used in the ART1 algorithm are derived in Section 5.2.4. The algorithm uses fast learning, which assumes that the input pattern is presented for a long enough period of time for the weights to reach equilibrium. Since none of the activations of the F_1 units change during resonance, it is not necessary actually to perform the iterations to solve the weight-update differential equations numerically.

Several values of ρ are illustrated in the examples of Section 5.2.3.

5.2.3 Applications

Simple examples

The following two examples show in detail the application of the algorithm in the previous section, for a simple case with input vectors that are ordered quadruples and three cluster units. The role of the vigilance parameter is illustrated by the differences between the examples: The first example uses a relatively low value of ρ, the second a somewhat higher value.

Example 5.1 An ART1 net to cluster four vectors: low vigilance

The values and a description of the parameters in this example are:

n	=	4	number of components in an input vector;
m	=	3	maximum number of clusters to be formed;
ρ	=	0.4	vigilance parameter;
L	=	2	parameter used in update of bottom-up weights;
$b_{ij}(0)$	=	$\dfrac{1}{1+n}$	initial bottom-up weights (one-half the maximum value allowed);
$t_{ji}(0)$	=	1	initial top-down weights.

The example uses the ART1 algorithm to cluster the vectors $(1, 1, 0, 0)$, $(0, 0, 0, 1)$, $(1, 0, 0, 0)$, and $(0, 0, 1, 1)$, in at most three clusters.

Application of the algorithm yields the following:

Step 0.   Initialize parameters:

$$L = 2,$$

$$\rho = 0.4;$$

Initialize weights:

$$b_{ij}(0) = 0.2,$$

$$t_{ji}(0) = 1.$$

Step 1.   Begin computation.

   Step 2.   For the first input vector, $(1, 1, 0, 0)$, do Steps 3–12.

      Step 3.   Set activations of all F_2 units to zero.
Set activations of $F_1(a)$ units to input vector
$$\mathbf{s} = (1, 1, 0, 0).$$

      Step 4.   Compute norm of $\mathbf{s}$:

$$\|\mathbf{s}\| = 2.$$

      Step 5.   Compute activations for each node in the F_1 layer:

$$\mathbf{x} = (1, 1, 0, 0).$$

      Step 6.   Compute net input to each node in the F_2 layer:

$$y_1 = .2(1) + .2(1) + .2(0) + .2(0) = 0.4,$$

$$y_2 = .2(1) + .2(1) + .2(0) + .2(0) = 0.4,$$

$$y_3 = .2(1) + .2(1) + .2(0) + .2(0) = 0.4.$$

   Step 7.   While reset is true, do Steps 8–11.

      Step 8.   Since all units have the same net input,

$$J = 1.$$

Step 9.   Recompute the F_1 activations:

$$x_i = s_i t_{1i}; \text{ currently, } t_1 = (1, 1, 1, 1);$$

therefore, $\mathbf{x} = (1, 1, 0, 0)$

Step 10.   Compute the norm of $\mathbf{x}$:

$$\|\mathbf{x}\| = 2.$$

Step 11.   Test for reset:

$$\frac{\|\mathbf{x}\|}{\|\mathbf{s}\|} = 1.0 \geq 0.4;$$

therefore, reset is false.
Proceed to Step 12.

Step 12.   Update b_1; for $L = 2$, the equilibrium weights are

$$b_{i1}(\text{new}) = \frac{2x_i}{1 + \|\mathbf{x}\|}.$$

Therefore, the bottom-up weight matrix becomes

$$\begin{bmatrix} .67 & .2 & .2 \\ .67 & .2 & .2 \\ 0 & .2 & .2 \\ 0 & .2 & .2 \end{bmatrix}$$

Update t_1; the fast learning weight values are

$$t_{Ji}(\text{new}) = x_i,$$

therefore, the top-down weight matrix becomes

$$\begin{bmatrix} 1 & 1 & 0 & 0 \\ 1 & 1 & 1 & 1 \\ 1 & 1 & 1 & 1 \end{bmatrix}$$

Step 2.   For the second input vector, $(0, 0, 0, 1)$, do Steps 3–12.

Step 3.   Set activations of all F_2 units to zero.
Set activations of $F_1(a)$ units to input vector
$$\mathbf{s} = (0, 0, 0, 1).$$

Step 4.   Compute norm of $\mathbf{s}$:

$$\|\mathbf{s}\| = 1.$$

Step 5.   Compute activations for each node in the F_1 layer:

$$\mathbf{x} = (0, 0, 0, 1).$$

Step 6.   Compute net input to each node in the F_2 layer:

$$y_1 = .67(0) + .67(0) + 0(0) + 0(1) = 0.0,$$

$$y_2 = .2(0) + .2(0) + .2(0) + .2(1) = 0.2,$$

$$y_3 = .2(0) + .2(0) + .2(0) + .2(1) = 0.2.$$

Step 7.   While reset is true, do Steps 8–11.

    Step 8.   Since units Y_2 and Y_3 have the same net input

$$J = 2.$$

    Step 9.   Recompute the activation of the F_1 layer:

$$x_i = s_i t_{2i};$$

currently $\mathbf{t}_2 = (1, 1, 1, 1)$; therefore,

$$\mathbf{x} = (0, 0, 0, 1).$$

    Step 10.   Compute the norm of $\mathbf{x}$:

$$\|\mathbf{x}\| = 1.$$

    Step 11.   Test for reset:

$$\frac{\|\mathbf{x}\|}{\|\mathbf{s}\|} = 1.0 \geq 0.4;$$

therefore, reset is false. Proceed to Step 12.

Step 12.   Update $\mathbf{b}_2$; the bottom-up weight matrix becomes

$$\begin{bmatrix} .67 & 0 & .2 \\ .67 & 0 & .2 \\ 0 & 0 & .2 \\ 0 & 1 & .2 \end{bmatrix}$$

Update $\mathbf{t}_2$; the top-down weight matrix becomes

$$\begin{bmatrix} 1 & 1 & 0 & 0 \\ 0 & 0 & 0 & 1 \\ 1 & 1 & 1 & 1 \end{bmatrix}$$

Step 2.   For the third input vector, $(1, 0, 0, 0)$, do Steps 3–12.

    Step 3.   Set activations of all F_2 units to zero.

Set activations of $F_1(a)$ units to input vector
$$\mathbf{s} = (1, 0, 0, 0).$$

    Step 4.   Compute norm of $\mathbf{s}$:

$$\|\mathbf{s}\| = 1.$$

    Step 5.   Compute activations for each node in the F_1 layer:

$$\mathbf{x} = (1, 0, 0, 0).$$

Step 6.   Compute net input to each node in the F_2 layer:

$$y_1 = .67(1) + .67(0) + \ 0(0) + \ 0(0) = 0.67,$$

$$y_2 = \ \ 0(1) + \ \ \ 0(0) + \ \ 0(0) + \ \ 1(0) = 0.0,$$

$$y_3 = \ \ .2(1) + \ \ .2(0) + .2(0) + .2(0) = 0.2.$$

Step 7.   While reset is true, do Steps 8–11.
        Step 8. Since unit Y_1 has the largest net input,

$$J = 1.$$

Step 9.   Recompute the activation of the F_1 layer:

$$x_i = s_i t_{1i};$$

current, $t_1 = (1, 1, 0, 0)$; therefore,

$$\mathbf{x} = (1, 0, 0, 0).$$

Step 10.   Compute the norm of $\mathbf{x}$:

$$\|\mathbf{x}\| = 1.$$

Step 12.   Update $\mathbf{b}_1$; the bottom-up weight matrix becomes

$$\begin{bmatrix} 1 & 0 & .2 \\ 0 & 0 & .2 \\ 0 & 0 & .2 \\ 0 & 1 & .2 \end{bmatrix}$$

Update $\mathbf{t}_1$; the top-down weight matrix becomes

$$\begin{bmatrix} 1 & 0 & 0 & 0 \\ 0 & 0 & 0 & 1 \\ 1 & 1 & 1 & 1 \end{bmatrix}$$

Step 2.   For the fourth input vector, $(0, 0, 1, 1)$, do Steps 3–12.
        Step 3.   Set activations of all F_2 units to zero.
                Set activations of $F_1(a)$ units to input vector
                $\mathbf{s} = (0, 0, 1, 1)$.
Step 4.   Compute norm of $\mathbf{s}$:

$$\|\mathbf{s}\| = 2.$$

Step 5.   Compute activations for each node in the F_1 layer:

$$\mathbf{x} = (0, 0, 1, 1).$$

Step 6.   Compute net input to each node in the F_2 layer:

$$y_1 = \ 1(0) + \ 0(0) + \ 0(1) + \ 0(1) = 0.0,$$

$$y_2 = \ 0(0) + \ 0(0) + \ 0(1) + \ 1(1) = 1.0,$$

$$y_3 = .2(0) + .2(0) + .2(1) + .2(1) = 0.4.$$

Step 7.   While reset is true, do Steps 8–11.

    Step 8.   Since unit Y_2 has the largest net input,

$$J = 2.$$

    Step 9.   Recompute the activation of the F_1 layer:

$$x_i = s_i t_{2i};$$

currently, $t_2 = (0, 0, 0, 1)$; therefore,

$$\mathbf{x} = (0, 0, 0, 1).$$

    Step 10.   Compute the norm of **x**:

$$\|\mathbf{x}\| = 1.$$

    Step 11.   Test for reset:

$$\frac{\|\mathbf{x}\|}{\|\mathbf{s}\|} = 0.5 \geq 0.4;$$

therefore, reset is false. Proceed to Step 12.

Step 12.   Update $\mathbf{b}_2$; however, there is no change in the bottom-up weight matrix:

$$\begin{bmatrix} 1 & 0 & .2 \\ 0 & 0 & .2 \\ 0 & 0 & .2 \\ 0 & 1 & .2 \end{bmatrix}$$

Similarly, the top-down weight matrix remains

$$\begin{bmatrix} 1 & 0 & 0 & 0 \\ 0 & 0 & 0 & 1 \\ 1 & 1 & 1 & 1 \end{bmatrix}$$

Step 13.   Test stopping condition.
(This completes one epoch of training.)

The reader can check that no further learning takes place on subsequent presentations of these vectors, regardless of the order in which they are presented. Depending on the order of presentation of the patterns, more than one epoch may be required, but typically, stable weight matrices are obtained very quickly.

Example 5.2 An ART1 net to cluster four vectors: moderate vigilance

The same vectors are presented to the ART1 net (in the same order) as in Example 5.1. The vigilance parameter is set at 0.7. The training for vectors $(1, 1, 0, 0)$, $(0, 0, 0, 1)$, and $(1, 0, 0, 0)$ proceeds as before, giving the bottom-up weight matrix

$$\begin{bmatrix} 1 & 0 & .2 \\ 0 & 0 & .2 \\ 0 & 0 & .2 \\ 0 & 1 & .2 \end{bmatrix}$$

and the top-down weight matrix

$$\begin{bmatrix} 1 & 0 & 0 & 0 \\ 0 & 0 & 0 & 1 \\ 1 & 1 & 1 & 1 \end{bmatrix}$$

However, for the fourth input vector, $\mathbf{s} = (0, 0, 1, 1)$, the results are different. We obtain:

Step 2.   For the fourth input vector, $(0, 0, 1, 1)$, do Steps 3–12.

         Step 3.   Set activations of all F_2 units to zero.

                               Set activations of $F_1(a)$ units to vector $\mathbf{s} = (0, 0, 1, 1)$.

         Step 4.   Compute norm of $\mathbf{s}$:

$$\|\mathbf{s}\| = 2.$$

         Step 5.   Compute activations for each node in the F_1 layer:

$$\mathbf{x} = (0, 0, 1, 1).$$

         Step 6.   Compute net input to each node in the F_2 layer:

$$y_1 = 1(0) + 0(0) + 0(1) + 0(1) = 0.0,$$

$$y_2 = 0(0) + 0(0) + 0(1) + 1(1) = 1.0,$$

$$y_3 = .2(0) + .2(0) + .2(1) + .2(1) = 0.4.$$

         Step 7.   While reset is true, do Steps 8–11.

            Step 8.   Since unit Y_2 has the largest net input,

$$J = 2.$$

            Step 9.   Recompute the activation of the F_1 layer:

$$x_i = s_i t_{2i};$$

currently, $\mathbf{t}_2 = (0, 0, 0, 1)$; therefore,

$$\mathbf{x} = (0, 0, 0, 1).$$

            Step 10.   Compute the norm of $\mathbf{x}$:

$$\|\mathbf{x}\| = 1.$$

            Step 11.   Check the vigilance criterion:

$$\frac{\|\mathbf{x}\|}{\|\mathbf{s}\|} = 0.5 < 0.7;$$

therefore, reset is true, so inhibit Y_2:

$$y_2 = -1.0.$$

Proceed with Step 7.

Step 7.   While reset is true, do Steps 8–11.

    Step 8.   Now the values for the F_2 layer are

$$y_1 = \ \ \ 0.0,$$

$$y_2 = -1.0,$$

$$y_3 = \ \ \ 0.4.$$

    So unit Y_3 has the largest net input, and

$$J = 3.$$

    Step 9.   Recompute the activation of the F_1 layer:

$$x_i = s_i t_{3i};$$

    currently, $t_3 = (1, 1, 1, 1)$; therefore,

$$x = (0, 0, 1, 1).$$

    Step 10.   Compute the norm of x:

$$\|x\| = 2.$$

    Step 11.   Test for reset:

$$\frac{\|x\|}{\|s\|} = 1.0 \geq 0.7;$$

    therefore, reset is false. Proceed with Step 12.

Step 12.   Update b_3; the bottom-up weight matrix becomes

$$\begin{bmatrix} 1 & 0 & 0 \\ 0 & 0 & 0 \\ 0 & 0 & .67 \\ 0 & 1 & .67 \end{bmatrix}$$

Update t_3; the top-down weight matrix becomes

$$\begin{bmatrix} 1 & 0 & 0 & 0 \\ 0 & 0 & 0 & 1 \\ 0 & 0 & 1 & 1 \end{bmatrix}$$

Step 13.   Test stopping condition.

Now, when the first vector is presented again, the vigilance criterion will not be satisfied for any of the F_2 nodes. The user may decide to add another F_2 unit, classify the first input vector as an outlier, or use a lower vigilance parameter. In contrast to some other neural networks, an ART net will not automatically force all input vectors onto a cluster if they are not sufficiently similar.

Character recognition

Examples 5.3–5.5 show the results of employing ART1 to cluster the patterns in Figure 5.3, using a representative selection of different values of the vigilance parameter, different input orders for the patterns, and different values for the maximum number of cluster units. Note that the weight vector for each cluster reflects all of the patterns placed on the cluster during training; the net does not

Input from
Font 1

Input from
Font 2

Input from
Font 3

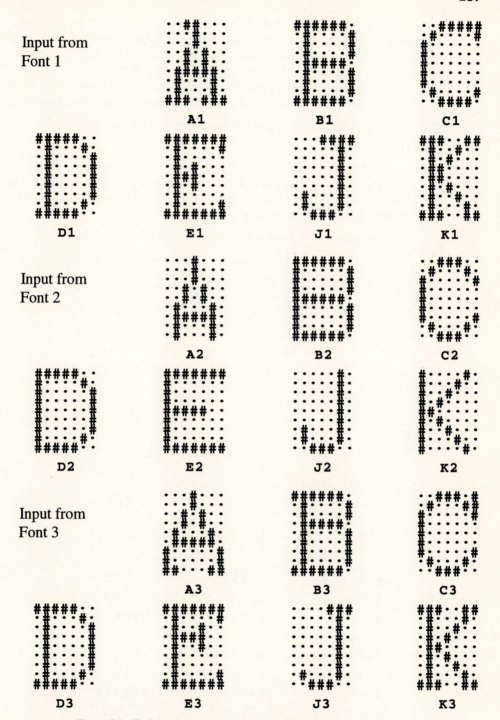

Figure 5.3   Training input patterns for character recognition examples.

forget patterns that were placed on the unit and later moved to another unit. Also, remember that there is no particular significance to patterns placed on cluster units whose indices are close together (in contrast to the topological interpretation of the indices for the Kohonen self-organizing feature maps).

The final weights associated with each cluster are shown in a two-dimensional array, since the input patterns represented two-dimensional patterns.

Example 5.3 An ART1 net to cluster letters from three fonts: low vigilance

Using the pattern input order

A1, A2, A3, B1, B2, B3, C1, C2, C3, . . . , J1, J2, J3, K1, K2, K3,

with a vigilance parameter of 0.30 and a maximum of 10 cluster units, results in stable cluster formation (and no weight changes) after three epochs of training. The placement of patterns during training is as follows:

CLUSTER	EPOCH 1	EPOCH 2	EPOCH 3
1	A1, A2, A3	A1, A2, A3	A1, A2, A3
2	B1, B2, B3 C1, C2, C3 J1		
3	D1, D2, D3 E1, E2, E3	B1, B2, B3 C1, C2, C3	C1, C2, C3
4	J2, J3	J1, J2, J3	J1, J2, J3
5	K1, K2	K1, K2	K1, K2
6	J3	D1, D2, D3, K3	D1, D2, D3, K3
7		E1, E2, E3	B1, B2, B3 E1, E2, E3

FINAL WEIGHTS

Cluster 1

Cluster 2

Cluster 3

Cluster 4

Cluster 5

Cluster 6

Cluster 7

Using the pattern input order

A1, B1, C1, D1, E1, J1, K1, A2, B2, C2, D2, E2, J2, K2, A3, B3, C3,
D3, E3, J3, K3,

with a vigilance parameter of 0.30 and 10 cluster units available, results in stable cluster formation (and no weight changes) after two epochs of training, shown as follows:

CLUSTER	EPOCH 1	EPOCH 2
1	A1, B1, C1	C1
2	D1, E1, J1 C2, J2	J2
3	K1, A2	A1, A2
4	B2, D2, E2, K2	B2, D2, E2, K2
5	A3, B3, E3	A3
6	C3, D3, J3	J1, C2, C3, J3
7	K3	B1, D1, E1, K1 B3, D3, E3, K3

FINAL WEIGHTS

Cluster 1   Cluster 2   Cluster 3   Cluster 4

Cluster 5   Cluster 6   Cluster 7

Example 5.4 An ART1 net to cluster letters from three fonts: higher vigilance and an insufficient number of clusters

Using a higher vigilance parameter of 0.70, but still allowing a maximum of only 10 cluster units, results in stable cluster formation (and no weight changes) after two epochs of training; however, some patterns cannot be placed on clusters. (These are shown as CNC, for "could not cluster," in the table that follows.) The pattern input order for this example is

A1, A2, A3, B1, B2, B3, C1, C2, C3, D1, D2, D3, E1, E2, E3,
J1, J2, J3, K1, K2, K3

We have:

CLUSTER	EPOCH 1	EPOCH 2
1	A1, A2	A2
2	A3	A3
3	B1, B2	B1, B2
4	B3, D1, D3	B3, D1, D3
5	C1, C2, K2	C1, C2
6	C3	C3
7	D2	D2
8	E1, E3, K1, K3	E1, E3
9	E2	E2
10	J1, J2, J3	J1, J2, J3
CNC	K2	A1, E1, E3, K2

FINAL WEIGHTS

Cluster 1   Cluster 2   Cluster 3   Cluster 4   Cluster 5

Cluster 6   Cluster 7   Cluster 8   Cluster 9   Cluster 10

Example 5.5 Higher vigilance (and more clusters) reduces sensitivity to order of input.

Using the same vigilance parameter as in Example 5.4 ($\rho = 0.7$), but allowing a maximum of 15 cluster units, results in stable cluster formation (and no weight changes) after two epochs of training. The clusters formed are less dependent on the pattern input order for the higher vigilance than were the results in Example 5.3

for the lower vigilance. Using the first pattern input order,

A1, A2, A3, B1, B2, B3, C1, C2, C3, D1, D2, D3, E1, E2, E3,
J1, J2, J3, K1, K2, K3,

we obtain the following results:

CLUSTER	EPOCH 1	EPOCH 2
1	A1, A2	A2
2	A3	A3
3	B1, B2	B1, B2
4	B3, D1, D3	B3, D1, D3
5	C1, C2	C1, C2
6	C3	C3
7	D2	D2
8	E1, E3, K1, K3	K1, K3
9	E2	E2
10	J1, J2, J3	J1, J2, J3
11	K2	K2
12		A1
13		E1, E3

FINAL WEIGHTS

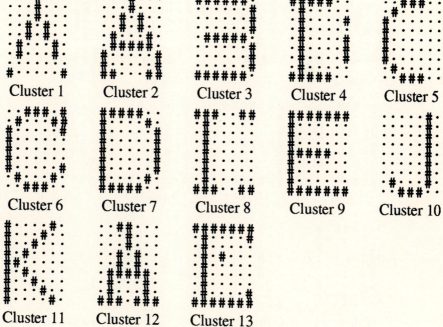

Cluster 1 Cluster 2 Cluster 3 Cluster 4 Cluster 5

Cluster 6 Cluster 7 Cluster 8 Cluster 9 Cluster 10

Cluster 11 Cluster 12 Cluster 13

The results for the second pattern input order,

A1, B1, C1, D1, E1, J1, K1, A2, B2, C2, D2, E2, J2, K2, A3,
B3, C3, D3, E3, J3, K3,

are quite similar (but not identical) to those for the first input order:

CLUSTER	EPOCH 1	EPOCH 2
1	A1, A2	A2
2	B1, D1, D3	B1, D1, D3
3	C1, C2	C1, C2
4	E1, K1, K3	E1, K1, K3
5	J1, J2, J3	J1, J2, J3
6	B2, D2	B2, D2
7	E2	E2
8	K2	K2
9	A3	A3
10	B3, E3	B3, E3
11	C3	C3
12		A1

FINAL WEIGHTS

Cluster 1   Cluster 2   Cluster 3   Cluster 4   Cluster 5

Cluster 6   Cluster 7   Cluster 8   Cluster 9   Cluster 10

Cluster 11   Cluster 12

5.2.4 Analysis

Learning in ART1

In this section, we present a simple derivation of the fast learning equations. In fast learning, it is assumed that the weights reach equilibrium during the learning trial (presentation of the training vector). The equilibrium values of these weights are easy to derive for ART1, since the activations of the F_1 units do not change while the weights are changing. Because only the weights on the winning F_2 unit (denoted by the index J) are modified, the differential equations that define the weight changes are given only for weights t_{Ji} and b_{iJ}. Vector **x** contains the activations of the F_1 units after the test for reset is performed (Step 11 in the algorithm in Section 5.2.2); therefore, x_i is 1 if unit X_i receives both a nonzero input signal s_i and a nonzero top-down signal t_{Ji} and x_i is 0 if either s_i or t_{Ji} is 0.

The differential equation for the top-down weights (on the winning F_2 unit J) is

$$\frac{d}{dt} t_{Ji} = K_2[-E_{ji}t_{Ji} + x_i].$$

Following the derivation of Carpenter and Grossberg [1987a] for ART1, we make the simple choices

$$K_2 = 1, \qquad E_{ji} = 1.$$

The differential equation then becomes

$$\frac{d}{dt} t_{Ji} = -t_{Ji} + x_i.$$

Since this is a linear differential equation with constant coefficients, for fast learning we set

$$\frac{d}{dt} t_{Ji} = 0,$$

in order to find the equilibrium values of the weights. In this manner, we find that

$$t_{Ji} = x_i.$$

The differential equation for the bottom-up weights (on the winning F_2 unit J) has essentially the same form as the equation for the top-down weights:

$$\frac{d}{dt} b_{iJ} = K_1[-E_{ij}b_{iJ} + x_i].$$

However, in order for ART1 to respond to relative differences in patterns (which may vary greatly in the number of components that are 1 rather than 0), it is important for the equilibrium bottom-up weights to be (approximately) inversely proportional to the norm of the vector of F_1 activations. This can be accomplished by taking

$$E_{ij} = x_i + L^{-1} \sum_{k \neq i} x_k \qquad \text{(for some positive constant } L)$$

and

$$K_1 = KL,$$

so that the differential equation becomes

$$\frac{d}{dt} b_{iJ} = KL[-b_{iJ}x_i - b_{iJ}L^{-1} \sum_{k \neq i} x_k + x_i]$$

$$= K[(1 - b_{iJ})Lx_i - b_{iJ} \sum_{k \neq i} x_k].$$

It is convenient to consider separately the cases when the F_1 unit X_i is inactive and when it is active:

(i) If the F_1 unit X_i is inactive, then

$$x_i = 0 \rightarrow$$

$$\sum_{k \neq i} x_k = \|\mathbf{x}\|$$

(because all active units are included in the summation). The differential equation becomes

$$\frac{d}{dt} b_{iJ} = K[-b_{iJ} \sum_{k \neq i} x_k] = K[-b_{iJ} \|\mathbf{x}\|].$$

As before, in order to find the equilibrium weights

$$b_{iJ} = 0,$$

we set the derivative to zero and solve for b_{iJ} in the resulting equation:

$$0 = K[-b_{iJ} \|\mathbf{x}\|].$$

(ii) If the F_1 unit X_i is active, then

$$x_i = 1 \rightarrow$$

$$\sum_{k \neq i} x_k = \|\mathbf{x}\| - 1$$

(because active unit X_i is not included in the summation). The differential equation becomes

$$\frac{d}{dt} b_{iJ} = K[(1 - b_{iJ})L - b_{iJ}(\|\mathbf{x}\| - 1)].$$

As before, to find the equilibrium weights

$$b_{iJ} = \frac{L}{L - 1 + \|\mathbf{x}\|},$$

we set the derivative to zero and solve for b_{iJ} in the resulting equation:

$$0 = K[(1 - b_{iJ})L - b_{iJ}(\|\mathbf{x}\| - 1)].$$

The formulas for the equilibrium bottom-up weights (in both cases) can be expressed by the single formula

$$b_{iJ} = \frac{Lx_i}{L - 1 + \|\mathbf{x}\|}$$

(since x_i is 1 if X_i is active and x_i is 0 if X_i is inactive).

To sum up, the equilibrium top-down weights are

$$t_{Ji} = x_i,$$

and the equilibrium bottom-up weights are

$$b_{iJ} = \frac{Lx_i}{L - 1 + \|\mathbf{x}\|}.$$

Activations in ART1

The activations of the F_1 (interface) units are controlled by a differential equation of the form

$$\epsilon \frac{d}{dt} x_i = -x_i + (1 - Ax_i)I^+ - (B + Cx_i)I^-,$$

where I^+ is the total excitatory input received by unit X_i (whose activation is x_i) and I^- is the total inhibitory input received by X_i.

It is assumed that activations change much more rapidly than signals are sent or weights change, so that each unit reaches its equilibrium activation virtually instantaneously.

When no F_2 unit is active, the equation for an F_1 (interface) unit becomes

$$\epsilon \frac{d}{dt} x_i = -x_i + (1 - Ax_i)s_i,$$

which gives

$$x_i = \frac{s_i}{1 + Ax_i}$$

at equilibrium. The algorithm in Section 5.2.2 uses the simplest form of this equation, namely, $A = 0$.

When one F_2 unit is active (say, unit J, with activation D and top-down weight vector $\mathbf{t}_J$), the equation for an F_1 (interface) unit becomes

$$\epsilon \frac{d}{dt} x_i = -x_i + (1 - Ax_i)(s_i + Dt_{Ji}) - (B + Cx_i),$$

which gives

$$x_i = \frac{s_i + Dt_{Ji} - B}{1 + A(s_i + Dt_{Ji}) + C}$$

at equilibrium. A threshold function is applied to x_i to obtain

$$x_i = 1 \text{ if } s_i \text{ and } t_{Ji} \text{ are both } 1; \ x_i = 0 \text{ otherwise.}$$

The algorithm in Section 5.2.2 is based on these results, with $A = 0$, $D = 1$, $B = 1.5$, and $C = 0$.

Initial weights for ART1

The following restrictions on the choice of initial weights for ART1 ensure that neither the reset mechanism nor the winner-take-all competition in the F_2 layer can interfere with the learning process in undesirable ways.

The initial top-down weights should be chosen so that when an uncommitted node (a cluster unit that has not learned any patterns yet) is first chosen as the candidate for learning, the reset mechanism does not reject it. This will be the case if the top-down weights are initialized to 1.

The initial bottom-up weights should be smaller than or equal to the equilibrium value

$$b_{iJ} = \frac{Lx_i}{L - 1 + \|\mathbf{x}\|}.$$

Otherwise, during learning, a vector could suddenly choose a new, uncommitted node. Larger initial bottom-up weights favor creation of new nodes over attempting to put a pattern on a previously trained cluster unit.

5.3 ART2

ART2 is designed to perform for continuous-valued input vectors the same type of tasks as ART1 does for binary-valued input vectors. The differences between ART2 and ART1 reflect the modifications needed to accommodate patterns with continuous-valued components. The more complex F_1 field of ART2 is necessary because continuous-valued input vectors may be arbitrarily close together. The F_1 field in ART2 includes a combination of normalization and noise suppression, in addition to the comparison of the bottom-up and top-down signals needed for the reset mechanism. The learning process for ART2 is summarized in Section 5.3.2 and discussed in more detail in 5.3.4.

There are two types of continuous-valued inputs for which ART2 may be used. The first might be called "noisy binary" signals. These consist of patterns whose information is conveyed primarily by which components are "on" or "virtually off," rather than by the differences in the magnitude of the components that are positive. The equilibrium weights found by the fast learning mode are suitable for this type of data. However, it is not as easy to find equilibrium weights in ART2 as it is for ART1, because the differential equations for the weight updates depend on activations in the F_1 layer, which are changing as learning progresses.

For patterns in which the range of values of the components carries significant information and the weight vector for a cluster unit is to be interpreted as an exemplar for the patterns placed on that unit, the slow learning mode may be more appropriate. We can think of this second type of data as "truly continuous." Both forms of learning are included in the algorithm in Section 5.3.2.

5.3.1 Architecture

A typical ART2 architecture [Carpenter & Grossberg, 1987b] is illustrated in Figure 5.4. The F_1 layer consists of six types of units (the W, X, U, V, P, and Q units). There are n units of each of these types (where n is the dimension of an input pattern). Only one unit of each type is shown in the figure. A supplemental unit between the W and X units receives signals from all of the W units, computes the norm of the vector $\mathbf{w}$, and sends this (inhibitory) signal to each of the X units. Each of these also receives an excitatory signal from the corresponding W unit. Detail of this portion of the net is shown in Figure 5.5. A similar supplemental unit performs the same role between the P and Q units, and another does so between the V and U units. Each X unit is connected to the corresponding V unit, and each Q unit is connected to the corresponding V unit also.

The symbols on the connection paths between the various units in the F_1 layer in Figure 5.4 indicate the transformation that occurs to the signal as it passes from one type of unit to the next; they do not indicate multiplication by the given quantity. However, the connections between units P_i (of the F_1 layer) and Y_j (of the F_2 layer) do show the weights that multiply the signal transmitted over those paths. The activation of the winning F_2 unit is d, where $0 < d < 1$. The activation function applied to $\mathbf{x}$ and $\mathbf{q}$ will be discussed later. The symbol $\cdots\rightarrow$ indicates normalization; i.e., the vector $\mathbf{q}$ of activations of the Q units is just the vector $\mathbf{p}$ of activations of the P units, normalized to approximately unit length.

The action of the F_2 layer (units Y_j in Figure 5.4) is essentially unchanged from the action in ART1. The units compete in a winner-take-all mode for the right to learn each input pattern. As in ART1, learning occurs only if the top-down weight vector for the winning unit is sufficiently similar to the input vector. The tests for reset in ART1 and ART2 differ. (See the algorithm in Section 5.3.2 for details of the test used in ART2.)

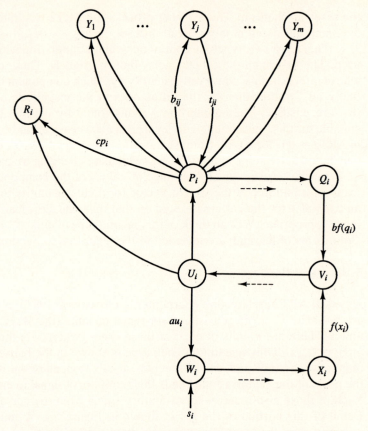

Figure 5.4 Typical ART2 architecture [Carpenter & Grossberg, 1987b].

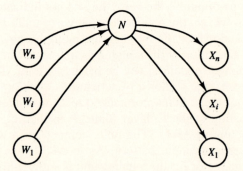

Figure 5.5 Detail of connections from W units to X units, showing the supplemental unit N to perform normalization.

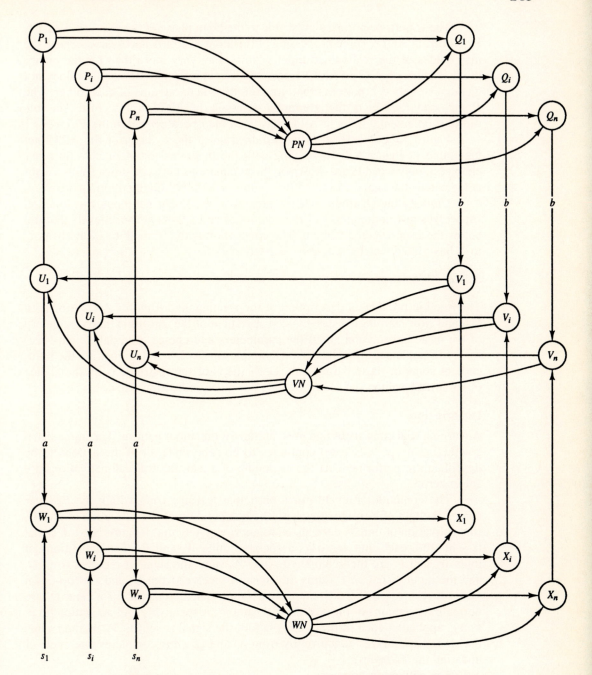

Figure 5.6   Expanded diagram of the F_1 layer of a typical ART2 architecture.

The U units perform a similar role to the input phase of the F_1 layer in ART1. However, in ART2, some processing of the input vector is necessary because the magnitudes of the real-valued input vectors may vary more than for the binary input vectors of ART1. ART2 treats small components as noise and does not distinguish between patterns that are merely scaled versions of each other. The P units play the role of the interface F_1 units in the ART1 architecture. The role of the supplemental units in ART1 has been incorporated within the F_1 layer.

Figure 5.6 presents an expanded diagram of the F_1 layer of the ART2 net illustrated in Figure 5.4. In keeping with the diagrams for other nets (in other chapters), only weights are shown on the connections between units. (If no weight is indicated, the signal traverses that pathway without modification.)

Units X_i and Q_i apply an activation function to their net input; this function suppresses any components of the vectors of activations at those levels that fall below the user-selected value θ. The connection paths from W to U and from Q to V have fixed weights a and b, respectively.

5.3.2 Algorithm

This section provides a description of the training algorithm for ART2 (for both fast and slow learning), a step-by-step statement of the algorithm, and a summary of the basic information about the parameters and choice of initial weights necessary to implement the algorithm. Examples in Section 5.3.3 illustrate the influence of some of these parameters. Derivations of the restrictions on the choice of parameter values and initial weights are given in Section 5.3.4.

Description

A learning trial consists of one presentation of one input pattern. The input signal $\mathbf{s} = (s_1, \ldots, s_i, \ldots, s_n)$ continues to be sent while all of the actions to be described are performed. At the beginning of a learning trial, all activations are set to zero.

The computation cycle (for a particular learning trial) within the F_1 layer can be thought of as originating with the computation of the activation of unit U_i (the activation of unit V_i normalized to approximately unit length). Next, a signal is sent from each unit U_i to its associated units W_i and P_i. The activations of units W_i and P_i are then computed. Unit W_i sums the signal it receives from U_i and the input signal s_i. P_i sums the signal it receives from U_i and the top-down signal it receives if there is an active F_2 unit. The activations of X_i and Q_i are normalized versions of the signals at W_i and P_i, respectively. An activation function is applied at each of these units before the signal is sent to V_i. V_i then sums the signals it receives concurrently from X_i and Q_i; this completes one cycle of updating the F_1 layer.

The activation function

$$f(x) = \begin{cases} x & \text{if } x \geq \theta \\ 0 & \text{if } x < \theta \end{cases}$$

is used in examples in Carpenter and Grossberg's (1987b) original paper and here. With this function, the activations of the U and P units reach equilibrium after two updates of the F_1 layer. This function treats any signal that is less than θ as noise and suppresses it (sets it to zero). The value of the parameter θ is specified by the user. Noise suppression helps the net achieve stable cluster formation. The net is stable when the first cluster unit chosen for each input pattern is accepted and no reset occurs. For slow learning, the weight vectors for the clusters will also converge to stable values.

After the activations of the F_1 units have reached equilibrium, the P units send their signals to the F_2 layer, where a winner-take-all competition chooses the candidate cluster unit to learn the input pattern. (It is not important how fast this competition takes place relative to iterations in the F_1 layer, since the activations of the U and P units there will not change until the top-down signal from the winning F_2 unit is received by the P units.)

The units U_i and P_i in the F_1 layer also send signals to the corresponding reset unit R_i. The reset mechanism can check for a reset each time it receives a signal from P_i, since the necessary computations are based on the value of that signal and the most recent signal the unit R_i had received from U_i. However, this needs to be done only when P_i first receives a top-down signal, since parameter values are chosen such that no reset will occur if no F_2 unit is active, or after learning has started. The reset mechanism will be considered in more detail in Section 5.3.4.

After the conditions for reset have been checked, the candidate cluster unit either will be rejected as not similar enough to the input pattern or will be accepted. If the cluster unit is rejected, it will become inhibited (prevented from further participation in the current learning trial), and the cluster unit with the next largest net input is chosen as the candidate. This process continues until an acceptable cluster unit is chosen (or the supply of available cluster units is exhausted). When a candidate cluster unit is chosen that passes the reset conditions, learning will occur.

In slow learning, only one iteration of the weight update equations occurs on each learning trial. A large number of presentations of each pattern is required, but relatively little computation is done on each trial. For convenience, these repeated presentations are treated as epochs in the algorithm that follows. However, there is no requirement that the patterns be presented in the same order or that exactly the same patterns be presented on each cycle through them.

In the fast learning mode, the weight updates (alternating with updates of the F_1 layer activations) continue until the weights reach equilibrium on each trial. Only a few epochs are required, but a large number of iterations through the weight-update–F_1-update portion of the algorithm must be performed on each learning trial (presentation of a pattern). In fast learning, the placement of the patterns on clusters stabilizes (no reset occurs), but the weights will change for each pattern presented.

Training algorithm

The algorithm that follows can be used for either fast learning or slow learning. In fast learning, iterations of weight change followed by updates of F_1 activations proceed until equilibrium is reached. In slow learning, only one iteration of the weight-update step is performed, but a large number of learning trials is required in order for the net to stabilize. [Carpenter & Grossberg, 1987b]. Fast learning and slow learning are compared in the examples in Section 5.3.3.

Calculations for Algorithm. The following calculations are repeated at several steps of the algorithm and will be referred to as "update F_1 activations." Unit J is the winning F_2 node after competition. If no winning unit has been chosen, d will be zero for all units. Note that the calculations for w_i and p_i can be done in parallel, as can the calculations for x_i and q_i.

The update F_1 activations are:

$$u_i = \frac{v_i}{e + \|\mathbf{v}\|},$$

$$w_i = s_i + au_i, \qquad p_i = u_i + dt_{Ji},$$

$$x_i = \frac{w_i}{e + \|\mathbf{w}\|}, \qquad q_i = \frac{p_i}{e + \|\mathbf{p}\|},$$

$$v_i = f(x_i) + bf(q_i).$$

The activation function is

$$f(x) = \begin{cases} x & \text{if } x \geq \theta \\ 0 & \text{if } x < \theta. \end{cases}$$

Algorithm

Step 0. Initialize parameters:

$$a, \ b, \ \theta, \ c, \ d, \ e, \ \alpha, \ \rho.$$

Step 1. Do Steps 2–12 N_EP times.
(Perform the specified number of epochs of training.)

Step 2. For each input vector s, do Steps 3–11.

Step 3. Update F_1 unit activations:

$$u_i = 0, \qquad\qquad x_i = \frac{s_i}{e + \|\mathbf{s}\|},$$

$$w_i = s_i,$$

$$\qquad\qquad q_i = 0,$$

$$p_i = 0,$$

$$\qquad\qquad v_i = f(x_i).$$

Update F_1 unit activations again:

$$u_i = \frac{v_i}{e + \|\mathbf{v}\|},$$

$$w_i = s_i + au_i,$$

$$p_i = u_i,$$

$$x_i = \frac{w_i}{e + \|\mathbf{w}\|},$$

$$q_i = \frac{p_i}{e + \|\mathbf{p}\|},$$

$$v_i = f(x_i) + bf(q_i).$$

Step 4.   Compute signals to F_2 units:

$$y_j = \sum_i b_{ij} p_i.$$

Step 5.   While reset is true, do Steps 6–7.

    Step 6.   Find F_2 unit Y_J with largest signal. (Define J such that $y_J \geq y_j$ for $j = 1, \ldots, m$.)

    Step 7.   Check for reset:

$$u_i = \frac{v_i}{e + \|\mathbf{v}\|},$$

$$p_i = u_i + dt_{Ji},$$

$$r_i = \frac{u_i + cp_i}{e + \|\mathbf{u}\| + c\|\mathbf{p}\|}.$$

If $\|\mathbf{r}\| < \rho - e$, then

$$y_J = -1 \quad \text{(inhibit } J\text{)}$$

(reset is true; repeat Step 5);
If $\|\mathbf{r}\| \geq \rho - e$, then

$$w_i = s_i + au_i,$$

$$x_i = \frac{w_i}{e + \|\mathbf{w}\|},$$

$$q_i = \frac{p_i}{e + \|\mathbf{p}\|},$$

$$v_i = f(x_i) + bf(q_i).$$

Reset is false; proceed to Step 8.

Step 8.   Do Steps 9–11 *N_IT* times.
               (Perform the specified number of learning iterations.)
    Step 9.   Update weights for winning unit *J*:

$$t_{Ji} = \alpha d u_i + \{1 + \alpha d (d - 1)\} t_{Ji},$$

$$b_{iJ} = \alpha d u_i + \{1 + \alpha d (d - 1)\} b_{iJ}.$$

    Step 10.   Update F_1 activations:

$$u_i = \frac{v_i}{e + \|\mathbf{v}\|},$$

$$w_i = s_i + a u_i,$$

$$p_i = u_i + d t_{Ji},$$

$$x_i = \frac{w_i}{e + \|\mathbf{w}\|},$$

$$q_i = \frac{p_i}{e + \|\mathbf{p}\|},$$

$$v_i = f(x_i) + b f(q_i).$$

    Step 11.   Test stopping condition for weight updates.
Step 12.   Test stopping condition for number of epochs.

Comments.   In the preceding algorithms, we have made use of the following facts:

1. Reset cannot occur during resonance (Step 8).
2. A new winning unit cannot be chosen during resonance.

We have not used the facts that:

1. Typically, in slow learning *N_IT* = 1, and Step 10 can be omitted.
2. In fast learning, for the first pattern learned by a cluster, *u* will be parallel to *t* throughout the training cycle and the equilibrium weights will be

$$t_{Ji} = \frac{1}{1 - d} u_i;$$

$$b_{iJ} = \frac{1}{1 - d} u_i.$$

(See Exercise 5.6.)

Other possible stopping conditions are as follows:

Repeat Step 8 until the weight changes are below some specified tolerance. For slow learning, repeat Step 1 until the weight changes are below some specified tolerance. For fast learning, repeat Step 1 until the placement of patterns on the cluster units does not change from one epoch to the next.

Steps 3 through 11 constitute one learning trial (one presentation of a pattern). It will be convenient (especially in describing the implementation of fast learning and slow learning using this algorithm) to refer to the performance of a learning trial for each input pattern as an *epoch*. It is not necessary that the patterns be presented in the same order on each epoch. Note that the action of updating the F_1 activations during the extended resonance phase of the fast learning mode (the computations in Step 8) causes the values of the u_i that appear in the weight update equations to change as learning progresses.

Choices

Parameters.   A summary of the role of the various parameters used in the algorithm is given here. Derivations of some of these restrictions are given in Section 5.3.4. The parameters and their roles are as follows:

n number of input units (F_1 layer);

m number of cluster units (F_2 layer);

a, b fixed weights in the F_1 layer; sample values are $a = 10$, $b = 10$. Setting either $a = 0$ or $b = 0$ produces instability in the net; other than that, the net is not particularly sensitive to the values chosen.

c fixed weight used in testing for reset; a sample value is $c = .1$. A small c gives a larger effective range of the vigilance parameter (see Section 5.3.4).

d activation of winning F_2 unit; a sample value is $d = .9$. Note that c and d must be chosen to satisfy the inequality

$$\frac{cd}{1 - d} \le 1$$

(in order to prevent a reset from occurring during a learning trial). The ratio should be chosen close to 1 to achieve a larger effective range for vigilance (see Section 5.3.4).

e a small parameter introduced to prevent division by zero when the norm of a vector is zero. This value prevents the normalization to unity from being exact. A value of zero is typically used in the hand computations and derivations that follow and may be used in the algorithm if the normalization step is skipped when the vector is zero.

θ noise suppression parameter, a sample value is $\theta = 1/\sqrt{n}$. The sample value may be larger than desired in some applications. Components of the normalized input vector (and other vectors in the F_1 loop) that are less than this value are set to zero.

α learning rate. A smaller value will slow the learning in either the fast or the slow learning mode. However, a smaller value will ensure that the weights (as well as the placement of patterns on clusters) eventually reach equilibrium in the slow learning mode.

ρ vigilance parameter. Along with the initial bottom-up weights, this parameter determines how many clusters will be formed. Although, theoretically, values from 0 to 1 are allowed, only values between approximately 0.7 and 1 perform any useful role in controlling the number of clusters. (Any value less than 0.7 will have the same effect as setting ρ to zero.) Some choices of values for c and d will restrict the effective range of values for ρ even further.

Initial Weights.   The initial weights for the ART2 net are as follows:

$t_{ji}(0)$ initial top-down weights (must be small to ensure that no reset will occur for the first pattern placed on a cluster unit);

$$t_{ji}(0) = 0.$$

$b_{ij}(0)$ initial bottom-up weights; must be chosen to satisfy the inequality

$$b_{ij}(0) \le \frac{1}{(1 - d)\sqrt{n}},$$

to prevent the possibility of a new winner being chosen during "resonance" as the weights change. Larger values of b_{ij} encourage the net to form more clusters.

Learning Mode.   Fast learning and slow learning differ not only in the theoretical assumptions on which they are based (in terms of the speed of learning relative to the duration of presentation of a pattern during any one learning trial), but also in the performance characteristics of the clusters and weight vectors formed when they are used. Some differences in the performance of fast and slow learning are summarized here and are illustrated in the examples of the next section.

Fast learning results in weight vectors that have some of the same characteristics as the weights found by ART1. Typically, a component of the weight vector for a particular cluster unit that is set to zero during a learning trial will not become nonzero during a subsequent learning trial. Furthermore, if a pattern being learned by a cluster has one or more components that are zero after noise suppression is applied, the corresponding components of the weight vector will be set to zero during learning. However, the weights of the nonzero components will change on each learning trial to reflect the relative magnitudes of the nonzero components of the current input vector. This suggests that fast learning may be more appropriate for data in which the primary information is contained in the pattern of components that are "small" or "large," rather than for data in which the relative sizes of the nonzero components is important.

Slow learning requires many epochs of training, with only one weight update iteration performed on each learning trial. The weight vector for each cluster is the average of the patterns it has learned, which may make the weight vector a more suitable exemplar for the cluster for certain types of applications. The weights reach equilibrium (to within a tolerance determined by the learning rate). In some applications, slow learning may also produce clustering that is less influenced by the order of presentation of the patterns than the clustering produced by fast learning.

5.3.3 Applications

Simple examples

We now consider several examples of the operation of ART2 for input with two components. In each of these examples, the parameter values are as follows:

$$a = 10,$$

$$b = 10,$$

$$c = 0.1,$$

$$d = 0.9,$$

$$e = 0 \quad \text{(not shown in subsequent formulas, for simplicity).}$$

Example 5.6 ART2 with fast learning: first pattern on a given cluster

This example illustrates that the first time a first cluster unit is chosen as the winner, it learns the noise-suppressed input pattern. No reset can occur for the first pattern learned by a cluster unit. The final weights are $1/(1 - d)$ times the noise-suppressed input vector.

The parameter values are:

$$\rho = 0.9, \qquad \theta = 0.7.$$

The initial bottom-up weight vector (approximately the largest permissible value) is

$$\mathbf{b}_j = (7.0, 7.0) \text{ for each cluster unit.}$$

The initial top-down weight vector is

$$\mathbf{t}_j = (0.0, 0.0) \text{ for each cluster unit.}$$

The input is

$$\mathbf{s} = (0.8, 0.6).$$

All other activations are initially zero. For the first F_1 loop, we have:

$$\mathbf{u} = \frac{\mathbf{v}}{\|\mathbf{v}\|} \qquad = (0.0, 0.0),$$

$$\mathbf{w} = \mathbf{s} + a\mathbf{u} \qquad = (0.8, 0.6),$$

$$\mathbf{p} = \mathbf{u} \qquad\qquad = (0.0, 0.0),$$

$$\mathbf{x} = \frac{\mathbf{w}}{\|\mathbf{w}\|} \qquad\qquad = (0.8, 0.6),$$

$$\mathbf{q} = \frac{\mathbf{p}}{\|\mathbf{p}\|} \qquad\qquad = (0.0, 0.0),$$

$$\mathbf{v} = f(\mathbf{x}) + bf(\mathbf{q}) = (0.8, 0.0).$$

For the second F_1 loop;

$$\mathbf{u} = \frac{\mathbf{v}}{\|\mathbf{v}\|} \qquad\qquad = (1.0, 0.0).$$

$$\mathbf{w} = \mathbf{s} + a\mathbf{u} \qquad\quad = (0.8, 0.6) + 10\,(1.0, 0.0)$$

$$\qquad\qquad\qquad\qquad = (10.8, 0.6),$$

$$\mathbf{p} = \mathbf{u} \qquad\qquad = (1.0, 0.0),$$

$$\mathbf{x} = \frac{\mathbf{w}}{\|\mathbf{w}\|} \qquad\qquad = (0.998, 0.055),$$

$$\mathbf{q} = \frac{\mathbf{p}}{\|\mathbf{p}\|} \qquad\qquad = (1.0, 0.0),$$

$$\mathbf{v} = f(\mathbf{x}) + bf(\mathbf{q}) = (0.998, 0.0) + 10\,(1.0, 0.0)$$

$$\qquad\qquad\qquad\qquad = (10.998, 0.0).$$

Further iterations will not change $\mathbf{u}$ or $\mathbf{p}$, so we now proceed to send a signal from the P units so that the F_2 layer can find a winner.

Since this is the first pattern presented, and the bottom-up weights for all cluster units are initialized to the same values, all F_2 units will receive the same input. Taking the usual tie breaker of letting the unit with the lowest index win, cluster unit 1 will be chosen as the winner.

In the loop that tests for a reset,

$$\mathbf{u} = \frac{\mathbf{v}}{\|\mathbf{v}\|} = (1.0, 0.0).$$

As soon as the P units receive a top-down signal from the winning cluster unit (unit J), the test for a reset is performed. However, since this unit has not learned any patterns previously (and the top-down weights are initialized to zero), the activation of the P units is unchanged by the top-down signal; i.e.,

$$\mathbf{p} = \mathbf{u} + d\mathbf{t}_J = (1.0, 0.0) + 0.9\,(0.0, 0.0).$$

Since

$$\mathbf{p} = \mathbf{u},$$

the check for a reset gives

$$\|\mathbf{r}\| = \frac{\|\mathbf{u} + c\mathbf{u}\|}{\|\mathbf{u}\| + c\,\|\mathbf{u}\|} = 1.$$

For this cluster unit to learn, we must have

$$\|\mathbf{r}\| \ge \rho.$$

However, $\|\mathbf{r}\| = 1 \ge \rho$ for any valid value of ρ (since $\rho \le 1$), so the winning cluster unit will be allowed to learn the current pattern. This example shows that a reset cannot occur for the first pattern on any cluster unit.

Now, we finish the F_1 loop calculations:

$$\mathbf{w} = \mathbf{s} + a\mathbf{u} \qquad = (0.8, 0.6) + 10\,(1.0, 0.0)$$

$$= (10.8, 0.6),$$

$$\mathbf{x} = \frac{\mathbf{w}}{\|\mathbf{w}\|} \qquad = (0.998, 0.055),$$

$$\mathbf{q} = \frac{\mathbf{p}}{\|\mathbf{p}\|} \qquad = (1.0, 0.0),$$

$$\mathbf{v} = f(\mathbf{x}) + bf(\mathbf{q}) = (0.998, 0.0) + 10\,(1.0, 0.0)$$

$$= (10.998, 0.0).$$

We update the weights, using a learning rate of 0.6:

$$\mathbf{t}_J(\text{new}) = 0.6\,(0.9)\mathbf{u} + [1.0 - 0.6\,(0.9)(0.1)]\mathbf{t}_J(\text{old})$$

$$= 0.54\mathbf{u} + 0.946\mathbf{t}_J(\text{old}),$$

$$\mathbf{t}_J = (0.54, 0.0).$$

$$\mathbf{b}_J(\text{new}) = 0.6\,(0.9)\mathbf{u} + [1.0 - 0.6\,(0.9)(0.1)]\mathbf{b}_J(\text{old})$$

$$= 0.54\mathbf{u} + 0.946\mathbf{b}_J(\text{old}),$$

$$\mathbf{b}_J = (0.54, 0.0) + (6.62, 6.62),$$

$$= (7.16, 6.62).$$

For the F_1 loop,

$$\mathbf{u} = \frac{\mathbf{v}}{e + \|\mathbf{v}\|} \qquad = (1.0, 0.0),$$

$$\mathbf{w} = \mathbf{s} + a\mathbf{u} \qquad = (0.8, 0.6) + 10\,(1.0, 0.0)$$

$$= (10.8, 0.6),$$

$$\mathbf{p} = \mathbf{u} + d\mathbf{t}_J \qquad = (1.0, 0.0) + .9\,(0.09, 0.0),$$

$$\mathbf{x} = \frac{\mathbf{w}}{\|\mathbf{w}\|} \qquad = (0.998, 0.055),$$

$$\mathbf{q} = \frac{\mathbf{p}}{\|\mathbf{p}\|} \qquad = (1.0, 0.0),$$

$$\mathbf{v} = f(\mathbf{x}) + bf(\mathbf{q}) = (0.998, 0.0) + 10(1.0, 0.0)$$

$$= (10.998, 0.0).$$

We update the weights again:

$$\mathbf{t}_J(\text{new}) = 0.54\mathbf{u} + 0.946\mathbf{t}_J(\text{old}),$$

$$\mathbf{t}_J = (0.54, 0.0) + 0.946\,(0.54, 0.0) = (1.05, 0.0).$$

$$\mathbf{b}_J(\text{new}) = 0.6\,(0.9)\mathbf{u} + [1.0 - 0.6\,(0.9)\,(0.1)]\,\mathbf{b}_J(\text{old})$$

$$= 0.54\mathbf{u} + 0.946\mathbf{b}_J(\text{old}),$$

$$\mathbf{b}_J = (0.54, 0.0) + (6.77, 6.26),$$

$$= (7.32, 6.26).$$

For the F_1 loop,

$$\mathbf{u} = \frac{\mathbf{v}}{\|\mathbf{v}\|} = (1.0, 0.0),$$

$$\mathbf{w} = (0.8, 0.6) + 10\,(1.0, 0.0) = (10.8, 0.6),$$

$$\mathbf{p} = (1.0, 0.0) + 0.9\,(0.18, 0.0),$$

$$\mathbf{x} = (0.998, 0.055),$$

$$\mathbf{q} = (1.0, 0.0),$$

$$\mathbf{v} = (0.998, 0.0) + 10\,(1.0, 0.0).$$

Thus, $\mathbf{p}$ never gets a contribution to the component that is zero, $\mathbf{q}$ does not change, $\mathbf{u}$ does not change, and $\mathbf{t}_J$ gradually grows to a multiple of $\mathbf{u}$. In fact, since $\mathbf{u}$ will not change during learning, the equilibrium values of the weights can be found immediately from the following formulas:

$$\frac{d}{dt}t_{Ji} = du_i + d(d-1)t_{Ji},$$

$$0 = du_i + d(d-1)t_{Ji},$$

$$t_{1i} = \frac{1}{1-d}u_i,$$

$$\mathbf{t}_1 = (10, 0).$$

Although the bottom-up weights start from different initial values than the top-down weights, the differential equations are the same, and the same analysis shows that they converge to the same values. Thus, we see that the equilibrium weights for the first pattern learned by any cluster unit can be found without iterative solution of the differential equations for the weights.

There are two special aspects to this example, namely, that the pattern is the first one learned by the cluster unit and some components of the input are suppressed. The original formulation of ART2 suggested $\theta = 1/\sqrt{n}$, which gives an approximate value of $\theta = 0.7$ for $n = 2$. However, it is easy to see that using this value of θ will drive any input (whose components are not exactly equal to each other to $\mathbf{u} = (1, 0)$ or $\mathbf{u} = (0, 1)$ on the first iteration. Thus, we see that the choice of the noise parameter θ can have a significant effect on the performance of the net. Noise suppression is considered in more detail in Section 5.3.4.

The preceding computations would be the same for slow learning, down to the point where the first weight update was performed. However, after that, a new pattern would be presented to the net, rather than an alternation of F_1 and weight updates. We shall consider the effect of slow and fast learning further in the examples that follow.

We now consider some examples with no effect from noise suppression. Except for the parameter values given for Examples 5.7 and 5.8, all other parameters are as specified in Example 5.6.

Example 5.7 Effect of initial bottom-up weights in ART2 cluster formation

We illustrate the effect of the initial bottom-up weights on the number of clusters formed using fast learning. The noise suppression parameter and input vector are:

$$\theta = 0.1, \qquad \mathbf{s} = (0.8, 0.6).$$

All other activations are initially zero.

For the first F_1 loop, we have:

$$\mathbf{u} \qquad\qquad = (0.0, 0.0),$$

$$\mathbf{w} = \mathbf{s} + a\mathbf{u} = (0.8, 0.6), \qquad \mathbf{p} = \mathbf{u} = (0.0, 0.0),$$

$$\mathbf{x} = \frac{\mathbf{w}}{\|\mathbf{w}\|} \qquad = (0.8, 0.6), \qquad \mathbf{q} = \frac{\mathbf{p}}{\|\mathbf{p}\|} = (0.0, 0.0),$$

$$\mathbf{v} \qquad\qquad = f(\mathbf{x}) + bf(\mathbf{q}) = (0.8, 0.6).$$

For the second F_1 loop,

$$\mathbf{u} = \frac{\mathbf{v}}{\|\mathbf{v}\|} \qquad\qquad = (0.8, 0.6),$$

$$\mathbf{w} = \mathbf{s} + a\mathbf{u} \qquad = (0.8, 0.6) + 10\,(0.8, 0.6) = (8.8, 6.6),$$

$$\mathbf{p} = \mathbf{u} \qquad\qquad = (0.8, 0.6),$$

$$\mathbf{x} = \frac{\mathbf{w}}{\|\mathbf{w}\|} \qquad\qquad = (0.8, 0.6),$$

$$\mathbf{q} = \frac{\mathbf{p}}{\|\mathbf{p}\|} \qquad\qquad = (0.8, 0.6),$$

$$\mathbf{v} = f(\mathbf{x}) + bf(\mathbf{q}) = (0.8, 0.6) + 10\,(0.8, 0.6) = (8.8, 6.6).$$

Further iterations will not change $\mathbf{u}$ or $\mathbf{p}$, so the F_1–F_2 iteration to find a winner can be started:

Signals are sent from the P units to the F_2 layer.

The F_2 units that are not inhibited compete to find the winning unit.

The winning F_2 unit sends a signal back down to the F_1 layer, but since this is the first pattern learned by this cluster (and the top-down weights are initialized to zero), the signal is zero.

In general, we would update the activations of the P units to incorporate the top-down signal from the winning F_2 unit and then check the reset condition. (If the condition is met, we would update the rest of the F_1 activations). However, as observed in Example 5.6, a reset will not occur for the first pattern on a cluster if the top-down weights are initialized to zero. We have:

$$\mathbf{u} = \frac{\mathbf{v}}{\|\mathbf{v}\|} \qquad = (0.8, 0.6).$$

$$\mathbf{p} = \mathbf{u} + d\mathbf{t}_J = (0.8, 0.6).$$

Test for a reset would occur at this point.

$$\mathbf{w} = \mathbf{s} + a\mathbf{u} \qquad = (0.8, 0.6) + 10\,(0.8, 0.6) = 8.8, 6.6),$$

$$\mathbf{x} = \frac{\mathbf{w}}{\|\mathbf{w}\|} \qquad = (0.8, 0.6),$$

$$\mathbf{q} = \frac{\mathbf{p}}{\|\mathbf{p}\|} \qquad = (0.8, 0.6),$$

$$\mathbf{v} = f(\mathbf{x}) + bf(\mathbf{q}) = (0.8, 0.6) + 10\,(0.8, 0.6).$$

We update the weights, using a learning rate of 0.6:

$$t_{Ji}(\text{new}) = 0.6\,(0.9)u_i + [1.0 - 0.6\,(0.9)(0.1)]t_{Ji}(\text{old})$$

$$= 0.54u_i + 0.946t_{Ji}(\text{old}),$$

$$\mathbf{t} = 0.54\,(0.8, 0.6) = (0.432, 0.324).$$

Next, we update F_1:

$$\mathbf{u} = (0.8, 0.6),$$

$$\mathbf{w} = (0.8, 0.6) + 10\,(0.8, 0.6) = 11\,(0.8, 0.6),$$

$$\mathbf{p} = (0.8, 0.6) + 0.9\,(0.432, 0.324) = 1.486\,(0.8, 0.6),$$

$$\mathbf{x} = (0.8, 0.6),$$

$$\mathbf{q} = (0.8, 0.6),$$

$$\mathbf{v} = (0.8, 0.6) + 10\,(0.8, 0.6) = 11\,(0.8, 0.6).$$

Now we update the top-down weights:

$$\mathbf{t} = 0.54\mathbf{u} + 0.946\mathbf{t}(\text{old})$$

$$= 0.54\,(0.8, 0.6) + 0.946\,(0.54)(0.8, 0.6),$$

$$= 1.05\,(0.8, 0.6).$$

Since all of the vectors are multiples of $(0.8, 0.6)$, it is easy to see that the top-down weights will converge to a multiple of $(0.8, 0.6)$. In fact, since $\mathbf{u}$ is constant, the equilibrium top-down and bottom-up weights are defined from the following formulas:

$$\frac{d}{dt}\, t_{Ji} = du_i + d(d-1)t_{Ji},$$

$$0 = du_i + d(d-1)t_{Ji},$$

$$t_{1i} = \frac{1}{1-d}\, u_i,$$

$$\mathbf{t}_1 = 10(0.8,\, 0.6) = (8.0,\, 6.0),$$

$$\mathbf{b}_1 = 10(0.8,\, 0.6) = (8.0,\, 6.0).$$

We continue the example by presenting a second pattern. The bottom-up weights are now

$$\mathbf{b}_1 = (8.0,\, 6.0), \qquad \mathbf{b}_2 = (7.0,\, 7.0).$$

The top-down weights are

$$\mathbf{t}_1 = (8.0,\, 6.0), \qquad \mathbf{t}_2 = (0.0,\, 0.0).$$

We present a second input pattern, namely,

$$\mathbf{s} = (0.6,\, 0.8).$$

All other activations are initially zero. For the first F_1 loop, we have:

$$\mathbf{u} = (0.0,\, 0.0),$$

$$\mathbf{w} = \mathbf{s} + a\mathbf{u} \qquad = (0.6,\, 0.8),$$

$$\mathbf{p} = \mathbf{u} \qquad = (0.0,\, 0.0),$$

$$\mathbf{x} = \frac{\mathbf{w}}{\|\mathbf{w}\|} \qquad = (0.6,\, 0.8),$$

$$\mathbf{q} = \frac{\mathbf{p}}{\|\mathbf{p}\|} \qquad = (0.0,\, 0.0),$$

$$\mathbf{v} = f(\mathbf{x}) + bf(\mathbf{q}) = (0.6,\, 0.8).$$

For the second F_1 loop,

$$\mathbf{u} = \frac{\mathbf{v}}{\|\mathbf{v}\|} \qquad = (0.6,\, 0.8),$$

$$\mathbf{w} = \mathbf{s} + a\mathbf{u} \qquad = (0.6,\, 0.8) + 10\,(0.6,\, 0.8) = (6.6,\, 8.8),$$

$$\mathbf{p} = \mathbf{u} \qquad = (0.6,\, 0.8),$$

$$\mathbf{x} = \frac{\mathbf{w}}{\|\mathbf{w}\|} \qquad = (0.6,\, 0.8),$$

$$\mathbf{q} = \frac{\mathbf{p}}{\|\mathbf{p}\|} \qquad = (0.6,\, 0.8),$$

$$\mathbf{v} = f(\mathbf{x}) + bf(\mathbf{q}) = (0.6,\, 0.8) + 10\,(0.6,\, 0.8) = (6.6,\, 8.8).$$

Further iterations will not change $\mathbf{u}$ or $\mathbf{p}$, so the F_1–F_2 iterations to find a winner can be started. Signals are sent from the P units to the F_2 layer. The net input to cluster unit 1 is

$$(0.6, 0.8)(8.0, 6.0) = 4.8 + 4.8 = 9.6.$$

The net input to cluster unit 2 is

$$(0.6, 0.8)(7.0, 7.0) = 4.2 + 5.6 = 9.8.$$

Thus, the winner is cluster unit 2.

Cluster unit 2 would learn this pattern in a manner similar to that already described for the first pattern being learned by cluster unit 1.

However, if the initial bottom-up weights are taken to be (5.0, 5.0), rather than the maximum permissible value of (7.0, 7.0), the second pattern will pick the first cluster as the winner. The value of the vigilance parameter will determine whether the first cluster will learn this pattern.

Example 5.8 Equilibrium weights for ART2: fast learning and no noise suppression

We continue Example 5.7, using a low enough value of the vigilance parameter so that the first cluster unit will learn the second pattern. Thus, pattern 1 = (0.8, 0.6) has been presented, and the top-down and bottom-up weights for cluster unit 1 are (8.0, 6.0). The second pattern (0.6, 0.8), is presented to the net, and the F_1 loop iterations are performed as in Example 5.7. The computations continue to determine the winning F_2 unit.

The net input to cluster unit 1 is

$$(0.6, 0.8)(8.0, 6.0) = 4.8 + 4.8 = 9.6.$$

The net input to cluster unit 2 is

$$(0.6, 0.8)(5.0, 5.0) = 3.0 + 4.0 = 7.0.$$

Now, cluster unit 1 is the winner.

The winning unit sends a top-down signal, and the reset condition is checked:

$$\mathbf{u} = \frac{\mathbf{v}}{\|\mathbf{v}\|} = (0.6, 0.8),$$

$$\mathbf{p} = \mathbf{u} + d\mathbf{t}_1 = (0.6, 0.8) + 0.9(8.0, 6.0)$$

$$= (7.8, 6.2).$$

The reset condition requires that, in order for this unit to be allowed to learn, we have

$$\|\mathbf{r}\| = \frac{\|\mathbf{u} + c\mathbf{p}\|}{\|\mathbf{u}\| + c\,\|\mathbf{p}\|} > \rho.$$

In this case,

$$\mathbf{u} + c\mathbf{p} = (0.6, 0.8) + (0.78, 0.62) = (1.38, 1.42),$$

$$\|\mathbf{u} + c\mathbf{p}\| = 1.98,$$

$$\|\mathbf{p}\| = 9.964,$$

$$\|\mathbf{u}\| = 1.0,$$

$$\|\mathbf{u}\| + .1 \|\mathbf{p}\| = 1.9964,$$

$$\|\mathbf{r}\| = 0.992 > 0.9 = \rho.$$

The winning unit is accepted, so the rest of the F_1 activations are updated:

$$\mathbf{w} = \mathbf{s} + a\mathbf{u} = (0.6, 0.8) + 10 (0.6, 0.8) = (6.6, 8.8),$$

$$\mathbf{x} = \frac{\mathbf{w}}{\|\mathbf{w}\|} = (0.6, 0.8),$$

$$\mathbf{q} = \frac{\mathbf{p}}{\|\mathbf{p}\|} = (0.78, 0.62),$$

Next, the weights are updated, using a learning rate of 0.6:

$$t_{1i}(\text{new}) = 0.6 (0.9)u_i + [1.0 - 0.6 (0.9)(0.1)]t_{1i}(\text{old})$$

$$= 0.54u_i + 0.946t_{1i}(\text{old}),$$

$$\mathbf{t} = 0.54 (0.6, 0.8) + .946 (8, 6) = (7.9, 6.1).$$

During learning, an F_1 loop follows each weight update. The evolution of the weights is shown in Figure 5.7.

Note that the final weights are essentially the second input vector; virtually all information about the first vector learned by this cluster unit has been lost.

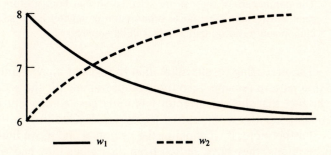

Figure 5.7   Weight changes for Example 5.8.

Example 5.9 Equilibrium weights for ART2: slow learning and no noise suppression

If we repeat Example 5.8 using slow learning, we see that the weight vector eventually becomes the average of the two patterns that were placed on the cluster. This is illustrated in Figure 5.8.

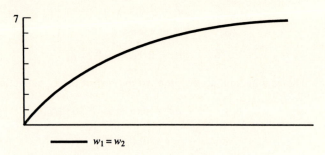

Figure 5.8   Weight changes for Example 5.9.

Example 5.10 Equilibrium weights for ART2: fast learning and moderate noise suppression

This example illustrates the fact that, after training a cluster that has learned previously, the weights are zero for any component for which either the previous weight vector or the noise-suppressed input vector is zero. The nonzero components are simply scaled versions of the corresponding components of the input vector.

Figure 5.9 shows the effect of first presenting the vector

$$(0.2, 0.4, 0.6, 0.8, 1.0)$$

followed by the vector

$$(1.0, 0.8, 0.6, 0.4, 0.2).$$

The parameters of the net are selected so that both vectors will be placed on the same cluster unit. The noise suppression parameter is set so that only the smallest component in each input vector will be suppressed.

The preceding results show that the weight component that was zero from the first pattern remains zero throughout training for the second pattern, in spite of the fact that the input vector has its largest component in that position (the first component).

Furthermore, if we monitor the vectors $\mathbf{u}$ and $\mathbf{p}$, we find that within the first 100 iterations, they have shifted from the values they had when learning started

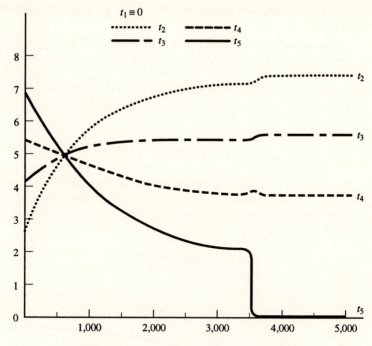

Figure 5.9   Weight changes for Example 5.10.

(representing the noise-suppressed input vector) to reflect quite accurately the weight vector for the winning cluster unit. The vectors **u** and **p** are very close to parallel at this stage of training; they will become virtually parallel by the 200th iteration and remain so throughout the rest of the learning trial. However, they move gradually toward the direction of the input vector. The last component of **u** and **p** (which is small enough in the input vector to be suppressed to zero) decreases gradually until about iteration 3,600, when it suddenly reaches the noise suppression level and is set to zero. Following this, the other components change so that $\|t\| = \dfrac{1}{1-d}$.

The relatively large value of d (0.9) forces the weight vector to have a large magnitude (10). This helps to ensure that weights once set to zero stay zero (a property of the equilibrium weights in ART1 that carries over to the fast learning mode for ART2). However, this large a value slows down the iterative solution of the weight update equations during fast learning. (See Section 5.3.4 for an alternative "shortcut.")

Example 5.11 Equilibrium weights for ART2: slow learning and moderate noise suppression

This example illustrates the fact that, using slow learning, after training a cluster that has learned previously, the weights are the average of the patterns placed on that cluster.

Figure 5.10 shows the effect of first presenting the vector

$$(0.2, 0.4, 0.6, 0.8, 1.0)$$

followed by the vector

$$(1.0, 0.8, 0.6, 0.4, 0.2).$$

The parameters of the net are selected so that both vectors will be placed on the same cluster unit (in this case by having only one cluster unit; the same effect could be achieved by choosing the initial bottom-up weights sufficiently small). The noise suppression parameter is set so that only the smallest component in each input vector will be suppressed.

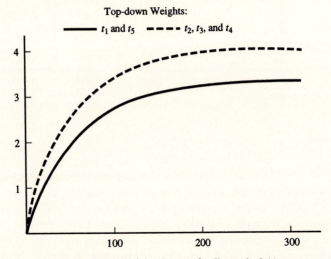

Figure 5.10 Weight changes for Example 5.11.

Spanning tree data

Sample data developed by Kohonen [1989a] can be used to illustrate the behavior of ART2 neural networks. The relationships between the patterns can be displayed graphically as in Figure 5.11 [Kohonen, 1989a]. Patterns that are displayed in a row or a column differ in only one component. Furthermore, the distance between patterns in the same row or column on the diagram in Figure 5.11 corresponds directly to the Euclidean distance between the vectors. For example, patterns X and Y are adjacent; they differ only in the fourth component and the Euclidean distance between $(3, 3, 6, 2, 0)$ and $(3, 3, 6, 3, 0)$ is one. The original data are given in Figure 5.12. Because of this nice structure, we shall refer to these test data as the spanning tree example.

```
A   B   C   D   E
        F
        G
        H   K   L   M   N   O   P   Q   R
        I           S           W
        J           T           X   1   2   3   4   5   6
                    U           Y
                    V           Z
```

Figure 5.11 Spanning tree test data structure [Kohonen, 1989a].

PATTERN	COMPONENTS				
A	1	0	0	0	0
B	2	0	0	0	0
C	3	0	0	0	0
D	4	0	0	0	0
E	5	0	0	0	0
F	3	1	0	0	0
G	3	2	0	0	0
H	3	3	0	0	0
I	3	4	0	0	0
J	3	5	0	0	0
K	3	3	1	0	0
L	3	3	2	0	0
M	3	3	3	0	0
N	3	3	4	0	0
O	3	3	5	0	0
P	3	3	6	0	0
Q	3	3	7	0	0
R	3	3	8	0	0
S	3	3	3	1	0
T	3	3	3	2	0
U	3	3	3	3	0
V	3	3	3	4	0
W	3	3	6	1	0
X	3	3	6	2	0
Y	3	3	6	3	0
Z	3	3	6	4	0
1	3	3	6	2	1
2	3	3	6	2	2
3	3	3	6	2	3
4	3	3	6	2	4
5	3	3	6	2	5
6	3	3	6	2	6

Figure 5.12 Spanning tree test data [Kohonen, 1989a].

Example 5.12 Spanning tree data clustered by ART2: high vigilance and noise suppression

The clusters formed using spanning tree data and fast learning are indicated in Figure 5.13; the results for slow learning are shown in Figure 5.14. The vigilance and noise suppression parameter values are relatively high, namely, $\rho = .99$ and $\theta = .447$.

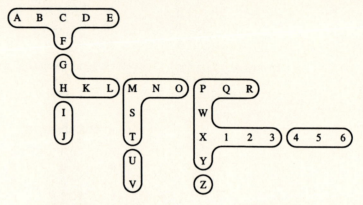

Figure 5.13   Spanning tree test data as clustered by ART2; fast learning, $\rho = .99$, $\theta = .447$.

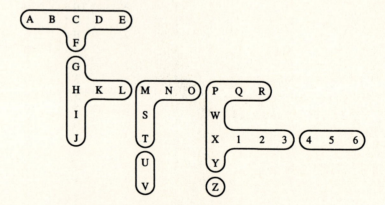

Figure 5.14   Spanning tree test data as clustered by ART2; slow learning, $\rho = .99$, $\theta = .447$.

Example 5.13 Spanning tree data clustered by ART2: moderate vigilance and high noise suppression

For the moderate vigilance and high noise suppression parameter values

$$\rho = .95, \qquad \theta = .447,$$

the clusters formed using fast and slow learning are indicated in Figures 5.15 and 5.16, respectively.

Note that the net is sensitive to fairly small changes in the vigilance parameter. As expected (and desired), more clusters are formed for the higher vigilance parameter:

CLUSTERS FORMED

	$\rho = .99$	$\rho = .95$
Fast	8	6
Slow	7	4

In each of these cases, 1,000 epochs with one weight update iteration per epoch were performed for the slow learning; three epochs of training were performed for the fast learning.

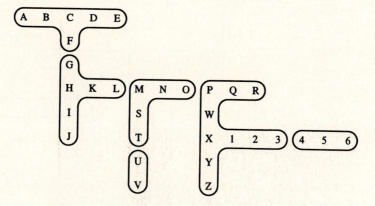

Figure 5.15   Spanning tree test data as clustered by ART2; fast learning, $\rho = .95$, $\theta = .447$.

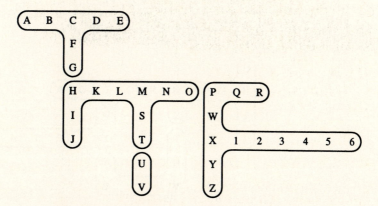

Figure 5.16   Spanning tree test data as clustered by ART2; slow learning, $\rho = .95$, $\theta = .447$.

Example 5.14 Spanning tree data clustered by ART2: high vigilance and moderate noise suppression

For the high vigilance and moderate noise suppression parameter values

$$\rho = .99, \qquad \theta = .2,$$

there is more difference in the number of clusters formed by fast and slow learning, as illustrated in Figures 5.17 and 5.18, than there was for the higher noise suppression (Example 5.12). In each of these examples the net was allowed a maximum of 10 cluster units. Fast learning used all 10 units, whereas slow learning only uses 6 units.

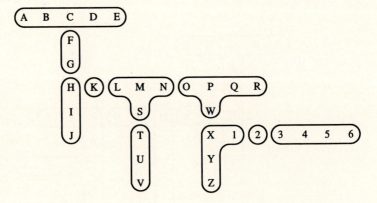

Figure 5.17 Spanning tree test data as clustered by ART2; fast learning, $\rho = .99$, $\theta + .2$.

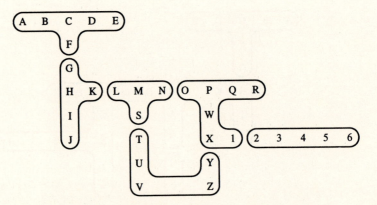

Figure 5.18 Spanning tree test data as clustered by ART2; slow learning, $\rho = .99$, $\theta = .2$.

Character recognition

Example 5.15 Clustering letters from different fonts using ART2 with slow learning

ART2 can be used for binary data. Using slow learning gives weights that may be more useful as exemplars than those formed by ART1 (see Example 5.3). The clusters formed are much less sensitive to the order of presentation than were those in that example.

Here, we take as input vectors the representations of seven letters from each of the three fonts presented in Figure 5.3. The results that follow use the standard values of $a = 10$, $b = 10$, $c = 0.1$, and $d = 0.9$, together with vigilance parameter $\rho = 0.8$ and noise suppression parameter $\theta = 0.126$.

If the order of presentation is A1, A2, A3, B1, B2, B3, . . . , the patterns are clustered as follows:

CLUSTER	PATTERNS
1	A1, A2
2	A3
3	C1, C2, C3, D2
4	B1, D1, E1, K1, B3, D3, E3, K3
5	K2
6	J1, J2, J3
7	B2, E2

If the data are input in the order A1, B1, C1, . . . , A2, B2, C2, . . . , the results are:

CLUSTER	PATTERNS
1	A1, A2
2	B1, D1, E1, K1, B3, D3, E3, K3
3	C1, C2, C3
4	J1, J2, J3
5	B2, D2, E2
6	K2
7	A3

Note that although the order of the clusters is different for the two orders of presentation, the patterns placed together are almost identical.

The top-down weights can be shown in a two-dimensional array (just as the original input data represented a two-dimensional pattern). The weights are either 0 (indicated by •) or the average of the input signals for the patterns placed on that cluster (indicated by #).

The weights for the clusters formed with the first order of presentation are:

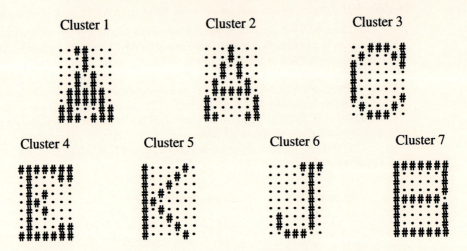

The top-down weights for the second order of presentation are:

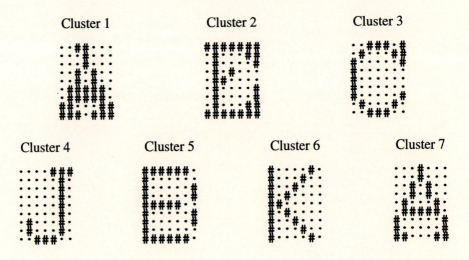

5.3.4 Analysis

Variations

The following computationally streamlined variations of ART2 produce essentially the same final clusters and weight vectors as the iterative (resonating) algorithms for fast and slow learning. In each variation, several iterative computations have been replaced by shortcut (nonneural) approaches.

Instant ART2.   Instant ART2 is computationally streamlined algorithm for ART2 that exhibits the same stability and clustering characteristics as the fast learning algorithm analyzed in the preceding sections.

There is no need to check for reset for the first pattern being learned by a cluster unit. (This can be determined by checking whether the top-down weights are zero, since they are initialized to zero, but at least some components will be nonzero after the unit has learned a pattern.)

Carpenter and Grossberg (1987b) indicate that ART2 is designed to have components that are set to zero—in either the previous weight vector or the noise-suppressed input vector—remain zero after fast learning. This is illustrated in Examples 5.8 and 5.11, and forms the basis for the Instant ART2 algorithm.

Note that in the following algorithm, the parameter e has been taken to be zero. Some protection against division by zero should therefore be included in any computer implementation of this algorithm.

Training Algorithm for Instant ART2

Step 0.   Initialize parameters:

$$a, b, \theta, c, d, \alpha, \rho.$$

Step 1.   Do Steps 2–9 *N_EP* times.
(Perform specified number of epochs of training.)

   Step 2.   For each input vector **s**, do Steps 3–8.

      Step 3.   Update F_1 unit activations:

$$w_i = s_i,$$

$$x_i = \frac{s_i}{\|\mathbf{s}\|},$$

$$v_i = f(x_i).$$

Update F_1 unit activations again:

$$u_i = \frac{v_i}{\|\mathbf{v}\|},$$

$$w_i = s_i + au_i,$$

$$p_i = u_i,$$

$$x_i = \frac{w_i}{\|\mathbf{w}\|},$$

$$q_i = \frac{p_i}{\|\mathbf{p}\|},$$

$$v_i = f(x_i) + bf(q_i).$$

Step 4.   Compute signals to F_2 units:

$$y_j = \sum_i b_{ij}p_i.$$

Step 5.   While reset is true, do Steps 6–7.

    Step 6.   Find F_2 unit with largest signal.

        (Define J such that $y_J \geq y_j$ for $j = 1, \ldots, m$.)

        If $y_J = -1$ all cluster units are inhibited; this pattern cannot be clustered.

    Step 7.   If $\mathbf{t}_J = \mathbf{0}$, proceed to Step 8.

        If $\mathbf{t}_J \neq \mathbf{0}$, then check for reset:

$$u_i = \frac{v_i}{\|\mathbf{v}\|},$$

$$p_i = u_i + dt_{Ji},$$

$$r_i = \frac{u_i + cp_i}{\|\mathbf{u}\| + c\,\|\mathbf{p}\|}.$$

If $\|\mathbf{r}\| < \rho$, then

$$y_J = -1 \qquad \text{(inhibit } J).$$

(Reset is true; repeat Step 5).

If $\|\mathbf{r}\| \geq \rho$, then

$$w_i = s_i + au_i,$$

$$x_i = \frac{w_i}{\|\mathbf{w}\|},$$

$$q_i = \frac{p_i}{\|\mathbf{p}\|},$$

$$v_i = f(x_i) + bf(q_i).$$

Reset is false; proceed to Step 8.

Step 8.   Update weights for winning unit J:
If $\mathbf{t}_J \neq \mathbf{0}$, then
If $t_{Ji} \neq 0$,

$$t_{Ji} = \frac{1}{1 - d} u_i;$$

$$b_{iJ} = \frac{1}{1 - d} u_i.$$

If $\mathbf{t}_J = \mathbf{0}$, then

$$\mathbf{t}_J = \frac{1}{1 - d} \mathbf{u};$$

$$\mathbf{b}_J = \frac{1}{1 - d} \mathbf{u}.$$

Step 9.   Test stopping condition for number of epochs.

Derivations

The derivations in this section follow the original development by Carpenter and Grossberg (1987b) closely.

Differential Equations for Activations.   The general form of the differential equation for the activation of an arbitrary unit Z in the F_1 layer is

$$\frac{dz}{dt} = -Az + J^+ - zJ^-,$$

where A is a positive constant, J^+ is the total excitatory input to the unit, and J^- is the total inhibitory input to the unit.

In ART nets, the assumption is made that activations change much more rapidly than any other process. Thus, the activations reach equilibrium before any input signals have time to change. This leads to the general form of the equation for the equilibrium activation:

$$z = \frac{J^+}{A + J^-}.$$

We have chosen the value of A to obtain the simplest form of the equation for the activations.

For the units that receive no inhibitory input, we take $A = 1$. Inhibitory input is indicated in Figure 5.4 by dashed arrows. Thus, the W units receive no inhibitory input; their excitatory input comes from the U units and the input signal. The equilibrium activation for the generic W unit, W_i, is

$$w_i = s_i + au_i.$$

Similarly, the P units receive excitatory input from the U units and the F_2 units. The activation of an F_2 unit is zero if it is not active and d if it is the current winner. We denote the winning unit's index by J, so that the equilibrium activation for the generic P unit, P_i, is

$$p_i = u_i + dt_{Ji},$$

where t_{Ji} is the top-down weight from unit Y_J to unit P_i. If no F_2 unit is active, the activation of unit P_i is simply

$$p_i = u_i.$$

The role of inhibition in ART2 is to normalize the activations at certain points in the computation cycle. For units that receive inhibitory inputs, we take A equal to a small parameter e. This protects against difficulties that would otherwise arise if the unit received no input signal, but allows the unit to normalize the vector to (approximately) unit length. Each X unit receives an excitatory signal from the corresponding W unit and an inhibitory signal equal to the norm of the vector of activations of the W units; thus, the activation is

$$x_i = \frac{w_i}{e + \|\mathbf{w}\|}.$$

Similarly, each Q unit receives an excitatory signal from the corresponding P unit and an inhibitory signal equal to the norm of the vector of activations of the P units, leading to the same form of an equation for the activation of a Q unit:

$$q_i = \frac{p_i}{e + \|\mathbf{p}\|}.$$

For the V units we take $A = 1$, and for the U units we take $A = e$, as described before for the other units of the F1 layer. We obtain

$$v_i = f(x_i) + bf(q_i),$$

$$u_i = \frac{v_i}{e + \|\mathbf{v}\|}.$$

Differential Equations for Weight Changes. The differential equations for the top-down weights (where J denotes the winning cluster unit) are

$$\frac{d}{dt} t_{Ji} = du_i + d(d - 1)t_{Ji}.$$

These can be approximated by the difference equations

$$t_{Ji}(\text{new}) - t_{Ji}(\text{old}) = \alpha[du_i + d(d - 1)t_{Ji}(\text{old})],$$

or

$$t_{Ji}(\text{new}) = \alpha du_i + [1 + \alpha d(d - 1)]t_{Ji}(\text{old}),$$

where α is the step size or learning rate.

In the same manner, the differential equations for the bottom-up weights,

$$\frac{d}{dt} b_{iJ} = du_i + d(d - 1)b_{iJ},$$

can be approximated by the difference equations

$$b_{iJ}(\text{new}) = \alpha du_i + [1 + \alpha d(d - 1)]b_{iJ}(\text{old}).$$

For fast learning, an input pattern is presented for a long enough period so that the weights reach equilibrium during each learning trial (presentation). However, unless we assume that learning progresses sufficiently rapidly that no iterations in the F_1 loop occur as the weights change, the activations of the U units will be changing as the top-down weights change. Thus, in general, an iterative process of learning is required for ART2 nets.

Reset Mechanism.   The test for reset is

if $\|\mathbf{r}\| \geq \rho$, then accept the winning F_2 unit and proceed with learning:
if $\|\mathbf{r}\| < \rho$, then reject (inhibit) the winning F_2 unit and reset all activations to 0;

where

$$\|\mathbf{r}\| = \frac{\|\mathbf{u} + c\mathbf{p}\|}{\|\mathbf{u}\| + c \|\mathbf{p}\|}.$$

Before a winner is chosen in the F_2 layer,

$$\mathbf{p} = \mathbf{u};$$

therefore,

$$\|\mathbf{r}\| = 1.$$

This shows that it is not necessary to check for a reset until an F_2 winner has been chosen.

It is desired that after a winner has been accepted to learn, no reset should suddenly cause the unit to stop being acceptable. In order to analyze the precautions that are necessary to ensure that this cannot happen, we now consider the possible changes in $\|\mathbf{r}\|$ during learning.

After a winning F_2 unit has been chosen,

$$\mathbf{p} = \mathbf{u} + d\mathbf{t};$$

therefore,

$$\|\mathbf{r}\| = \frac{\|\mathbf{u} + c\mathbf{p}\|}{\|\mathbf{u}\| + c \|\mathbf{p}\|}$$

can be written as

$$\|\mathbf{r}\| = \frac{\|\{1 + c\}\, \mathbf{u} + c d\mathbf{t}\|}{1 + c\, \|\mathbf{u} + d\mathbf{t}\|}.$$

Using the law of cosines, we obtain

$$\|(1 + c)\mathbf{u} + c d\mathbf{t}\|^2 = \|(1 + c)\mathbf{u}\|^2 + \|c d\mathbf{t}\|^2 + 2\, \|(1 + c)\mathbf{u}\|\, \|c d\mathbf{t}\|\, \cos(\mathbf{u}, \mathbf{t})$$

$$= (1 + c)^2 + \|c d\mathbf{t}\|^2 + 2(1 + c)\, \|c d\mathbf{t}\|\, \cos(\mathbf{u}, \mathbf{t})$$

and

$$\|\mathbf{u} + d\mathbf{t}\|^2 = \|\mathbf{u}\|^2 + \|d\mathbf{t}\|^2 + 2\, \|\mathbf{u}\|\, \|d\mathbf{t}\|\, \cos(\mathbf{u}, \mathbf{t})$$

$$= 1 + \|d\mathbf{t}\|^2 + 2\, \|d\mathbf{t}\|\, \cos(\mathbf{u}, \mathbf{t}),$$

from which we find that

$$\|\mathbf{r}\| = \frac{\sqrt{(1 + c)^2 + \|c d\mathbf{t}\|^2 + 2\,(1 + c)\, \|c d\mathbf{t}\|\, \cos(\mathbf{u}, \mathbf{t})}}{1 + c\, \sqrt{1 + \|d\mathbf{t}\|^2 + 2\, \|d\mathbf{t}\|\, \cos(\mathbf{u}, \mathbf{t})}}.$$

If we let $X = \|c d\mathbf{t}\|$ and $\beta = \cos(\mathbf{u}, \mathbf{t})$, then we can write $\|\mathbf{r}\|$ as a function of X and the parameter β as follows:

$$\|\mathbf{r}\| = \frac{\sqrt{(1 + c)^2 + 2\,(1 + c)\, \beta X + X^2}}{1 + \sqrt{c^2 + 2\, c\beta X + X^2}}.$$

The minimum value of $\|\mathbf{r}\|$ occurs at $X_{\min} = \sqrt{3c^2 + 4c + 1}$.

Figure 5.21 shows a sketch of $\|\mathbf{r}\|$ as a function of X for three values of the parameter β: 0, .5, and 1. These values correspond to the cases in which the vectors $\mathbf{u}$ and $\mathbf{t}$ are orthogonal, at an angle of $\pi/3$, and parallel, respectively. Clearly, the minimum value of $\|\mathbf{r}\|$ occurs when $\beta = 0$. In this case, the minimum is

$$\|\mathbf{r}\|_{\min} = \frac{\sqrt{(1 + c)^2 + 3c^2 + 4c + 1}}{1 + \sqrt{c^2 + 3c^2 + 4c + 1}}$$

$$= \frac{\sqrt{4c^2 + 6c + 2}}{1 + \sqrt{4c^2 + 4c + 1}} = \frac{\sqrt{2c + 1}}{\sqrt{2c + 2}}.$$

Since $0 < c < 1$, the minimum value of $\|\mathbf{r}\|$ ranges between $1/\sqrt{2}$ and $\sqrt{3}/2$. Thus, although the vigilance parameter can be set to any value between 0 and 1, values below approximately 0.7 will have no effect; the results will be the same as if the vigilance parameter were set to 0. However, it is not likely to have a pattern $\mathbf{u}$ assigned to a cluster with top-down weight vector $\mathbf{t}$ such that $\beta = \cos(\mathbf{u}, \mathbf{t}) = 0$. Therefore, larger values of ρ are necessary before the reset mechanism will have any effect.

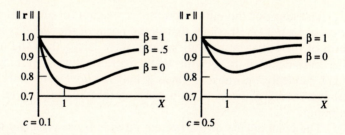

Figure 5.19   The effect of parameter c on the relation between $\|\mathbf{r}\|$ and $X = \|c d\mathbf{t}\|$
for selected values of $\beta = \cos(\mathbf{u}, \mathbf{t})$.
(a) c = 0.1
(b) c = 0.5

Figure 5.19 also shows why a small value of c is preferable to larger value:
For small c, the minimum value of $\|\mathbf{r}\|$ is closer to $1/\sqrt{2}$, and the effective range
for the vigilance parameter will be as wide as possible.

It is desired that the winning unit, once accepted by the reset mechanism,
should not be rejected during learning. We know that $\|\mathbf{t}\| = 1/(1 - d)$ if the weights
have been trained on a previous pattern (and we know that the reset mechanism
will never reject a cluster unit the first time it is chosen as the winner). However,
$\|\mathbf{t}\|$ may decrease during training, before returning to $1/(1 - d)$. Since the minimum
value of $\|\mathbf{r}\|$ occurs for $X_{\min}$ (which is greater than 1), we can ensure that $\|\mathbf{r}\|$ does
not decrease during training as long as the initial value of X is less than 1. If this
is guaranteed, then $\|\mathbf{r}\|$ will increase if $\|\mathbf{t}\|$ decreases. The initial value of X will be
less than 1 if

$$\|c d\mathbf{t}\| < 1,$$

or

$$\frac{cd}{1 - d} < 1.$$

This analysis shows that as long as the parameters are chosen to satisfy these
constraints, it is not necessary to check for a reset during training.

Again, the graph illustrates the advantage in choosing the parameters c and
d so that the ratio $cd/(1 - d)$ is close to 1. The initial value of X (i.e., $\|c d\mathbf{t}\|$) will
determine the effective range of the vigilance parameter, since X can move only
to the left during training.

Initialization of Weights.   The top-down weights are initialized to zero to
prevent a reset when a cluster unit that has not learned any patterns previously
(an uncommitted node) is first selected as the winner. (If the top-down weights
were nonzero, the combination of $\mathbf{u}$ and $\mathbf{t}$ might not be sufficiently similar to $\mathbf{u}$
to pass the reset check.)

The norm of the initial bottom-up weights for any unit should be less than or equal to $1/(1 - d)$, since that is the norm of the weights after training. If the norm of the initial bottom-up weights were larger than the norm of the weights after learning, then the weights would decrease during learning, and a new winner might suddenly be chosen. Once a unit is allowed to start learning a pattern, nothing should cause it to stop learning (before equilibrium is reached for fast learning).

The components of the bottom-up vector are typically taken to be equal to each other, and bottom-up weights are often chosen to be the same for all cluster units.

Choosing initial bottom-up weights (with equal components) equal to the maximum possible norm would result in

$$b_{ij}(0) = \frac{1}{(1 - d)\sqrt{n}} .$$

This will give new cluster units (units not previously selected to learn) the best chance of being selected the winner. However, if there are input vectors that are very similar to each other, they may be assigned to several different clusters, rather than being grouped together. Choosing smaller values for the initial bottom-up weights may reduce the number of clusters formed, since units that have learned previously will be selected preferentially to new units. In this case, the vigilance will play a stronger role in determining the nature of the clusters formed. If the vigilance parameter is relatively low, the net may change the weights on the clusters that are already formed (rather than forming a new cluster) more often than is desirable.

5.4 SUGGESTIONS FOR FURTHER STUDY

5.4.1 Readings

The standard reference for ART1 is

CARPENTER, G. A., & S. GROSSBERG. (1987a). "A Massively Parallel Architecture for a Self-Organizing Neural Pattern Recognition Machine." *Computer Vision, Graphics, and Image Processing,* 37:54–115.

The presentation of ART2 is given in

CARPENTER, G. A., & S. GROSSBERG. (1987b). "ART2: Self-organization of Stable Category Recognition Codes for Analog Input Patterns." *Applied Optics,* 26:4919–4930. Reprinted in Anderson, Pellionisz, & Rosenfeld (1990), pp. 151–162.

5.4.2 Exercises

ART1

5.1 Consider an ART1 neural net with four F_1 units and three F_2 units. After some training, the weights are as follows:

Bottom-up weights b_{ij}

0.67	0.0	0.2
0.0	0.0	0.2
0.0	0.0	0.2
0.0	0.67	0.2

Top-down weights t_{ji}

1	0	0	0
0	0	0	1
1	1	1	1

Determine the new weight matrices after the vector (0, 0, 1, 1) is presented if
a. the vigilance parameter is 0.3;
b. the vigilance parameter is 0.7.

5.2 Consider an ART1 network with nine input (F_1) units and two cluster (F_2) units. After some training, the bottom-up weights (b_{ij}) and the top-down weights (t_{ji}) have the following values:

Bottom-up weights b_{ij}

1/3	1/10
0	1/10
1/3	1/10
0	1/10
1/3	1/10
0	1/10
1/3	1/10
0	1/10
1/3	1/10

Top-down weights t_{ji}

1	0	1	0	1	0	1	0	1
1	1	1	1	1	1	1	1	1

The pattern (1, 1, 1, 1, 0, 1, 1, 1, 1) is presented to the network. Compute the action of the network if
a. the vigilance parameter is 0.5;
b. the vigilance parameter is 0.8.

ART2

5.3 Consider an ART2 network with two input units ($n = 2$). Show that using $\theta = 0.07$ will force the input patterns (0.71, 0.69) and (0.69, 0.71) to different clusters. What role does the vigilance parameter play in this situation?

5.4 Consider an ART2 network intended to cluster the input vectors

$$(0.6, 0.8, 0.0), (0.8, 0.6, 0.0), (0.0, 1.0, 0.0), (1.0, 0.0, 0.0)$$

$$(0.0, 0.6, 0.8), (0.0, 0.8, 0.6), (0.0, 0.0, 1.0).$$

Under what circumstances will the net place the first two vectors listed, namely, (0.6, 0.8, 0.0) and (0.8, 0.6, 0.0), together? When will it place (0.6, 0.8, 0.0) together with (0.0, 1.0, 0.0)? Use the noise suppression parameter value $\theta = 1/\sqrt{3} \approx 0.577$, and consider different values of the vigilance and different initial weights. Assume fast learning. (Several of the vectors listed do not enter into the computations for this problem. They are included simply to illustrate a situation in which it might be reasonable to expect the first four vectors to be viewed as ordered triples rather than using $n = 2$.)

5.5 Continue Exercise 5.4, assuming that the net has placed the vectors (0.6, 0.8, 0.0) and (0.8, 0.6, 0.0) together. What will happen if the vector (0.55, 0.84, 0.0) is presented to the net? Does the value of the vigilance parameter or the initial weights on the cluster units that have not learned any patterns affect the action of the net?

5.6 Show that the vector **u** does not change during the iterations in the F_1 layer for the first pattern being learned by a cluster unit. Therefore, the equilibrium values of the weights in the fast learning mode *for the first pattern placed on a cluster* can be found as we did for ART1:

$$\frac{d}{dt} t_{Ji} = du_i + d(d - 1)t_{Ji},$$

$$0 = du_i + d(d - 1)t_{Ji},$$

$$t_{Ji} = \frac{1}{1 - d} u_i;$$

and in a similar manner,

$$\frac{d}{dt} b_{iJ} = du_i + d(d - 1)b_{iJ},$$

$$0 = du_i + d(d - 1)b_{iJ},$$

$$b_{iJ} = \frac{1}{1 - d} u_i.$$

The value of **u** when the winning F_2 unit is chosen is simply the input vector normalized and with any components that are less than the noise suppression parameter θ set to zero.

5.7 Show that if the formulas in Exercise 5.6 were used for a cluster that had learned previously, the results would depend on precisely when in the F_1–F_2 iterations they were applied. Consider the results in the following cases:

   a. if they were applied immediately after the winning unit was accepted for learning, but before the new values for the activations of the P units had reached the U units;

   b. if they were applied after the top-down signal was incorporated into **u**. Will components of the input vector that have been noise suppressed be set to zero in the new weight vector? Will components of the weight vector that were zero remain zero (if the corresponding component of the input vector is nonzero)?

5.8 Show that two iterations of signals through the F_1 layer are sufficient to suppress all noise. (Activations of some units may change after this time, but not activations of the units that determine the winning F_2 units or that determine a reset or acceptance of the winning unit.)

     Start with all activations zero, and take the parameter e to be zero. Assume that the first component of the vector **x** (the input vector after it is normalized to unit length) falls below θ (on the first iteration) and the other components do not. Define the vector $\mathbf{ss} = (0, s_2, \ldots, s_n)$ as the input vector, with the component that is "noise suppressed" by the activation function set to zero.

   a. Compute the activations of the F_1 units for the first iteration. Compute the activations for **u** and **w** for the second iteration.

   b. Show that $\|\mathbf{ss}\| + a \le \|\mathbf{w}\| \le \|\mathbf{s}\| + a$.

   c. Using the results from part b, show that the norm of **w** increases from the first iteration to the second. On the first iteration, $\mathbf{w} = (s_1, s_2, \ldots, s_n)$; on the second iteration, $\mathbf{w} = (s_1, s_2 + au_2, \ldots, s_n + au_n)$.

   d. Show that the components of **x** that were set to zero on the first iteration will be set to zero again on the next iteration and that the components that were not set to zero on the first iteration will not be set to zero on subsequent iterations.

5.9 Using fast learning, show that noise suppression can help to prevent instability in pattern clustering by considering the performance of an ART2 net for the following input patterns:

$$
\begin{aligned}
\text{Pat } 10 &= (0.984798,\ 0.173648), \\
\text{Pat } 20 &= (0.939683,\ 0.342017), \\
\text{Pat } 30 &= (0.866018,\ 0.499993), \\
\text{Pat } 40 &= (0.766034,\ 0.642785), \\
\text{Pat } 50 &= (0.642785,\ 0.766034), \\
\text{Pat } 60 &= (0.499993,\ 0.866018), \\
\text{Pat } 70 &= (0.342017,\ 0.939683), \\
\text{Pat } 80 &= (0.173648,\ 0.984798).
\end{aligned}
$$

Use the standard parameter values ($a = 10$, $b = 10$, $c = 0.10$, $d = 0.90$), together with a vigilance of 0.99 and initial bottom-up weights of (6.50, 6.50). Use the fact that for fast learning, each cluster unit learns the current input pattern perfectly.

Present the patterns in the following order: Pat 40, Pat 30, Pat 20, Pat 10, Pat 40, Pat 50, Pat 60, Pat 70, Pat 80, Pat 20, Pat 30, Pat 40.

a. Use $\theta = 0.0$.

b. Use $\theta = 0.2$.

5.4.3 Projects

ART1

5.1 Write a computer program to implement the ART1 neural network. Explore the performance of the net for various input orders of the training patterns used in the examples in the text.

ART2

5.2 Write a computer program to implement the ART2 neural network, allowing for either fast or slow learning, depending on the number of epochs of training and the number of weight update iterations performed on each learning trial. Use this program to explore the relationships between fast learning and slow learning for various input patterns.

5.3 Because the ART2 net normalizes its input, it is sometimes advisable to create an additional component for each of the vectors before presenting the data to the net. This extra component is constructed so that the new vectors have the same first components as the original vectors, but the new vectors will have the same norm N [Dayhoff, 1990]. N can be chosen to be any number larger than the norm of the largest of the original vectors.

Applying this process to the spanning tree patterns and using $N = 10$ gives the patterns shown in the table that follows. The sixth component of each vector is the square root of the quantity N minus the norm of the original vector.

Using this form of the data, repeat the spanning tree example. Compare and discuss your results.

PATTERN	COMPONENTS					
A	1.0	0.0	0.0	0.0	0.0	9.9498
B	2.0	0.0	0.0	0.0	0.0	9.7979
C	3.0	0.0	0.0	0.0	0.0	9.5393
D	4.0	0.0	0.0	0.0	0.0	9.1651
E	5.0	0.0	0.0	0.0	0.0	8.6602
F	3.0	1.0	0.0	0.0	0.0	9.4868
G	3.0	2.0	0.0	0.0	0.0	9.3273
H	3.0	3.0	0.0	0.0	0.0	9.0553
I	3.0	4.0	0.0	0.0	0.0	8.6602
J	3.0	5.0	0.0	0.0	0.0	8.1240
K	3.0	3.0	1.0	0.0	0.0	9.0000
L	3.0	3.0	2.0	0.0	0.0	8.8317

PATTERN	COMPONENTS					
M	3.0	3.0	3.0	0.0	0.0	8.5440
N	3.0	3.0	4.0	0.0	0.0	8.1240
O	3.0	3.0	5.0	0.0	0.0	7.5498
P	3.0	3.0	6.0	0.0	0.0	6.7823
Q	3.0	3.0	7.0	0.0	0.0	5.7445
R	3.0	3.0	8.0	0.0	0.0	4.2426
S	3.0	3.0	3.0	1.0	0.0	8.4852
T	3.0	3.0	3.0	2.0	0.0	8.3066
U	3.0	3.0	3.0	3.0	0.0	8.0000
V	3.0	3.0	3.0	4.0	0.0	7.5498
W	3.0	3.0	6.0	1.0	0.0	6.7087
X	3.0	3.0	6.0	2.0	0.0	6.4807
Y	3.0	3.0	6.0	3.0	0.0	6.0827
Z	3.0	3.0	6.0	4.0	0.0	5.4772
1	3.0	3.0	6.0	2.0	1.0	6.4031
2	3.0	3.0	6.0	2.0	2.0	6.1644
3	3.0	3.0	6.0	2.0	3.0	5.7445
5	3.0	3.0	6.0	2.0	5.0	4.1231
4	3.0	3.0	6.0	2.0	4.0	5.0990
6	3.0	3.0	6.0	2.0	6.0	2.4494

CHAPTER 6

Backpropagation Neural Net

6.1 STANDARD BACKPROPAGATION

The demonstration of the limitations of single-layer neural networks was a significant factor in the decline of interest in neural networks in the 1970s. The discovery (by several researchers independently) and widespread dissemination of an effective general method of training a multilayer neural network [Rumelhart, Hinton, & Williams, 1986a, 1986b; McClelland & Rumelhart, 1988] played a major role in the reemergence of neural networks as a tool for solving a wide variety of problems. In this chapter, we shall discuss this training method, known as *backpropagation* (of errors) or the *generalized delta rule*. It is simply a gradient descent method to minimize the total squared error of the output computed by the net.

The very general nature of the backpropagation training method means that a backpropagation net (a multilayer, feedforward net trained by backpropagation) can be used to solve problems in many areas. Several of the applications mentioned in Chapter 1—for example, NETtalk, which learned to read English aloud—were based on some variation of the backpropagation nets we shall describe in the sections that follow. Applications using such nets can be found in virtually every field that uses neural nets for problems that involve mapping a given set of inputs to a specified set of target outputs (that is, nets that use supervised training). As is the case with most neural networks, the aim is to train the net to achieve a balance between the ability to respond correctly to the input patterns that are used for training (memorization) and the ability to give reasonable (good) responses to input that is similar, but not identical, to that used in training (generalization).

The training of a network by backpropagation involves three stages: the feedforward of the input training pattern, the calculation and backpropagation of the associated error, and the adjustment of the weights. After training, application of the net involves only the computations of the feedforward phase. Even if training is slow, a trained net can produce its output very rapidly. Numerous variations of backpropagation have been developed to improve the speed of the training process.

Although a single-layer net is severely limited in the mappings it can learn, a multilayer net (with one or more hidden layers) can learn any continuous mapping to an arbitrary accuracy. More than one hidden layer may be beneficial for some applications, but one hidden layer is sufficient.

In Section 6.1, we shall describe standard backpropagation, including a few of the choices that must be made in designing a net with this feature. In the next section, we mention a few of the many variations of backpropagation that have been developed. Finally, the mathematical derivation of the training algorithm and a brief summary of some of the theorems dealing with the ability of multilayer nets to approximate arbitrary (continuous) functions are given.

6.1.1 Architecture

A multilayer neural network with one layer of hidden units (the Z units) is shown in Figure 6.1. The output units (the Y units) and the hidden units also may have biases (as shown). The bias on a typical output unit Y_k is denoted by w_{0k}; the bias on a typical hidden unit Z_j is denoted v_{0j}. These bias terms act like weights on connections from units whose output is always 1. (These units are shown in Figure 6.1 but are usually not displayed explicitly.) Only the direction of information flow for the feedforward phase of operation is shown. During the backpropagation phase of learning, signals are sent in the reverse direction.

The algorithm in Section 6.1.2 is presented for one hidden layer, which is adequate for a large number of applications. The architecture and algorithm for two hidden layers are given in Section 6.2.4.

6.1.2 Algorithm

As mentioned earlier, training a network by backpropagation involves three stages: the feedforward of the input training pattern, the backpropagation of the associated error, and the adjustment of the weights.

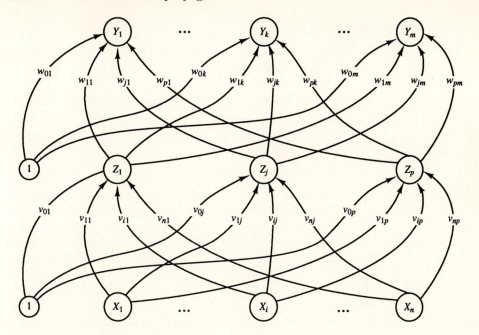

Figure 6.1 Backpropagation neural network with one hidden layer.

During feedforward, each input unit (X_i) receives an input signal and broadcasts this signal to the each of the hidden units $Z_1, \ldots, Z_p$. Each hidden unit then computes its activation and sends its signal (z_j) to each output unit. Each output unit (Y_k) computes its activation (y_k) to form the response of the net for the given input pattern.

During training, each output unit compares its computed activation y_k with its target value t_k to determine the associated error for that pattern with that unit. Based on this error, the factor δ_k ($k = 1, \ldots, m$) is computed. δ_k is used to distribute the error at output unit Y_k back to all units in the previous layer (the hidden units that are connected to Y_k). It is also used (later) to update the weights between the output and the hidden layer. In a similar manner, the factor δ_j ($j = 1, \ldots, p$) is computed for each hidden unit Z_j. It is not necessary to propagate the error back to the input layer, but δ_j is used to update the weights between the hidden layer and the input layer.

After all of the δ factors have been determined, the weights for all layers are adjusted simultaneously. The adjustment to the weight w_{jk} (from hidden unit Z_j to output unit Y_k) is based on the factor δ_k and the activation z_j of the hidden unit Z_j. The adjustment to the weight v_{ij} (from input unit X_i to hidden unit Z_j) is based on the factor δ_j and the activation x_i of the input unit.

Nomenclature

The nomenclature we use in the training algorithm for the backpropagation net is as follows:

x   Input training vector:

$$\mathbf{x} = (x_1, \ldots, x_i, \ldots, x_n).$$

t   Output target vector:

$$\mathbf{t} = (t_1, \ldots, t_k, \ldots, t_m).$$

δ_k Portion of error correction weight adjustment for w_{jk} that is due to an error at output unit Y_k; also, the information about the error at unit Y_k that is propagated back to the hidden units that feed into unit Y_k.

δ_j Portion of error correction weight adjustment for v_{ij} that is due to the backpropagation of error information from the output layer to the hidden unit Z_j.

α Learning rate.

X_i Input unit i:
For an input unit, the input signal and output signal are the same, namely, x_i.

v_{0j} Bias on hidden unit j.

Z_j Hidden unit j:
The net input to Z_j is denoted z_in_j:

$$z\_in_j = v_{0j} + \sum_i x_i v_{ij}.$$

The output signal (activation) of Z_j is denoted z_j:

$$z_j = f(z\_in_j).$$

w_{0k} Bias on output unit k.

Y_k Output unit k:
The net input to Y_k is denoted y_in_k:

$$y\_in_k = w_{0k} + \sum_j z_j w_{jk}.$$

The output signal (activation) of Y_k is denoted y_k:

$$y_k = f(y\_in_k).$$

Activation function

An activation function for a backpropagation net should have several important characteristics: It should be continuous, differentiable, and monotonically non-decreasing. Furthermore, for computational efficiency, it is desirable that its derivative be easy to compute. For the most commonly used activation functions, the value of the derivative (at a particular value of the independent variable) can be expressed in terms of the value of the function (at that value of the independent variable). Usually, the function is expected to *saturate*, i.e., approach finite maximum and minimum values asymptotically.

One of the most typical activation functions is the binary sigmoid function, which has range of (0, 1) and is defined as

$$f_1(x) = \frac{1}{1 + \exp(-x)},$$

with

$$f_1'(x) = f_1(x)[1 - f_1(x)].$$

This function is illustrated in Figure 6.2.

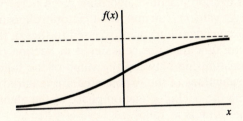

Figure 6.2   Binary sigmoid, range (0, 1).

Another common activation function is bipolar sigmoid, which has range of $(-1, 1)$ and is defined as

$$f_2(x) = \frac{2}{1 + \exp(-x)} - 1,$$

with

$$f_2'(x) = \frac{1}{2}[1 + f_2(x)][1 - f_2(x)].$$

This function is illustrated in Figure 6.3. Note that the bipolar sigmoid function is closely related to the function

$$\tanh(x) = \frac{e^x - e^{-x}}{e^x + e^{-x}}.$$

(See Section 1.4.3.)

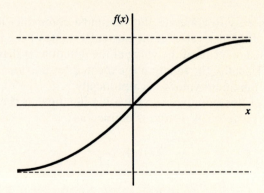

Figure 6.3   Bipolar sigmoid, range $(-1, 1)$.

Training algorithm

Either of the activation functions defined in the previous section can be used in the standard backpropagation algorithm given here. The form of the data (especially the target values) is an important factor in choosing the appropriate function. The relevant considerations are discussed further in the next section. Other suitable activation functions are considered in Section 6.2.2. Note that because of the simple relationship between the value of the function and its derivative, no additional evaluations of the exponential are required to compute the derivatives needed during the backpropagation phase of the algorithm.

  The algorithm is as follows:

Step 0. Initialize weights.
     (Set to small random values).
Step 1. While stopping condition is false, do Steps 2–9.
    Step 2. For each training pair, do Steps 3–8.
      Feedforward:
      Step 3. Each input unit ($X_i, i = 1, \ldots, n$) receives input signal x_i and broadcasts this signal to all units in the layer above (the hidden units).
      Step 4. Each hidden unit ($Z_j, j = 1, \ldots, p$) sums its weighted input signals,

$$z\_in_j = v_{0j} + \sum_{i=1}^{n} x_i v_{ij},$$

        applies its activation function to compute its output signal,

$$z_j = f(z\_in_j),$$

        and sends this signal to all units in the layer above (output units).

Step 5.   Each output unit (Y_k, $k = 1, \ldots, m$) sums its weighted input signals,

$$y\_in_k = w_{0k} + \sum_{j=1}^{p} z_j w_{jk}$$

and applies its activation function to compute its output signal,

$$y_k = f(y\_in_k).$$

Backpropagation of error:

Step 6.   Each output unit (Y_k, $k = 1, \ldots, m$) receives a target pattern corresponding to the input training pattern, computes its error information term,

$$\delta_k = (t_k - y_k) f'(y\_in_k),$$

calculates its weight correction term (used to update w_{jk} later),

$$\Delta w_{jk} = \alpha \delta_k z_j,$$

calculates its bias correction term (used to update w_{0k} later),

$$\Delta w_{0k} = \alpha \delta_k,$$

and sends δ_k to units in the layer below.

Step 7.   Each hidden unit (Z_j, $j = 1, \ldots, p$) sums its delta inputs (from units in the layer above),

$$\delta\_in_j = \sum_{k=1}^{m} \delta_k w_{jk},$$

multiplies by the derivative of its activation function to calculate its error information term,

$$\delta_j = \delta\_in_j \, f'(z\_in_j),$$

calculates its weight correction term (used to update v_{ij} later),

$$\Delta v_{ij} = \alpha \delta_j x_i,$$

and calculates its bias correction term (used to update v_{0j} later),

$$\Delta v_{0j} = \alpha \delta_j.$$

Update weights and biases:

Step 8.   Each output unit ($Y_k, k = 1, \ldots, m$) updates its bias and weights ($j = 0, \ldots, p$):

$$w_{jk}(\text{new}) = w_{jk}(\text{old}) + \Delta w_{jk}.$$

Each hidden unit ($Z_j, j = 1, \ldots, p$) updates its bias and weights ($i = 0, \ldots, n$):

$$v_{ij}(\text{new}) = v_{ij}(\text{old}) + \Delta v_{ij}.$$

Step 9.   Test stopping condition.

Note that in implementing this algorithm, separate arrays should be used for the deltas for the output units (Step 6, δ_k) and the deltas for the hidden units (Step 7, δ_j).

An epoch is one cycle through the entire set of training vectors. Typically, many epochs are required for training a backpropagation neural net. The foregoing algorithm updates the weights after each training pattern is presented. A common variation is batch updating, in which weight updates are accumulated over an entire epoch (or some other number of presentations of patterns) before being applied.

Note that $f'(y\_in_k)$ and $f'(z\_in_j)$ can be expressed in terms of y_k and z_j, respectively, using the appropriate formulas on page 293 (depending on the choice of activation function).

The mathematical basis for the backpropagation algorithm is the optimization technique known as *gradient descent*. The gradient of a function (in this case, the function is the error and the variables are the weights of the net) gives the direction in which the function increases more rapidly; the negative of the gradient gives the direction in which the function decreases most rapidly. A derivation of the weight update rules is given in Section 6.3.1. The derivation clarifies the reason why the weight updates should be done after all of the δ_k and δ_j expressions have been calculated, rather than during backpropagation.

Choices

Choice of initial weights and biases.

Random Initialization.   The choice of initial weights will influence whether the net reaches a global (or only a local) minimum of the error and, if so, how quickly it converges. The update of the weight between two units depends on both the derivative of the upper unit's activation function and the activation of the lower unit. For this reason, it is important to avoid choices of initial weights that would make it likely that either activations or derivatives of activations are zero. The values for the initial weights must not be too large, or the initial input signals to each hidden or output unit will be likely to fall in the region where the derivative of the sigmoid function has a very small value (the so-called saturation region). On the other hand, if the initial weights are too small, the net input to a hidden or output unit will be close to zero, which also causes extremely slow learning.

A common procedure is to initialize the weights (and biases) to random values between -0.5 and 0.5 (or between -1 and 1 or some other suitable interval). The values may be positive or negative because the final weights after training may be of either sign also. Section 6.2.2 presents some possible modifications to the logistic sigmoid function described before that can customize the function to help prevent difficulties caused by very small activations or derivatives. A simple modification of random initialization, developed by Nguyen and Widrow [1990], is given here.

Nguyen-Widrow Initialization.   The following simple modification of the common random weight initialization just presented typically gives much faster learning. The approach is based on a geometrical analysis of the response of the hidden neurons to a single input; the analysis is extended to the case of several inputs by using Fourier transforms. Weights from the hidden units to the output units (and biases on the output units) are initialized to random values between -0.5 and 0.5, as is commonly the case.

The initialization of the weights from the input units to the hidden units is designed to improve the ability of the hidden units to learn. This is accomplished by distributing the initial weights and biases so that, for each input pattern, it is likely that the net input to one of the hidden units will be in the range in which that hidden neuron will learn most readily. The definitions we use are as follows:

n number of input units
p number of hidden units
β scale factor:

$$\beta = 0.7 \, (p)^{1/n} = 0.7 \, \sqrt[n]{p}$$

The procedure consists of the following simple steps:
for each hidden unit ($j = 1, \ldots, p$):
    Initialize its weight vector (from the input units):

$v_{ij}(\text{old}) = $ random number between -0.5 and 0.5 (or between $-\gamma$ and γ).

    Compute $\|\mathbf{v}_j(\text{old})\|$.
    Reinitialize weights:

$$v_{ij} = \frac{\beta v_{ij}(\text{old})}{\|\mathbf{v}_j(\text{old})\|} \, .$$

    Set bias:

$$v_{0j} = \text{random number between } -\beta \text{ and } \beta.$$

The Nguyen-Widrow analysis is based on the activation function

$$\tanh(x) = \frac{e^x - e^{-x}}{e^x + e^{-x}},$$

which is closely related to the bipolar sigmoid activation function in Section 6.1.2. In Example 6.4, we demonstrate that using Nguyen-Widrow initialization gives improved training for the XOR problems considered in Examples 6.1–6.3.

How long to train the net.   Since the usual motivation for applying a back-propagation net is to achieve a balance between correct responses to training patterns and good responses to new input patterns (i.e., a balance between memorization and generalization), it is not necessarily advantageous to continue training until the total squared error actually reaches a minimum. Hecht-Nielsen (1990) suggests using two sets of data during training: a set of training patterns and a set of training-testing patterns. These two sets are disjoint. Weight adjustments are based on the training patterns; however, at intervals during training, the error is computed using the training-testing patterns. As long as the error for the training-testing patterns decreases, training continues. When the error begins to increase, the net is starting to memorize the training patterns too specifically (and starting to lose its ability to generalize). At this point, training is terminated.

How many training pairs there should be.   A relationship among the number of training patterns available, P, the number of weights to be trained, W, and the accuracy of classification expected, e, is summarized in the following rule of thumb. For a more precise statement, with proofs, see Baum and Haussler, (1989). The question to be answered is "Under what circumstances can I be assured that a net which is trained to classify a given percentage of the training patterns correctly will also classify correctly testing patterns drawn from the same sample space?" Specifically, if the net is trained to classify the fraction $1 - (e/2)$ of the training patterns correctly, where $0 < e \leq 1/8$, can I be sure that it will classify $1 - e$ of the testing patterns correctly? The answer is that if there are enough training patterns, the net will be able to generalize as desired (classify unknown testing patterns correctly). Enough training patterns is determined by the condition

$$\frac{W}{P} = e,$$

or

$$P = \frac{W}{e}.$$

For example, with $e = 0.1$, a net with 80 weights will require 800 training patterns to be assured of classifying 90% of the testing patterns correctly, assuming that the net was trained to classify 95% of the training patterns correctly.

Data Representation.   In many problems, input vectors and output vectors have components in the same range of values. Because one factor in the weight correction expression is the activation of the lower unit, units whose activations are zero will not learn. This suggests that learning may be improved if the input is represented in bipolar form and the bipolar sigmoid is used for the activation function.

In many neural network applications, the data (input or target patterns) may be given by either a continuous-valued variable or a "set or ranges". For example, the temperature of food could be represented by the actual temperature (a continuous-valued variable) or one of the four states (ranges of temperature): frozen, chilled, room temperature, or hot. In the later case, four neurons, each with bipolar values, would be appropriate; in the former case a single neuron would be used. In general, it is easier for a neural net to learn a set of distinct responses than a continuous-valued response. However, breaking truly continuous data into artificial distinct categories can make it more difficult for the net to learn examples that occur on, or near, the boundaries of the groups. Continuous-valued inputs or targets should not be used to represent distinct quantities, such as letters of the alphabet [Ahmad & Tesauro, 1989; Lawrence, 1993].

Number of Hidden Layers.   For a neural net with more than one layer of hidden units, only minor modifications of the algorithm on page 294 are required. The calculation of the δ's is repeated for each additional hidden layer in turn, summing over the δ's for the units in the previous layer that feed into the current layer for which δ is being calculated. With reference to the algorithm, Step 4 is repeated for each hidden layer in the feedforward phase, and Step 7 is repeated for each hidden layer in the backpropagation phase. The algorithm and architecture for a backpropagation net with two hidden layers are given in Section 6.2.4. The theoretical results presented in Section 6.3 show that one hidden layer is sufficient for a backpropagation net to approximate any continuous mapping from the input patterns to the output patterns to an arbitrary degree of accuracy. However, two hidden layers may make training easier in some situations.

Application procedure

After training, a backpropagation neural net is applied by using only the feedforward phase of the training algorithm. The application procedure is as follows:

Step 0.   Initialize weights (from training algorithm).

Step 1.   For each input vector, do Steps 2–4.

    Step 2.   For $i = 1, \ldots, n$: set activation of input unit

$$x_i;$$

    Step 3.   For $j = 1, \ldots, p$:

$$z\_in_j = v_{0j} + \sum_{i=1}^{n} x_i v_{ij};$$

$$z_j = f(z\_in_j).$$

Step 4.   For $k = 1, \ldots, m$:

$$y\_in_k = w_{0k} + \sum_{j=1}^{p} z_j w_{jk};$$

$$y_k = f(y\_in_k).$$

6.1.3 Applications

Simple examples

The simple examples given here illustrate the training of a 2–4–1 backprop net (a net with two input units, four hidden units in one hidden layer, and one output unit) to solve the XOR problem. Example 6.1 uses binary data representation for this problem, with the binary sigmoid for the activation function on all hidden and output units. Example 6.2 uses bipolar data representation and the bipolar sigmoid function.

In each of these examples, the same set of initial weights is used; random values were chosen between -0.5 and $+0.5$.

The initial weights to the hidden layer are:

-0.3378 0.2771 0.2859 -0.3329 (biases to the four hidden units)
0.1970 0.3191 -0.1448 0.3594 (weights from the first input unit)
0.3099 0.1904 -0.0347 -0.4861 (weights from the second input unit).

The initial weights from the hidden units to the output unit are:

-0.1401 (bias on the output unit)
0.4919 (weight from the first hidden unit)
-0.2913 (weight from the second hidden unit)
-0.3979 (weight from the third hidden unit)
0.3581 (weight from the fourth hidden unit)

Although the training speed varies somewhat for different choices of the initial weights, the relative speeds shown here for these variations are typical. The learning rate for each of the examples is 0.02. Training continued until the total squared error for the four training patterns was less than 0.05.

Example 6.1 A backprop net for the XOR function: binary representation

Training using binary input is relatively slow; any unit that receives an input of zero for a particular pattern cannot learn that pattern. For the binary case, training took almost 3,000 epochs (see Figure 6.4).

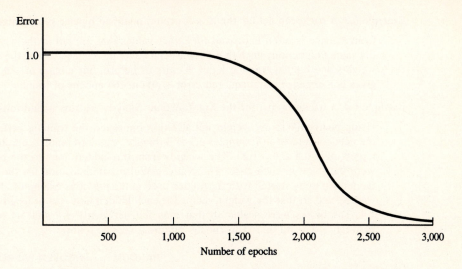

Figure 6.4   Total squared error for binary representation of XOR problem.

Example 6.2 A backprop net for the XOR function: bipolar representation

Using bipolar representation for the training data for the XOR problem and the bipolar sigmoid for the activation function for the hidden and output units gives much faster training than for the binary case illustrated in Example 6.1. Now, training takes only 387 epochs (see Figure 6.5).

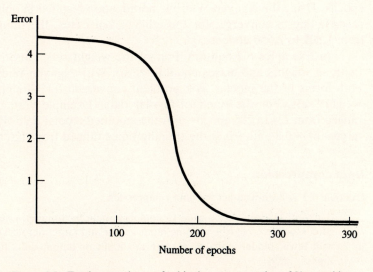

Figure 6.5   Total squared error for bipolar representation of XOR problem.

Example 6.3 A backprop net for the X$_{OR}$ function: modified bipolar representation

Convergence is often improved for target values that are not at the asymptotes of the sigmoid function, since those values can never be reached. Using the same initial weights and bipolar training input vectors as before, but targets of $+0.8$ or -0.8, gives convergence (total squared error $<.05$) in 264 epochs of training.

Example 6.4 A backprop net for the X$_{OR}$ function: Nguyen-Widrow weight initialization

Using Nguyen-Widrow weight initialization improved the training performance of the nets considered in Examples 6.1–6.3 (binary, standard bipolar, and bipolar with targets of $+0.8$ and -0.8). The weights from the hidden units to the output units were the same in each case. The Nguyen-Widrow initialization for the weights to the hidden units started with the values used in the previous examples. The weights were scaled so that the weight vector for each hidden unit was of length $0.7 \sqrt{4} = 1.4$. The biases were scaled so that they fell between -1.4 and 1.4 (rather than between -0.5 and 0.5). The epochs required were as follows:

	RANDOM	NGUYEN-WIDROW
Binary X$_{OR}$	2,891	1,935
Bipolar X$_{OR}$	387	224
Modified bipolar X$_{OR}$ (targets $= +0.8$ and -0.8)	264	127

In these examples, the use of Nguyen-Widrow initialization not only improved the training speed, but also greatly reduced the chances of generating initial weights for which the net fails to converge. To study the sensitivity of these simple backpropagation nets to the choice of initial weights, eight different sets of random weights were used in the examples. For the binary representation, with random initial weights, only one of the eight cases converged in fewer than 3,000 epochs. Using the Nguyen-Widrow modification resulted in only one of the eight cases failing to converge; for the other seven cases, the training times ranged from 1,605 to 2,556 epochs.

In Examples 6.2 and 6.3, four of the weight sets caused the net to freeze (with all weights and biases equal to zero). With Nguyen-Widrow initialization, both forms of the bipolar X$_{OR}$ problem converged for all eight of the modified weights sets. For the standard bipolar data (Example 6.2), the training times ranged from 224 to 285 epochs. For the modified bipolar data (Example 6.3), with targets of $+0.8$ and -0.8, the training times ranged from 120 to 174 epochs.

Data compression

Example 6.5 A backprop net for data compression

Backpropagation can be used to compress data by training a net to function as an autoassociative net (the training input vector and target output vector are the same) with fewer hidden units than there are input or output units [Cottrell, Munro, & Zipser, 1989].

In this simple example, the image data set chosen was the set of characters A, B, . . . , J shown in Figure 6.6. Each character was defined in terms of binary values on a grid of size 7×9 pixels.

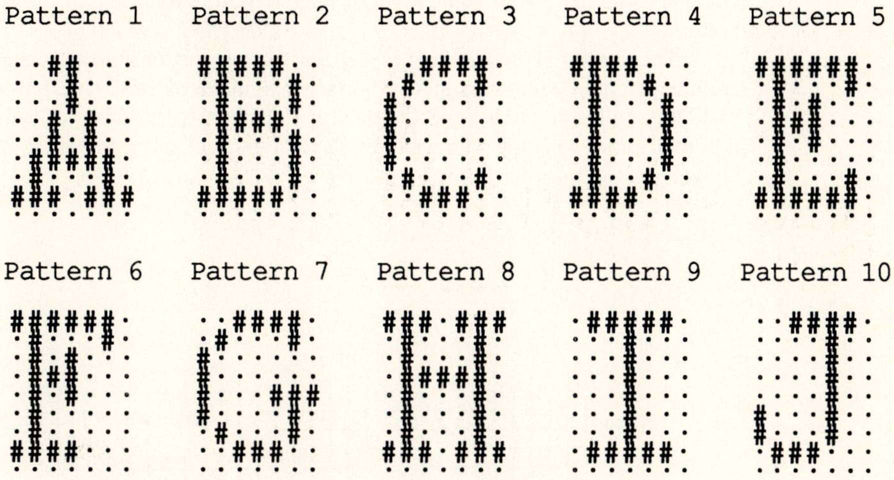

```
Pattern 1      Pattern 2      Pattern 3      Pattern 4      Pattern 5

. . # # . . .   # # # # # . .   . . # # # # .   # # # # . . .   # # # # # # .
. . # # . . .   . . . . . # .   . # . . . # .   . # . . # . .   . # . . . # .
. # . # . . .   . . . . . # .   # . . . . # .   . # . . . # .   . # . . . # .
. # . # . . .   . # # # . . .   # . . . . . .   . # . . . # .   . # . . # # .
# # # # # . .   . # # # . . .   # . . . . . .   . # . . . # .   # # # # # . .
. # . . . # .   . . . . . # .   # . . . . # .   . # . . . # .   . # . . # # .
# # # . # # #   # # # # # . .   . . # # # . .   # # # # . . .   # # # # # . .

Pattern 6      Pattern 7      Pattern 8      Pattern 9      Pattern 10

# # # # # # .   . . # # # # .   # # # . # # #   . # # # # # .   . . # # # # .
. # . . . # .   . # . . . # .   . # . . . # .   . . # . . . .   . . . . # . .
. # . . . . .   # . . . . . .   . # # # # # .   . . # . . . .   . . . . # . .
. # # # . . .   # . . . . . .   . # # # # . .   . . # . . . .   . . . . # . .
. # . . . . .   # . . # # # .   . # . . . . .   . . # . . . .   # . . . # . .
. # . . . . .   . # . . . # .   . # . . . . .   . . # . . . .   # . . . # . .
# # # # . . .   . . # # # . .   # # # . # # #   . # # # # # .   . # # # . . .
```

Figure 6.6   Total patterns for Example 6.5.

Because each character is represented in the input data as a vector with 63 binary components, the input layer of the neural network has 63 input units. The hidden layer is given a smaller number of units (for compression). It is known [Rumelhart, McClelland, & PDP Research Group, 1986] that a set of N orthogonal input patterns can be mapped onto $\log_2 N$ hidden units to form a binary code with a distinct pattern for each of the N input patterns. Because the characters in the set of input patterns are not orthogonal, the value of $\log_2 N$ can be taken as a theoretical lower bound for the number of hidden units that can be used for compression if perfect reconstruction of the characters is required. (This is called *lossless reconstruction*.) The number of hidden units was varied as part of the investigation. The output layer had 63 units (for restoration).

The net is considered to have learned a pattern if all computed output values are within a specified tolerance of the desired values (0 or 1). The results shown in Figure 6.7 give the number of epochs required for the net to learn the 10 input patterns for two values of the tolerance. Points marked by an "x" are based on a tolerance of 0.2. In other words, the response of a unit was considered correct if its activation was no greater than 0.2 and the pixel (in the training pattern) corresponding to that unit was "off." Similarly, a unit corresponding to a pixel than was "on" had to have an activation that was no less than 0.8 to be considered correct. For a tolerance of 0.1, the corresponding values are "no more than 0.1" and "no less than 0.9". All units in the net had to have the correct activation (to within the specified tolerance) for all training patterns before learning was considered successful.

It is important to note that the accuracy of these results was evaluated in terms of 100% correctness of the reconstructed characters in the training set. In some applications, (e.g., some types of communications) this requirement for lossless reconstruction is intrinsic. In other applications, (e.g., some types of image sets [Arozullah & Namphol, 1990; Cottrell, Munro, & Zipser, 1989; Sonehara, Kawato, Miyake, & Nakane, 1989]) some level of degradation is tolerable in the reconstructed pattern.

The results shown in Figure 6.7 are for one set of starting weights for each architecture and each tolerance. The larger number of epochs required for 21 and 24 hidden units, using a larger tolerance, reflects the variation in training time for different initial weights. The case for 18 hidden units and a tolerance of 0.2 did not converge.

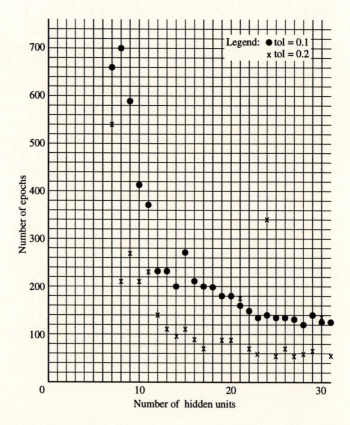

Figure 6.7 Number of epochs required as a function of number of hidden units.

6.2 VARIATIONS

Several modifications can be made to the backpropagation algorithm presented on page 294 which may improve its performance in some situations. The modifications we discuss involve changes to the weight update procedure, alternatives to the sigmoid activation functions presented previously, a variation to improve biological plausibility and computational power, and finally an explicit statement of the backpropagation algorithm for 2 hidden layers.

6.2.1 Alternative Weight Update Procedures

Momentum

In backpropagation with momentum, the weight change is in a direction that is a combination of the current gradient and the previous gradient. This is a modification of gradient descent whose advantages arise chiefly when some training data are very different from the majority of the data (and possibly even incorrect). It is desirable to use a small learning rate to avoid a major disruption of the direction of learning when a very unusual pair of training patterns is presented. However, it is also preferable to maintain training at a fairly rapid pace as long as the training data are relative similar.

Convergence is sometimes faster if a momentum term is added to the weight update formulas. In order to use momentum, weights (or weight updates) from one or more previous training patterns must be saved. For example, in the simplest form of backpropagation with momentum, the new weights for training step $t + 1$ are based on the weights at training steps t and $t - 1$. The weight update formulas for backpropagation with momentum are

$$w_{jk}(t + 1) = w_{jk}(t) + \alpha \delta_k z_j + \mu[w_{jk}(t) - w_{jk}(t - 1)],$$

or

$$\Delta w_{jk}(t + 1) = \alpha \delta_k z_j + \mu \Delta w_{jk}(t),$$

and

$$v_{ij}(t + 1) = v_{ij}(t) + \alpha \delta_j x_i + \mu[v_{ij}(t) - v_{ij}(t - 1)],$$

or

$$\Delta v_{ij}(t + 1) = \alpha \delta_j x_i + \mu \Delta v_{ij}(t),$$

where the momentum parameter μ is constrained to be in the range from 0 to 1, exclusive of the end points.

Momentum allows the net to make reasonably large weight adjustments as long as the corrections are in the same general direction for several patterns, while using a smaller learning rate to prevent a large response to the error from any one training pattern. It also reduces the likelihood that the net will find weights that are a local, but not global, minimum. When using momentum, the net is proceeding not in the direction of the gradient, but in the direction of a combination of the current gradient and the previous direction of weight correction.

As in the case of delta-bar-delta updates, momentum forms an exponentially weighted sum (with μ as the base and time as the exponent) of the past and present weight changes. Limitations to the effectiveness of momentum include the fact that the learning rate places an upper limit on the amount by which a weight can be changed and the fact that momentum can cause the weight to be changed in a direction that would increase the error [Jacobs, 1988].

Example 6.6 A backprop net with momentum for the XOR function: bipolar representation

> Using the same initial weights and architecture as in Examples 6.1–6.3 and target values of $+1.0$ and -1.0, adding momentum (0.9) with a learning rate as before (0.2) reduces the training requirements from 387 epochs to 38 epochs.

Batch updating of weights

In some cases it is advantageous to accumulate the weight correction terms for several patterns (or even an entire epoch if there are not too many patterns) and make a single weight adjustment (equal to the average of the weight correction terms) for each weight rather than updating the weights after each pattern is presented. This procedure has a smoothing effect on the correction terms. In some cases, this smoothing may increase the chances of convergence to a local minimum.

Adaptive learning rates

The standard backpropagation algorithm modifies weights in the direction of most rapid decrease of the error surface for the current weights. In general, this does not move the weights directly toward the optimal weight vector. For further elaboration, see a standard work on adaptive filter theory, such as Widrow and Stearns (1985). A variety of methods to modify the direction of the weight adjustment have been proposed and studied.

One way in which researchers have attempted to improve the speed of training for backpropagation is by changing the learning rate during training. Some adjustable learning rate algorithms are designed for specific problems, such as classification problems in which there are significantly fewer training patterns from some classes than from others. If the traditional approach, duplication or creating noisy copies of the training patterns from the underrepresented classes, is not practical, the learning rate may be increased when training patterns from the underrepresented classes are presented [DeRouin, Brown, Beck, Fausett, & Schneider, 1991].

Another type of adjustable learning rate algorithm is based on determination of the maximum safe step size at each stage of training [Weir, 1991]. Although this algorithm requires additional computations of gradients that are not calculated in standard backpropagation, it provides protection against the overshoot of the minimum error that can occur in other forms of backpropagation.

Perhaps the most extensive work in adjustable learning rate algorithms deals with "risk-taking" algorithms. Algorithms of this type have been developed by many researchers, among them Cater (1987), Fahlman (1988) and Silva and Almeida (1990). The method described here, delta-bar-delta [Jacobs, 1988], is an extension of previous work by a number of other researchers [Kesten, 1958; Saridis, 1970; Sutton, 1986; and Barto and Sutton, 1981].

Delta-Bar-Delta.   The general approach of the delta-bar-delta algorithm is to allow each weight to have its own learning rate, and to let the learning rates vary with time as training progresses. In addition to the assumptions that each weight has its own learning rate and that the learning rates vary with time, two heuristics are used to determine the appropriate changes in the learning rate for each weight. If the weight change is in the same direction (increase or decrease) for several time steps, the learning rate for that weight should be increased. (The weight change will be in the same direction if the partial derivative of the error with respect to that weight has the same sign for several time steps.) However, if the direction of the weight change (i.e., sign of the partial derivative) alternates, the learning rate should be decreased. Note that no claim is made that these heuristics will always improve the performance of the net, although in many examples they do.

The delta-bar-delta rule consists of a weight update rule and a learning rate update rule. Let $w_{ij}(t)$ denote an arbitrary weight at time t, let $\alpha_{ij}(t)$ be the learning rate for that weight at time t, and let E represent the squared error for the pattern presented at time t.

The delta-bar-delta rule changes the weights as follows:

$$w_{jk}(t + 1) = w_{jk}(t) - \alpha_{jk}(t + 1) \frac{\partial E}{\partial w_{jk}}$$

$$= w_{jk}(t) + \alpha_{jk}(t + 1)\delta_k z_j.$$

This is the standard weight change for the backpropagation, with the modification that each weight may change by a different proportion of the partial derivative of the error with respect to that weight. Thus the direction of change of the weight vector is no longer in the direction of the negative gradient.

For each output unit, we define a "delta":

$$\Delta_{jk} = \frac{\partial E}{\partial w_{jk}} = -\delta_k z_j;$$

and for each hidden unit:

$$\Delta_{ij} = \frac{\partial E}{\partial v_{ij}} = -\delta_j x_i.$$

The delta-bar-delta rule uses a combination of information about the current and past derivative to form a "delta-bar" for each output unit:

$$\overline{\Delta}_{jk}(t) = (1 - \beta)\Delta_{jk}(t) + \beta\overline{\Delta}_{jk}(t - 1),$$

and each hidden unit:

$$\overline{\Delta}_{ij}(t) = (1 - \beta)\Delta_{ij}(t) + \beta\overline{\Delta}_{ij}(t - 1);$$

the value of the parameter β must be specified by the user ($0 < \beta < 1$).

The heuristic that the learning rate should be increased if the weight changes are in the same direction on successive steps is implemented by increasing the learning rate (by a constant amount) if $\overline{\Delta}_{jk}(t - 1)$ and $\Delta_{jk}(t)$ are of the same sign. The learning rate is decreased (by a proportion γ of its current value) if $\overline{\Delta}_{jk}(t - 1)$ and $\Delta_{jk}(t)$ are of the opposite sign.

The new learning rate is given by:

$$\alpha_{jk}(t + 1) \begin{cases} \alpha_{jk}(t) + \kappa & \text{if } \overline{\Delta}_{jk}(t - 1)\Delta_{jk}(t) > 0, \\ (1 - \gamma)\alpha_{jk}(t) & \text{if } \overline{\Delta}_{jk}(t - 1)\Delta_{jk}(t) < 0, \\ \alpha_{jk}(t) & \text{otherwise.} \end{cases}$$

The values of parameters κ and γ must be specified by the user.

The comparison of methods summarized here, is presented in Jacobs (1988). Twenty-five simulations of the XOR problem (with different sets of initial weights) were performed, using a net with two input units, two hidden units, and one output unit. The XOR problem was formulated with binary input, and target values of 0.1 and 0.9. The simulations used batch updating of the weights. Successful completion was based on attaining a total squared error (per epoch) of less than 0.04, averaged over 50 epochs. The following parameter values are used.

	α	μ	κ	γ	β
Backpropagation	0.1				
Backpropagation with momentum	0.75	0.9			
Delta-bar-delta	0.8		0.035	0.333	0.7

The results summarized here show that, although the delta-bar-delta modification of backpropagation training may not always converge (22 successes for 25 simulations), when it does succeed, it does so very rapidly.

METHOD	SIMULATIONS	SUCCESSES	MEAN EPOCHS
Backpropagation	25	24	16,859.8
Backpropagation with momentum	25	25	2,056.3
Delta-bar-delta	25	22	447.3

6.2.2 Alternative Activation Functions

Customized sigmoid function for training patterns

As suggested earlier, the range of the activation function should be appropriate for the range of target values for a particular problem. The binary sigmoid function presented in Section 6.1.2, viz.,

$$f(x) = \frac{1}{1 + \exp(-x)},$$

with

$$f'(x) = f(x)[1 - f(x)],$$

can be modified to cover any range desired, to be centered at any desired value of x, and to have any desired slope at its center.

The binary sigmoid can have its range expanded and shifted so that it maps the real numbers into the interval $[a, b]$ for any a and b. To do so, given an interval $[a, b]$, we define the parameters

$$\gamma = b - a,$$

$$\eta = -a.$$

Then the sigmoid function

$$g(x) = \gamma f(x) - \eta$$

has the desired property, namely, its range is (a, b). Furthermore, its derivative also can be expressed conveniently in terms of the function value as

$$g'(x) = \frac{1}{\gamma} [\eta + g(x)][\gamma - \eta - g(x)].$$

For example, for a problem with bipolar target output, the appropriate activation function would be

$$g(x) = 2f(x) - 1,$$

with

$$g'(x) = \frac{1}{2} [1 + g(x)][1 - g(x)].$$

However, it may be preferable to scale the range of the activation function so that it extends somewhat beyond the largest and smallest target values.

The logistic sigmoid (or any other function) can be translated to the right or left by the use of an additive constant on the independent variable. However, this is not necessary, since the trainable bias serves the same role.

The steepness of the logistic sigmoid can be modified by a slope parameter σ. This more general sigmoid function (with range between 0 and 1) is

$$f(x) = \frac{1}{1 + \exp(-\sigma x)},$$

with

$$f'(x) = \sigma f(x)[1 - f(x)].$$

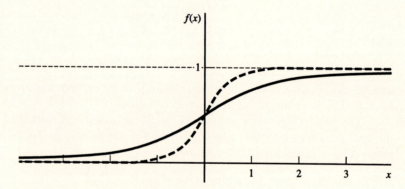

Figure 6.8   Binary sigmoid with $\sigma = 1$ and $\sigma = 3$.

The function f is illustrated in Figure 6.8 for $\sigma = 1$ and $\sigma = 3$. The slope may be determined so that the sigmoid function achieves a particular desired value for a given value of x. Combining the variations just defined gives

$$f(x) = \frac{1}{1 + \exp(-\sigma x)},$$

$$f'(x) = \sigma f(x)[1 - f(x)],$$

$$g(x) = \gamma f(x) - \eta$$

$$= \frac{\gamma}{1 + \exp(-\sigma x)} - \eta,$$

and

$$g'(x) = \frac{\sigma}{\gamma} [\eta + g(x)][\gamma - \eta - g(x)].$$

Suppose we have a logistic sigmoid with the following properties:

INPUT DATA:

centered at $x = 0,$

domain $(x_{min}, x_{max}),$

$x_{min} = -x_{max}.$

FUNCTION VALUES:

range $(a, b).$

We define

$$d = .5(x_{max} - x_{min}).$$

Then

$$c = .5(x_{max} + x_{min})   \text{(and } x_{max} = c + d).$$

Further, we define

$$\gamma = b - a,$$

$$\eta = -a.$$

Then $g(x) \to \gamma - \eta = b$ as $x \to \infty$, and $g(x) \to -\eta = a$ as $x \to -\infty$.

The binary sigmoid $f(x) = 1/(1 + \exp(-x))$ often used for input values between -1 and 1, has a value of approximately 0.75 when $x = 1$ and approximately 0.25 when $x = -1$. Therefore, it may be reasonable to choose the slope parameter σ for a more general sigmoid so that the function value when x has its largest value is approximately three-fourths of the distance from its smallest value, a, to its largest value, b. For

$$g(x) = \frac{\gamma}{1 + \exp(-\sigma x)} - \eta,$$

this condition becomes

$$g(x_{max}) = \frac{3b + a}{4}.$$

Solving for σ gives

$$\sigma = -\frac{1}{x_{max}} \ln \left[\frac{4\gamma}{3b + a + 4\eta} - 1 \right]$$

$$= -\frac{1}{x_{max}} \ln \left[\frac{4(b - a)}{3(b - a)} - 1 \right]$$

$$= -\frac{1}{x_{max}} \ln \left(\frac{1}{3} \right)$$

$$= \frac{\ln(3)}{x_{max}}.$$

Adaptive slope for sigmoid

The preceding algorithms have been presented in terms of a single activation function for the entire net (except for the input units that use the identity function as their activation function). The variations on the "original" sigmoid function involve setting one or more parameters to adjust the shape of the sigmoid to the range of input and output values for a particular application. Perhaps the parameter for which the appropriate value is most difficult to determine *a priori* is the slope parameter σ. In this section, we show that the slope can be adjusted during training, in a manner very similar to that used for adjusting the weights. The process is illustrated for a general activation function $f(x)$, where we consider the net input x to be of the form

$$x = \sigma_k y\_in_k$$

for an output unit Y_k, or

$$x = \sigma_j x\_in_j$$

for the hidden unit Z_j. However, since

$$y\_in_k = \sum_j z_j w_{jk},$$

the activation function for an output unit depends on both weights on connections coming into the unit and on the slope σ_k for that unit. Similar expressions apply for the hidden units. Note that each unit can have its own slope parameter, but we are assuming for simplicity that all units have the same form of the activation function.

Letting the net adjust the slopes allows a different value of σ to be used for each of the hidden units and for each of the output units. This will often improve the performance of the net. In fact, the optimal value of the slope for any unit may vary as training progresses.

With the abbreviations

$$y_k = f(\sigma_k y\_in_k)$$

and

$$z_j = f(\sigma_j z\_in_j)$$

to simplify the notation, the derivation is essentially the same as for standard backpropagation; it is given in Section 6.3.1. As in the standard backpropagation algorithm, it is convenient to define

$$\delta_k = [t_k - y_k]f'(y_k)$$

and

$$\delta_j = -\sum_k \delta_k \sigma_k w_{jk} f'(z_j).$$

The update for the weights to the output units are

$$\Delta w_{jk} = \alpha \delta_k \sigma_k z_j$$

and for the weights to the hidden units are

$$\Delta v_{ij} = \alpha \delta_j \sigma_j x_i.$$

Similarly, the updates for the slopes on the output units are

$$\Delta \sigma_k = \alpha \delta_k y\_in_k$$

and for the slopes on the hidden units are

$$\Delta \sigma_j = \alpha \delta_j z\_in_j.$$

See Tepedelenliogu, Rezgui, Scalero, and Rosario, 1991 for a related discussion (using the bipolar sigmoid) and sample results.

Another sigmoid function

The arctangent function is also used as an activation function for backpropagation nets. It saturates (approaches it asymptotic values) more slowly than the hyperbolic tangent function, $\tanh(x)$, or bipolar sigmoid. Scaled so that the function values range between -1 and $+1$, the function is

$$f(x) = \frac{2}{\pi} \arctan(x),$$

with derivative

$$f'(x) = \frac{2}{\pi} \frac{1}{1 + x^2}.$$

Nonsaturating activation functions

Hecht-Nielsen [1990] uses the identity function as the activation function on output units, especially if the target values are continuous rather than binary or bipolar. For some applications, where saturation is not especially beneficial, a nonsaturating activation function may be used. One suitable example is

$$f(x) = \begin{cases} \log(1 + x) & \text{for } x > 0 \\ -\log(1 - x) & \text{for } x < 0. \end{cases}$$

Note that the derivative is continuous at $x = 0$:

$$f'(x) = \begin{cases} \dfrac{1}{1 + x} & \text{for } x > 0 \\[2ex] \dfrac{1}{1 - x} & \text{for } x < 0. \end{cases}$$

This function can be combined with the identity function on the output units in some applications.

Example 6.7 A backprop net for the Xor function: log activation function

Fewer epochs of training are required for the Xor problem (with either standard bipolar or modified bipolar representation) when we use the logarithmic activation function in place of the bipolar sigmoid (see Examples 6.2 and 6.3). The following table compares the two functions with respect to the number of epochs of they require:

PROBLEM	LOGARITHMIC	BIPOLAR SIGMOID
standard bipolar Xor	144 epochs	387 epochs
modified bipolar Xor (targets of $+0.8$ or -0.8)	77 epochs	264 epochs

Example 6.8 A backprop net for the product of sine functions

A neural net with one hidden layer can be trained to map input vectors (x_1, x_2) to the corresponding output value y as follows:

Input points (x_1, x_2) range between 0 and 1 in steps of 0.2;
the corresponding target output is given by $y = \sin(2\,\pi x_1) \cdot \sin(2\,\pi x_2)$.

This is a surprisingly difficult problem for standard backpropagation. In this example, we used the logarithmic activation function for the hidden units and the identity function for the output units. With a learning rate of .05, the net achieved a mean squared error of 0.024 in 5,000 epochs. The results are shown in the following table, with target values given in italics and actual results from the net in bold (the example used one hidden layer with 10 hidden units):

x_2						
1.0	0.00	0.00	0.00	0.00	0.00	0.00
	−0.01	**−0.05**	**−0.12**	**−0.09**	**−0.01**	**−0.00**
0.8	0.00	−0.90	−0.56	0.56	0.90	0.00
	−0.03	**−0.91**	**−0.55**	**−0.53**	**−0.97**	**−0.02**
0.6	0.00	−0.56	−0.36	0.36	0.56	0.00
	−0.00	**−0.59**	**−0.32**	**−0.37**	**−0.51**	**−0.03**
0.4	0.00	0.56	0.36	−0.36	−0.56	0.00
	−0.01	**−0.57**	**−0.33**	**−0.35**	**−0.55**	**−0.00**
0.2	0.00	0.90	0.56	−0.56	−0.90	0.00
	−0.02	**−0.84**	**−0.57**	**−0.57**	**−0.89**	**−0.01**
0.0	0.00	0.00	0.00	0.00	0.00	0.00
	−0.02	**−0.02**	**−0.02**	**−0.02**	**−0.03**	**−0.02**
x_1	0.0	0.2	0.4	0.6	0.8	1.0

Nonsigmoid activation functions

Radial basis functions, activation functions with a local field of response, are also used in backpropagation neural nets. The response of such a function is non-negative for all values of x; the response decreases to 0 as $|x - c| \to \infty$. A common example is the Gaussian function illustrated in Figure 6.9. The function is defined as

$$f(x) = \exp(-x^2);$$

its derivative is given by

$$f'(x) = -2x \exp(-x^2) = -2xf(x).$$

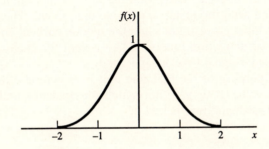

Figure 6.9   Gaussian activation function.

Radial basis function networks (RBFN) can be used for approximating functions and recognizing patterns [Park & Sandberg, 1991; Moody & Darken, 1989; Leonard, Kramer, & Ungar, 1992]. *Gaussian potential functions* [Lee & Kil, 1991] are also used in networks known as *regularization networks* [Poggio, 1990]. Regularization theory deals with techniques for transforming ill-posed problems, in which there is insufficient information in the data, into well-posed problems by introducing constraints [Hertz, Krogh, & Palmer, 1991]. The *probabilistic neural net,* which we discuss in Chapter 7, uses Gaussian potential functions for its activation functions.

6.2.3 Strictly Local Backpropagation

Backpropagation has been criticized by some as inappropriate for neurological simulation because it lacks biological plausibility. One of the arguments has been that the backpropagation algorithm requires sharing of information among processors, which is in violation of accepted theories on the functioning of biological neurons. The modified version [D. Fausett, 1990] described here alleviates this deficiency.

In this version of backpropagation, the net is viewed as consisting of three types of units: cortical units, synaptic units, and thalamic units. Each type performs certain calculations with information that is strictly local to it. The action of each during the feedforward phase of the strictly local backpropagation training algorithm is described next.

A cortical unit sums its inputs and sends the resulting value as a signal to the next unit above it. By contrast, input cortical units receive only one input signal, so no summation is necessary. Hidden cortical units sum their input signals and broadcast the resulting signal to each synaptic unit connected to them above. Output cortical units also sum their input signals, but each output cortical unit is connected to only one synaptic unit above it.

The function of the synaptic units is to receive a single input signal (from a cortical unit), apply an activation function to that input, multiply the result by a weight, and send the result to a single unit above. The input synaptic units (between the input cortical units and the hidden cortical units) use the identity function as their activation function. The weight for each output synaptic unit is 1; each sends its signal to a thalamic unit.

The purpose of the thalamic unit is to compare the computed output with the target value. If they do not match, the thalamic unit sends an error signal to the output synaptic unit below it.

It is during the backpropagation phase of the traditional backpropagation algorithm that information must be shared between units. The difficulty occurs in the calculation of the weight update terms

$$\Delta w_{jk} = \alpha \delta_k z_j$$

and

$$\Delta v_{ij} = \alpha \delta_j x_i.$$

The weight update term Δw_{jk} requires information from both output unit k and hidden unit j, thus violating the requirement for local computation. A similar comment applies to Δv_{ij}, which requires information from hidden unit j and input unit i.

The action of the thalamic, synaptic, and cortical units in strictly local backpropagation avoids this criticism of the traditional algorithm. Each output synaptic unit receives the error signal from the thalamic unit above it, multiplies this signal by its weight which is 1, and multiplies again by the derivative of its activation function. The result, δ_k, is sent to the output cortical unit below.

Each output cortical unit sends its lone input signal construed as a sum δ_k to the hidden synaptic units below it. Each hidden synaptic unit computes the weight update term to be used later (the product of its input signal δ_k, its activation, and a learning rate). It then multiplies its input signal by its weight and by the derivative of its activation function and sends the resulting value to the hidden cortical unit below it.

Next, the hidden cortical unit sums its input signals and sends the resulting value to the input synaptic unit below it. The input synaptic unit then computes its weight update term to be used later (the product of its input signal δ_k, its activation, and a learning rate).

In addition to addressing some of the biological implausibility objections to the traditional backpropagation algorithm, the strictly local backpropagation algorithm expands the computational power of the net by allowing even more variation in the activation functions used. Since the activation function now "lives" on a synaptic unit, there may be as many different functions as there are weights in the net. In the case in which the activation functions differ only in the value of their slope parameters, these parameters can be adjusted (trained) by a process similar to that used for the weights. The derivation is essentially the same as that for adjusting the slope parameters when each hidden or output unit can have a different form of the activation function. In the strictly local algorithm, each slope parameter would be double indexed to correspond to a particular synaptic unit.

Architecture

The architecture of a strictly local backpropagation neural net is illustrated in Figure 6.10.

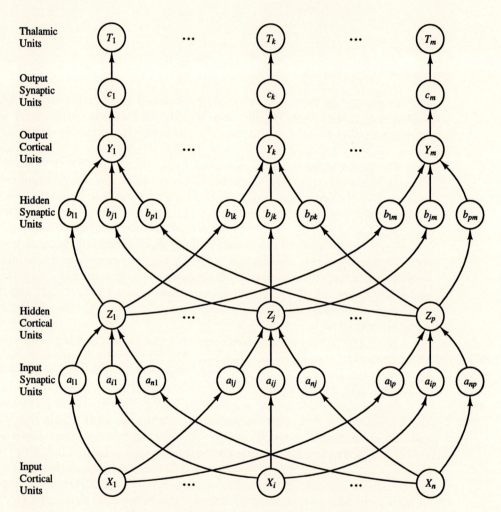

Figure 6.10 Strictly local backpropagation neural net.

Algorithm

The following outline of the computations needed to train a feedforward backpropagation neural net is presented to facilitate comparison between the standard and strictly local backpropagation algorithms. It illustrates the fact that the computations are the same; they are just arranged differently.

STANDARD BACKPROPAGATION	**STRICTLY LOCAL BACKPROPAGATION**
Feedforward	*Feedforward*
Input unit:	Input cortical unit:
Receives input signal	Receives input signal
	Input synaptic unit:
Path from input to hidden unit:	
Multiples by weights	Multiplies by weight
Hidden unit:	Hidden cortical unit:
Sums input signals	Sums input signals
Applies activation function	
	Hidden synaptic unit:
	Applies activation function
Path from hidden to output unit:	Multiplies by weight
Multiplies by weights	Output cortical unit:
Output unit:	Sums input signals
Sums input signals	
Applies activation function	
	Output synaptic unit:
	Applies activation function
Backpropagation of error	*Backpropagation of error*
Output unit:	Thalamic unit:
Calculates error	Calculates error
	Output synaptic unit:
Multiplies by f'	
	Multiplies by f'
Path from output to hidden unit:	Output cortical unit:
Multiplies by weights	Sends input δ_k to units below
	Hidden synaptic unit:
Calculates weight correction	Calculates weight correction
	Multiplies δ_k by weight
Hidden unit:	Multiplies by f'
Sums input from units above	Hidden cortical unit:
Multiplies by f'	Sums input from units above
Path from hidden to input unit:	
Calculates weight correction	Input synaptic unit:
	Calculates weight correction
Weight update	*Weight update*

6.2.4 Number of Hidden Layers

Although a single hidden layer is sufficient to solve any function approximation problem, some problems may be easier to solve using a net with two hidden layers. For example, bounds on the number of samples needed for successful classification of M clusters have been found for a net with two hidden layers [Mehrotra, Mohan, & Ranka, 1991]. In such a net, the first hidden layer often serves to partition the input space into regions and the units in the second hidden layer represent a cluster of points. If these clusters are separable, the output units can easily make the final classification. In this scenario, the number of boundary samples is of the order $\min(n, p) \cdot M$, where n is the dimension of the input space (the number of input units) and p is the number of hidden nodes.

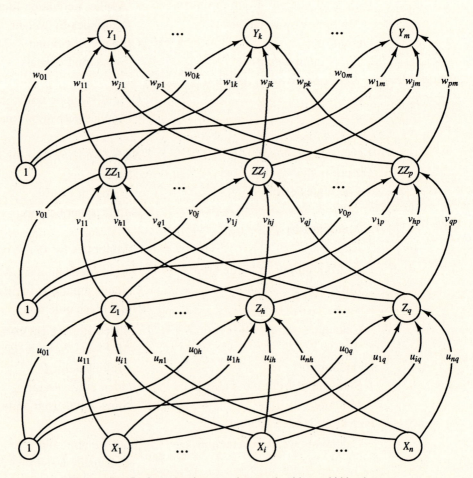

Figure 6.11 Backpropagation neural network with two hidden layers.

Architecture—two hidden layers

A multilayer neural network with two layers of hidden units (the Z units and the ZZ units) is shown in Figure 6.11. The output units (Y units) and hidden units may also have biases (as shown). The bias on a typical output unit Y_k is denoted w_{0k}; the bias on a typical hidden unit Z_j is denoted v_{0j}. These bias terms act like weights on connections from "units" whose output is always 1. (These "units" are not shown.) The main difficulty in generalizing the algorithm for this net is in bookkeeping (naming the units, weights, etc.). It is possible to use multiple indexing for the quantities dealing with the hidden layers, with the additional index denoting to which hidden layer an index refers.

Algorithm: two hidden layers

During feedforward, each input unit receives an input signal and broadcasts this signal to the each of the hidden units, $Z_1, \ldots, Z_q$, in the first layer. (X_i is a typical input unit.) Each of these hidden units then computes its activation and sends its signal to the hidden units, $ZZ_1, \ldots, ZZ_p$, in the second layer. (Z_h is a typical unit in the first hidden layer.) Next, each hidden unit in the second layer computes its activation and sends its signal to the output units. Finally, each output unit computes its activation (y_k is the activation of a typical output unit Y_k) to form the response of the net for the given input pattern.

   During training, each output unit compares its computed activation y_k with its target value t_k to determine the error associated with that unit. Then, based on this error, the factor δ_k is computed ($k = 1, \ldots, m$). δ_k is used to distribute the information on the error at output unit Y_k back to all units in the next lower layer. It is also used later to update the weights between the output and the second hidden layer. The factor δ_j ($j = 1, \ldots, p$) is computed for hidden unit ZZ_j and is then used to distribute the information on the error back to all units in the previous layer (units $Z_1, \ldots, Z_h, \ldots, Z_q$). It is also used later to update the weights between the second hidden layer and the first hidden layer. The factor δ_h ($h = 1, \ldots, q$) is computed for hidden unit Z_h. It is not necessary to propagate the error back to the input layer, but δ_h is used to update the weights between the first hidden layer (with units $Z_1, \ldots, Z_h, \ldots, Z_q$) and the input layer.

   After all of the δ factors have been determined, the weights for all layers are adjusted simultaneously. The adjustment to the weight w_{jk} (from hidden unit ZZ_j to output unit Y_k) is based on the factor δ_k and the activation of the hidden unit ZZ_j. The adjustment to the weight v_{hj} (from hidden unit Z_h to hidden unit ZZ_j) is based on the factor δ_j and the activation of unit Z_h. The adjustment to the weight u_{ih} (from input unit X_i to hidden unit Z_h) is based on the factor δ_h and the activation of the input unit.

   The steps for standard backpropagation for a net with two hidden layers are summarized next. The form of the activation function and its derivative are not explicitly specified; appropriate choices are as discussed for standard backpropagation.

Feedforward.

Each input unit $(X_i, i = 1, \ldots, n)$:

    broadcasts input signal to hidden units.

Each hidden unit $(Z_h, h = 1, \ldots, q)$:

    computes input signal

$$z\_in_h = u_{0h} + \sum_{i=1}^{n} x_i u_{ih},$$

applies activation function to compute output signal

$$z_h = f(z\_in_h),$$

and sends its output signal to the units in the second hidden layer.

Each hidden unit $(ZZ_j, j = 1, \ldots, p)$:

    computes input signal

$$zz\_in_j = v_{0j} + \sum_{h=1}^{n} z_h v_{hj},$$

applies activation function to compute output signal

$$zz_j = f(zz\_in_j).$$

and sends its output signal to output units.

Each output unit $(Y_k, k = 1, \ldots, m)$:

    sums weighted input signal

$$y\_in_k = w_{0k} + \sum_{j=1}^{p} zz_j w_{jk}$$

and applies activation function to compute its output signal

$$y_k = f(y\_in_k).$$

Backpropagation of error.

Each output unit $(Y_k, k = 1, \ldots, m)$:

    calculates its error

$$e_k = (t_k - y_k).$$

for the current training pattern, multiplies by derivative of activation function (expressed in terms of y_k) to get

$$\delta_k = e_k f'(y\_in_k),$$

calculates weight correction term (used to update w_{jk} later)

$$\Delta w_{jk} = \alpha \delta_k zz_j,$$

calculates bias correction term (used to update w_{0k} later)

$$\Delta w_{0k} = \alpha \delta_k,$$

and sends δ_k to hidden units $(ZZ_j, j = 1, \dots, p)$.
Each hidden unit $(ZZ_j, j = 1, \dots, p)$:
sums weighted input from units in layer above to get

$$\delta\_in_j = \sum_{k=1}^{m} \delta_k w_{jk},$$

multiplies by derivative of its activation function (expressed in terms of zz_j) to get

$$\delta_j = \delta\_in_j f'(zz\_in_j),$$

calculates weight correction term (used to update v_{ij} later)

$$\Delta v_{ij} = \alpha \delta_j x_i,$$

calculates bias correction term (used to update v_{0j} later)

$$\Delta v_{0j} = \alpha \delta_j,$$

and sends δ_j to hidden units $(Z_h, h = 1, \dots, q)$.
Each hidden unit $(Z_h, h = 1, \dots, q)$:
sums weighted input from units in layer above to get

$$\delta\_in_h = \sum_{j=1}^{p} \delta_j v_{hj},$$

multiplies by derivative of its activation function (expressed in terms of z_h) to get

$$\delta_h = \delta\_in_h f'(z\_in_h),$$

calculates weight correction term (used to update v_{ij} later)

$$\Delta u_{ih} = \alpha \delta_h x_i,$$

and calculates bias correction term (used to update v_{0j} later)

$$\Delta v_{0j} = \alpha \delta_j.$$

Update Weights and Biases.
For each output unit $(j = 0, \dots, p; k = 1, \dots, m)$:

$$w_{jk}(\text{new}) = w_{jk}(\text{old}) + \Delta w_{jk}.$$

For each hidden unit ZZ_j $(h = 0, \dots, q; j = 1, \dots, p)$:

$$v_{hj}(\text{new}) = v_{hj}(\text{old}) + \Delta v_{hj}.$$

For each hidden unit Z_h $(i = 0, \dots, n; h = 1, \dots, q)$:

$$u_{ih}(\text{new}) = u_{ih}(\text{old}) + \Delta u_{ih}.$$

6.3 THEORETICAL RESULTS

6.3.1 Derivation of Learning Rules

Standard backpropagation

As in the derivation of the delta rule in Chapter 3, we shall denote by w_{JK} the weight between the hidden unit Z_J and the output unit Y_K; these units are considered arbitrary, but fixed. The subscripts IJ are used analogously for the weight between input unit X_I and hidden unit, Z_J. With this notation, the corresponding lowercase letters can serve as summation indices in the derivation of the weight update rules. We shall return to the more common lowercase indices following the derivation.

The derivation is given here for an arbitrary activation function $f(x)$. The derivative of the activation function is denoted by f'. The dependence of the activation on the weights results from applying the activation function f to the net input

$$y\_in_K = \sum_j z_j w_{jK}$$

to find $f(y\_in_K)$.

The error (a function of the weights) to be minimized is

$$E = .5 \sum_k [t_k - y_k]^2. \qquad \text{\textit{summed over outputs}}$$

By use of the chain rule, we have

$$\frac{\partial E}{\partial w_{JK}} = \frac{\partial}{\partial w_{JK}} .5 \sum_k [t_k - y_k]^2$$

$$= \frac{\partial}{\partial w_{JK}} .5[t_K - f(y\_in_K)]^2$$

$$= -[t_K - y_K] \frac{\partial}{\partial w_{JK}} f(y\_in_K)$$

$$= -[t_K - y_K] f'(y\_in_K) \frac{\partial}{\partial w_{JK}} (y\_in_K)$$

$$= -[t_K - y_K] f'(y\_in_K) z_J.$$

It is convenient to define δ_K:

$$\delta_K = [t_K - y_K] f'(y\_in_K).$$

For weights on connections to the hidden unit Z_J:

$$\frac{\partial E}{\partial v_{IJ}} = -\sum_k [t_k - y_k] \frac{\partial}{\partial v_{IJ}} y_k$$

$$= -\sum_k [t_k - y_k] f'(y\_in_k) \frac{\partial}{\partial v_{IJ}} y\_in_k$$

$$= -\sum_k \delta_k \frac{\partial}{\partial v_{IJ}} y\_in_k$$

$$= -\sum_k \delta_k w_{Jk} \frac{\partial}{\partial v_{IJ}} z_J$$

$$= -\sum_k \delta_k w_{Jk} f'(z\_in_J)[x_I].$$

Define:

$$\delta_J = -\sum_k \delta_k w_{Jk} f'(z\_in_J)$$

Thus, the updates for the weights to the output units (returning to the more common lower case subscripts) are given by:

$$\Delta w_{jk} = -\alpha \frac{\partial E}{\partial w_{jk}}$$

$$= \alpha [t_k - y_k] f'(y\_in_k) z_j$$

$$= \alpha \delta_k z_j;$$

and for the weights to the hidden units:

$$\Delta v_{ij} = -\alpha \frac{\partial E}{\partial v_{ij}}$$

$$= \alpha f'(z\_in_j) x_i \sum_k \delta_k w_{jk},$$

$$= \alpha \delta_j x_i.$$

Generalized backpropagation—adaptive slope parameters

The derivation of the weight update rules for backpropagation with adaptive slope parameters is similar to that given in the previous section for standard back-propagation. As explained there, we use capitol letters for subscripts on the fixed, but arbitrary units and weights during the derivation to distinguish them from the corresponding indices of summation, returning to the more common notation (lowercase subscripts) for the final formulas. This derivation uses an arbitrary activation function $f(x)$; we consider the net input, x, to be of the form

$$x = \sigma_K y\_in_K$$

for an output unit, Y_K or

$$x = \sigma_J z\_in_J$$

for the hidden unit Z_J. Thus, since

$$y\_in_K = \sum_j z_j w_{jK},$$

the activation function for an output unit depends on both weights on connections coming into the unit, and on the slope parameter, σ_K, for that unit. Similar expressions apply for the hidden units. Note that each unit can have its own slope parameter, but we are assuming, for simplicity, that all units have the same form of the activation function. It is easy to generalize the derivation that follows to remove this assumption.

The error (a function of both the slope parameters and the weights) to be minimized is

$$E = .5 \sum_k [t_k - y_k]^2$$

By the use of the chain rule, we find that

$$\frac{\partial E}{\partial w_{JK}} = \frac{\partial}{\partial w_{JK}} .5 \sum_k [t_k - y_k]^2$$

$$= \frac{\partial}{\partial w_{JK}} .5[t_K - f(\sigma_K y\_in_K)]^2$$

$$= -[t_K - y_K] \frac{\partial}{\partial w_{JK}} f(\sigma_K y\_in_K)$$

$$= -[t_K - y_K]f'(\sigma_K y\_in_K) \frac{\partial}{\partial w_{JK}} (\sigma_K y\_in_K)$$

$$= -[t_K - y_K]f'(\sigma_K y\_in_K)\sigma_K z_J.$$

Similarly,

$$\frac{\partial E}{\partial \sigma_K} = -[t_K - y_K]f'(\sigma_K y\_in_K) \frac{\partial}{\partial \sigma_K} (\sigma_K y\_in_K)$$

$$= -[t_K - y_K]f'(\sigma_K y\_in_K)y\_in_K.$$

As in the standard backpropagation algorithm, it is convenient to define

$$\delta_K = [t_K - y_K]f'(\sigma_K y\_in_K).$$

For weights on connections to a hidden unit (Z_J),

$$\frac{\partial E}{\partial v_{IJ}} = -\sum_k [t_k - y_k] \frac{\partial}{\partial v_{IJ}} y_k$$

$$= -\sum_k [t_k - y_k] f'(\sigma_k y\_in_k) \frac{\partial}{\partial v_{IJ}} \sigma_k y\_in_k$$

$$= -\sum_k \delta_k \sigma_k \frac{\partial}{\partial v_{IJ}} y\_in_k$$

$$= -\sum_k \delta_k \sigma_k w_{Jk} \frac{\partial}{\partial v_{IJ}} z_J$$

$$= -\sum_k \delta_k \sigma_k w_{Jk} f'(\sigma_J z\_in_J) \sigma_J [x_I].$$

Similarly,

$$\frac{\partial E}{\partial \sigma_J} = -\sum_k [t_k - y_k] \frac{\partial}{\partial \sigma_J} y_k$$

$$= -\sum_k [t_k - y_k] f'(\sigma_k y\_in_k) \frac{\partial}{\partial \sigma_J} \sigma_k y\_in_k$$

$$= -\sum_k \delta_k \sigma_k \frac{\partial}{\partial \sigma_J} y\_in_k$$

$$= -\sum_k \delta_k \sigma_k w_{Jk} \frac{\partial}{\partial \sigma_J} z_J$$

$$= -\sum_k \delta_k \sigma_k w_{Jk} f'(\sigma_J z\_in_J) z\_in_J.$$

Now we define

$$\delta_J = -\sum_k \delta_k \sigma_k w_{Jk} f'(\sigma_J z\_in_J).$$

Returning to the usual lowercase subscripts, we have, for the updates for the weights to the output units,

$$\Delta w_{jk} = -\alpha \frac{\partial E}{\partial w_{jk}}$$

$$= \alpha [t_k - y_k] f'(\sigma_k y\_in_k) \sigma_k z_j$$

$$= \alpha \delta_k \sigma_k z_j;$$

and for the weights to the hidden units;

$$\Delta v_{ij} = -\alpha \frac{\partial E}{\partial v_{ij}}$$

$$= \alpha \sigma_j f'(\sigma_j z\_in_j) x_i \sum_k \delta_k \sigma_k w_{jk},$$

$$= \alpha \delta_j \sigma_j x_i.$$

Similarly, the updates for the slope parameters on the output units are

$$\Delta \sigma_k = -\alpha \frac{\partial E}{\partial \sigma_k}$$

$$= -\alpha [t_k - y_k] f'(\sigma_k y\_in_k) y\_in_k$$

$$= \alpha \delta_k y\_in_k;$$

and for the slope parameters on the hidden units,

$$\Delta \sigma_j = -\alpha \frac{\partial E}{\partial \sigma_j}$$

$$= -\alpha \sum_k \delta_k \sigma_k w_{jk} f'(\sigma_j z\_in_j) z\_in_j,$$

$$= \alpha \delta_j z\_in_j.$$

6.3.2 Multilayer Neural Nets as Universal Approximators

One use of a neural network is to approximate a continuous mapping f. Since there are very simple mappings that a single-layer net cannot represent, it is natural to ask how well a multilayer net can do. The answer is given by the "Kolmogorov mapping neural network existence theorem," which states that a feedforward neural network with three layers of neurons (input units, hidden units, and output units) can represent any continuous function exactly [Kolmogorov, 1957; Sprecher, 1965]. The following statements of the Kolmogorov and Sprecher theorems are based on the presentation by Funahashi (1989). The Hecht-Nielsen theorem, casting the Sprecher theorem in the terminology of neural nets, is as presented in Hecht-Nielsen (1987c).

Kolmogorov theorem

Any continuous function $f(x_1, \ldots, x_n)$ of several variables defined on I^n ($n \geq 2$), where $I = [0, 1]$, can be represented in the form

$$f(x) = \sum_{j=1}^{2n+1} \chi_j \left(\sum_{i=1}^{n} \psi_{ij}(x_i) \right),$$

where χ_j and ψ_{ij} are continuous functions of one variable and ψ_{ij} are monotonic functions that do not depend on f.

Sprecher theorem

For each integer $n \geq 2$, there exists a real, monotonically increasing function $\psi(x)$, $\psi:[0, 1] \rightarrow [0, 1]$, depending on n and having the following property: For each preassigned number $\delta > 0$, there is a rational number ϵ, $0 < \epsilon < \delta$, such that every real continuous function of n variables, $f(x)$, defined on I^n, can be represented as

$$f(x) = \sum_{j=1}^{2n+1} \chi \left(\sum_{i=1}^{n} \lambda^i \psi(x_i + \epsilon(j - 1)) + j - 1 \right),$$

where the function χ is real and continuous and λ is a constant that is independent of f.

Hecht-Nielsen theorem

Given any continuous function $f:I^n \rightarrow R^m$, where I is the closed unit interval $[0, 1]$, f can be represented exactly by a feedforward neural network having n input units, $2n + 1$ hidden units, and m output units.

The input units broadcast the input signal to the hidden units. The activation function for the jth hidden unit is $z_j = \left(\sum_{i=1}^{n} \lambda^i \psi(x_i + \epsilon j) + j \right)$ where the real constant λ and the continuous, real, monotonically increasing function ψ are independent of f (although they do depend on n) and the constant ϵ satisfies the conditions of the Sprecher theorem. The activation function for the output units is $y_k = \sum_{j=1}^{2n+1} g_k z_j$, where the functions g_k are real and continuous (and depend on f and ϵ).

Hornik, Stinchcombe, and White (1989) extended the foregoing results, in which the activation functions of at least some of the units depend on the function being approximated, to show that multilayer feedforward networks with arbitrary squashing functions can approximate virtually any function of interest (specifically, any Borel measurable function from one finite dimensional space to another finite dimensional space). A squashing function is simply a nondecreasing function $f(x)$ such that $0 \leq f(x) \leq 1$ for all x, $f(x) \rightarrow 0$ as $x \rightarrow -\infty$, and $f(x) \rightarrow 1$ as $x \rightarrow \infty$. These results require a sufficiently large number of hidden units; the authors do not address the number of units needed. Hornik, Stinchcombe, and White (1990) have also shown that with fairly mild assumptions and little additional work, a neural network can approximate both a function and its derivative (or generalized derivative). This is useful for applications such as a robot learning smooth movement [Jordan, 1989].

White (1990) has shown that the weights needed to achieve the approximation can be learned; i.e., the probability of the network error exceeding any specified level goes to zero as the size of the training set increases. In addition, the complexity of the net increases with the size of the training set.

It is not surprising that there has been a great deal of interest in determining the types of activation functions required to be assured that a multilayer neural net can approximate an arbitrary function to a specified accuracy. Kreinovich (1991) has shown that a neural network consisting of linear neurons and neurons with a single arbitrary (smooth) nonlinear activation function can represent any function to any specified (nonzero) precision. However, he assumes an unlimited number of hidden layers. Geva and Sitte (1992) have demonstrated a constructive method for approximating multivariate functions using multilayer neural networks. They combine two sigmoid functions to produce an activation function that is similar to a Gaussian potential function [Lee & Kil, 1991] or radial basis function [Chen, Cowan, & Grant, 1991] (which only respond to local information).

6.4 SUGGESTIONS FOR FURTHER STUDY

6.4.1 Readings

HECHT-NIELSEN, R. (1989). "Theory of the Backpropagation Neural Network." *International Joint Conference on Neural Networks, Washington, DC*, I-593:605.

McCLELLAND, J. L., & D. E. RUMELHART. (1988). *Explorations in Parallel Distributed Processing*, Cambridge, MA: MIT Press.

NGUYEN, D., & B. WIDROW. (1990). "Improving the Learning Speed of Two-Layer Neural Networks by Choosing Initial Values of the Adaptive Weights." *International Joint Conference on Neural Networks, San Diego, CA*, III:21–26.

RUMELHART, D. E., G. E. HINTON, & R. J. WILLIAMS. (1986b). "Learning Representations by Back-Propagating Error." *Nature*, 323:533–536. Reprinted in Anderson & Rosenfeld (1988), pp. 696–699.

RUMELHART, D. E., J. L. MCCLELLAND, & the PDP RESEARCH GROUP. (1986). *Parallel Distributed Processing, Explorations in the Microstructure of Cognition; Vol. 1: Foundations*, Cambridge, MA: MIT Press.

6.4.2 Exercises

Exercises 6.1–6.5 use the neural net illustrated in Figure 6.12.

6.1 Find the new weights when the net illustrated in Figure 6.12 is presented the input pattern (0, 1) and the target output is 1. Use a learning rate of $\alpha = 0.25$, and the binary sigmoid activation function.

6.2 Find the new weights when the net illustrated in Figure 6.12 is presented the input pattern $(-1, 1)$ and the target output is 1. Use a learning rate of $\alpha = 0.25$, and the bipolar sigmoid activation function.

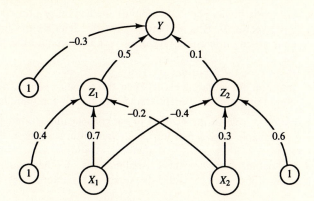

Figure 6.12   Neural network for Exercises 6.1–6.5.

6.3 Find the new weights when the net illustrated in Figure 6.12 is presented the input pattern (0, 1) and the target output is 0.8. Use a learning rate of $\alpha = 0.25$, and the binary sigmoid activation function.

6.4 Find the new weights when the net illustrated in Figure 6.12 is presented the input pattern (−1, 1) and the target output is 0.8. Use a learning rate of $\alpha = 0.25$, and the bipolar sigmoid activation function.

6.5 Repeat Exercises 6.1–6.4 using a slope parameter of $\sigma = 3.0$. Does this increase, or decrease the amount of learning (size of the weight changes)?

6.6 A neural network is being trained on the data for XOR problem. The architecture and the values of the weights and biases are shown in Figure 6.13.

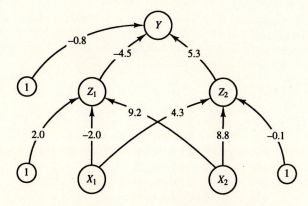

Figure 6.13   Neural network for Exercise 6.6.

a. Using the binary sigmoid, compute the activations for each of the units when the input vector $(0, 1)$ is presented. Find the delta factors for the output and hidden units. Using a learning rate of $\alpha = 0.25$, compute the weight corrections. Find the new weights (and biases).

b. Repeat for the input vector $(1, 0)$.

c. Interpret the differences between the weight changes on the connection to the output unit and the weight changes to the hidden units in parts a and b.

6.7 Explore the role of the weights in backpropagation training by finding weights that are reasonable in size, but for which very little learning will occur. For example, in Exercise 6.2, if $v_{02} + v_{22} \approx 0$, then $z_2 \approx 0$ so that $\Delta w_{21} \approx 0$, even if an error occurs at the output unit. Are there combinations of other weights for which very little learning will occur? Consider the situation for Exercise 6.1.

6.4.3 Projects

6.1 Code a computer program to implement a backpropagation neural network with one hidden layer. Use a bias on each hidden unit and each output unit. Use the bipolar sigmoid activation function. For each test case, print the initial weights, final weights, learning rate, number of training epochs, and network response to each input pattern at the end of training. The training data are given in the following table:

BIPOLAR XOR

$s(1) = (1, -1)$ $t(1) = 1$
$s(2) = (-1, 1)$ $t(2) = 1$
$s(3) = (1, 1)$ $t(3) = -1$
$s(4) = (-1, -1)$ $t(4) = -1$

Use initial weights distributed randomly on $(-0.5, 0.5)$, and a learning rate of (i) 0.05, (ii) 0.25, and (iii) 0.5. For each learning rate, perform 1,000, 10,000, and 25,000 epochs of training (using the same initial weights in each case). Use two input units, two hidden units, and one output unit.

6.2 Code a computer program to implement a backpropagation neural network with one hidden layer. Use a bias on each hidden unit and each output unit; use 10 input units, 4 hidden units, and 2 output units. Use the bipolar sigmoid activation function.

The input patterns are:

$s(1) =$ 0	1	2	3	4	5	6	7	8	9
$s(2) =$ 9	8	7	6	5	4	3	2	1	0
$s(3) =$ 0	9	1	8	2	7	3	6	4	5
$s(4) =$ 4	5	6	3	2	7	1	8	0	9
$s(5) =$ 3	8	2	7	1	6	0	5	9	4
$s(6) =$ 1	6	0	7	4	8	3	9	2	5
$s(7) =$ 2	1	3	0	4	9	5	8	6	7
$s(8) =$ 9	4	0	5	1	6	2	7	3	8

The corresponding ouput (target) patterns are:

t(1) =	−1	−1	t(5) =	1	1
t(2) =	1	1	t(6) =	1	−1
t(3) =	−1	1	t(7) =	−1	1
t(4) =	1	−1	t(8) =	−1	−1

Use random initial weights distributed on $(-0.5, 0.5)$ and a learning rate of (i) 0.05 and (ii) 0.5. For each learning rate, perform 5,000 and 50,000 epochs of training (using the same initial weights in each case).

6.3 Code a backpropagation network to store the following patterns. The input patterns are the "letters" given in the 5×3 arrays, and the associated target patterns are given below each input pattern:

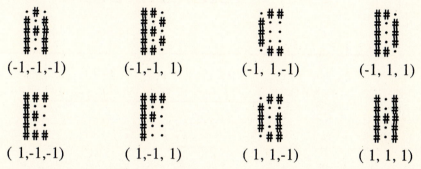

Experiment with the number of hidden units (no more than 15), the learning rate, and the initial weight values. Compare your results with those from the BAM network project.

6.4 Code a computer program to implement a backpropagation neural network with one hidden layer. Use a bias on each hidden unit and each output unit. Use a bipolar sigmoid activation function. The net is to be trained to learn the function

$$y = f(x_1, x_2) = \sin(2\pi x_1) \sin(2\pi x_2)$$

for $0 \le x_1 \le 1, 0 \le x_2 \le 1$. The number of hidden units may be varied as part of the experiment.

a. Try equally spaced training points $(i/5, j/5)$ for $i = 0, \ldots, 5, j = 0, \ldots, 5$. Scramble the order of presentation of the training points, and compute the correct target value for each point. Test the net on the points $(i/10, j/10)$ for $i = 0, \ldots, 10, j = 0, \ldots, 10$. Display your results in a form similar to that of Example 6.8. Is the response better after 10,000 epochs than it was after 1,000 epochs? This example is (perhaps surprisingly) difficult.

b. Try using more points for which the target value is non-zero and few, if any, points for which the target is zero.

c. Try using randomly generated training points.

6.5 Write a computer program to implement a backpropagation net for the data compression problem in Example 6.5. Use bipolar representation of the patterns. The target pattern is the same as the input pattern. (It is only necessary to use 56 input units and output units.)

## CHAPTER 7

A Sampler of Other Neural Nets

In this final chapter, we consider a variety of neural networks, each somewhat more specialized than those in the previous chapters. The level of detail will not be as great as before, since the intent is to suggest the ways in which the basic structures have been modified and adapted to form nets for particular applications.

The first group of nets are designed for constrained optimization problems, such as the traveling salesman problem. These nets have fixed weights that incorporate information concerning the constraints and the quantity to be optimized. The nets iterate to find a pattern of output signals that represents a solution to the problem. The Boltzmann machine (without learning), the continuous Hopfield net, and several variations (Gaussian and Cauchy nets) are described in Section 7.1.

In Section 7.2, we explore several nets that learn by means of extensions of the learning algorithms introduced in previous chapters. First, we consider two self-organizing nets that do not use competition. Oja has developed single-layer feedforward nets with linear neurons to extract information about the principal and minor components of data. These nets are trained with modified forms of Hebb learning. Then we describe the learning algorithm that Ackley, Hinton, and Sejnowski included in their presentation of the Boltzmann machine. This can be used for problems such as the encoder problem, in which the activations of some units in the net (input and output units) are known, but the correct activations of other (hidden) units are unknown. The section concludes with a discussion of three ways in which backpropagation (generalized delta rule learning) has been applied to recurrent nets.

Two examples of nets that adapt their architectures during training are presented in Section 7.3. The probabilistic neural net uses results from probability theory to classify input data in a Bayes-optimal manner. The cascade correlation algorithm constructs a net with a hierarchical arrangement of the hidden units. One hidden unit is added to the net at each stage of training, and the process is terminated as soon as the specified error tolerance is achieved. At each stage of training, only one layer of weights is adjusted (using the delta rule or a variation known as quickprop).

The final net in our sampler is the neocognitron. This net has been developed specifically for the task of recognizing handwritten digits. It has several layers of units, with very limited connections between units in successive layers. Weights between certain pairs of layers are fixed; the adaptive weights are trained one layer at a time.

There are many interesting and important neural nets that could not be included for lack of space. It is hoped that the nets we have chosen will serve to suggest the wide variety of directions in which neural network development is proceeding.

7.1 FIXED-WEIGHT NETS FOR CONSTRAINED OPTIMIZATION

In addition to solving mapping problems (including pattern classification and association) and clustering problems, neural nets can be used for constrained optimization problems. In this section, we discuss several nets designed for applications such as the traveling salesman problem, job shop scheduling, space allocation, prediction of RNA secondary structure, and map coloring, to name just a few. We will use the traveling salesman problem as our example application for these nets. (See Takefuji, 1992, for a discussion of many other applications.)

Description of the Traveling Salesman Problem.   In the classic constrained optimization problem known as the traveling salesman problem, the salesman is required to visit each of a given set of cities once and only once, returning to the starting city at the end of his trip (or tour). The tour of minimum distance is desired. The difficulty of finding a solution increases rapidly as the number of cities increases. (There is an extensive literature on solution techniques for this problem; see Lawler, Lenstra, Rinooy Kan, and Shmoys, 1985, for a discussion of approaches other than using neural networks.

We illustrate the operation of several nets in terms of their ability to find solutions for the 10-city problem, which has been used for comparison by several authors [Wilson & Pawley, 1988; Szu, 1988]. The positions of the cities are illustrated in Figure 7.1. The coordinates of the cities are as follows:

	x_1	x_2
A	0.4000	0.4439
B	0.2439	0.1463
C	0.1707	0.2293
D	0.2293	0.7610
E	0.5171	0.9414
F	0.8732	0.6536
G	0.6878	0.5219
H	0.8488	0.3609
I	0.6683	0.2536
J	0.6195	0.2634

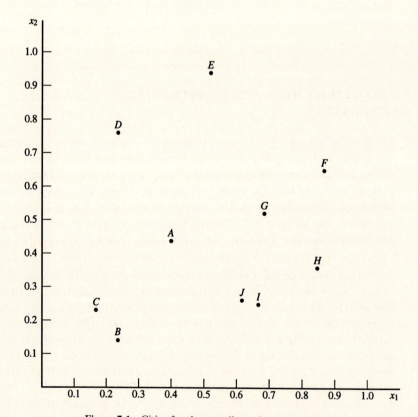

Figure 7.1   Cities for the traveling salesman problem.

The distances between the cities are given in the following symmetric distance matrix:

	A	B	C	D	E	F	G	H	I	J
A	.0000	.3361	.3141	.3601	.5111	.5176	.2982	.4564	.3289	.2842
B	.3361	.0000	.1107	.6149	.8407	.8083	.5815	.6418	.4378	.3934
C	.3141	.1107	.0000	.5349	.7919	.8207	.5941	.6908	.4982	.4501
D	.3601	.6149	.5349	.0000	.3397	.6528	.5171	.7375	.6710	.6323
E	.5111	.8407	.7919	.3397	.0000	.4579	.4529	.6686	.7042	.6857
F	.5176	.8083	.8207	.6528	.4579	.0000	.2274	.2937	.4494	.4654
G	.2982	.5815	.5941	.5171	.4529	.2274	.0000	.2277	.2690	.2674
H	.4564	.6418	.6908	.7375	.6686	.2937	.2277	.0000	.2100	.2492
I	.3289	.4378	.4982	.6710	.7042	.4494	.2690	.2100	.0000	.0498
J	.2842	.3934	.4501	.6323	.6857	.4654	.2674	.2492	.0498	.0000

Neural Net Approach to Constrained Optimization.   The neural nets described in this section have several characteristics in common. Each unit represents a hypothesis, with the unit "on" if the hypothesis is true, "off" if the hypothesis is false. The weights are fixed to represent both the constraints of the problem and the function to be optimized. The solution of the problem corresponds to the minimum of an energy function or the maximum of a consensus function for the net. The activity level of each unit is adjusted so that the net will find the desired maximum or minimum value.

Our discussion of the solution to the traveling salesman problem using the Boltzmann machine follows the formulation in Aarts and Korst (1989). This is mathematically equivalent to the original presentation of the Boltzmann machine [Ackley, Hinton, & Sejnowski, 1985], based on minimizing an energy function. Solutions using the Hopfield net use an energy function approach.

Neural nets have several potential advantages over traditional techniques for certain types of optimization problems. They can find near optimal solutions quickly for large problems. They can also handle situations in which some constraints are weak (desirable, but not absolutely required). For example, in the traveling salesman problem, it is physically impossible to visit two cities simultaneously, but it may be desirable to visit each city only once. The difference in these types of constraints could be reflected by making the penalty for having two units in the same column "on" simultaneously larger than the penalty for having two units in the same row "on" simultaneously. If it is more important to visit some cities than others, these cities can be given larger self-connection weights.

Position

City	1	2	3	4	5	6	7	8	9	10
A	$U_{A,1}$	$U_{A,2}$	$U_{A,3}$	$U_{A,4}$	$U_{A,5}$	$U_{A,6}$	$U_{A,7}$	$U_{A,8}$	$U_{A,9}$	$U_{A,10}$
B	$U_{B,1}$	$U_{B,2}$	$U_{B,3}$	$U_{B,4}$	$U_{B,5}$	$U_{B,6}$	$U_{B,7}$	$U_{B,8}$	$U_{B,9}$	$U_{B,10}$
C	$U_{C,1}$	$U_{C,2}$	$U_{C,3}$	$U_{C,4}$	$U_{C,5}$	$U_{C,6}$	$U_{C,7}$	$U_{C,8}$	$U_{C,9}$	$U_{C,10}$
D	$U_{D,1}$	$U_{D,2}$	$U_{D,3}$	$U_{D,4}$	$U_{D,5}$	$U_{D,6}$	$U_{D,7}$	$U_{D,8}$	$U_{D,9}$	$U_{D,10}$
E	$U_{E,1}$	$U_{E,2}$	$U_{E,3}$	$U_{E,4}$	$U_{E,5}$	$U_{E,6}$	$U_{E,7}$	$U_{E,8}$	$U_{E,9}$	$U_{E,10}$
F	$U_{F,1}$	$U_{F,2}$	$U_{F,3}$	$U_{F,4}$	$U_{F,5}$	$U_{F,6}$	$U_{F,7}$	$U_{F,8}$	$U_{F,9}$	$U_{F,10}$
G	$U_{G,1}$	$U_{G,2}$	$U_{G,3}$	$U_{G,4}$	$U_{G,5}$	$U_{G,6}$	$U_{G,7}$	$U_{G,8}$	$U_{G,9}$	$U_{G,10}$
H	$U_{H,1}$	$U_{H,2}$	$U_{H,3}$	$U_{H,4}$	$U_{H,5}$	$U_{H,6}$	$U_{H,7}$	$U_{H,8}$	$U_{H,9}$	$U_{H,10}$
I	$U_{I,1}$	$U_{I,2}$	$U_{I,3}$	$U_{I,4}$	$U_{I,5}$	$U_{I,6}$	$U_{I,7}$	$U_{I,8}$	$U_{I,9}$	$U_{I,10}$
J	$U_{J,1}$	$U_{J,2}$	$U_{J,3}$	$U_{J,4}$	$U_{J,5}$	$U_{J,6}$	$U_{J,7}$	$U_{J,8}$	$U_{J,9}$	$U_{J,10}$

Figure 7.2   Architecture for the 10-city traveling salesman problem.

Neural Net Architecture for the Traveling Salesman Problem.   For n cities, we use n^2 units, arranged in a square array, as illustrated in Figure 7.2. A valid tour is represented by exactly one unit being "on" in each row and in each column. Two units being "on" in a row indicates that the corresponding city was visited twice; two units being "on" in a column shows that the salesman was in two cities at the same time.

The units in each row are fully interconnected; similarly, the units in each column are fully interconnected. The weights are set so that units within the same row (or the same column) will tend not to be "on" at the same time. In addition, there are connections between units in adjacent columns and between units in the first and last columns, corresponding to the distances between cities. This will be discussed in more detail for the Boltzmann machine in Section 7.1.1 and for the Hopfield net in Section 7.1.2.

7.1.1 Boltzmann machine

Boltzmann machine neural nets were introduced by Hinton and Sejnowski (1983). The states of the units are binary valued, with probabilistic state transitions. The configuration of the network is the vector of the states of the units. The Boltzmann machine described in this section has fixed weights w_{ij}, which express the degree of desirability that units X_i and X_j both be "on."

In applying Boltzmann machines to constrained optimization problems, the weights represent the constraints of the problem and the quantity to be optimized. The description presented here is based on the maximization of a consensus function [Aarts & Korst, 1989].

The architecture of a Boltzmann machine is quite general, consisting of a set of units (X_i and X_j are two representative units) and a set of bidirectional connections between pairs of units. If units X_i and X_j are connected, $w_{ij} \neq 0$.

The bidirectional nature of the connection is often represented as $w_{ij} = w_{ji}$. A unit may also have a self-connection w_{ii}. (Or equivalently, there may be a bias unit, which is always "on" and connected to every other unit; in this interpretation, the self-connection weight would be replaced by the bias weight).

The state x_i of unit X_i is either 1 ("on") or 0 ("off"). The objective of the neural net is to maximize the consensus function

$$C = \sum_i [\sum_{j \leq i} w_{ij}\, x_i x_j].$$

The sum runs over all units of the net.

The net finds this maximum (or at least a local maximum) by letting each unit attempt to change its state (from "on" to "off" or vice versa). The attempts may be made either sequentially (one unit at a time) or in parallel (several units simultaneously). Only the sequential Boltzmann machine will be discussed here. The change in consensus if unit X_i were to change its state (from 1 to 0 or from 0 to 1) is

$$\Delta C(i) = [1 - 2x_i][w_{ii} + \sum_{j \neq i} w_{ij} x_j],$$

where x_i is the current state of unit X_i. The coefficient $[1 - 2x_i]$ will be $+1$ if unit X_i is currently "off" and -1 if unit X_i is currently "on."

Note that if unit X_i were to change its activation the resulting change in consensus can be computed from information that is local to unit X_i, i.e., from weights on connections and activations of units to which unit X_i is connected (with $w_{ij} = 0$ if unit X_j is not connected to unit X_i).

However, unit X_i does not necessarily change its state, even if doing so would increase the consensus of the net. The probability of the net accepting a change in state for unit X_i is

$$A(i, T) = \cfrac{1}{1 + \exp\left(-\cfrac{\Delta C(i)}{T}\right)}.$$

The control parameter T (called the temperature) is gradually reduced as the net searches for a maximal consensus. Lower values of T make it more likely that the net will accept a change of state that increases its consensus and less likely that it will accept a change that reduces its consensus. The use of a probabilistic update procedure for the activations, with the control parameter decreasing as the net searches for the optimal solution to the problem represented by its weights, reduces the chances of the net getting stuck in a local maximum.

This process of gradually reducing the temperature is called *simulated annealing* [Aarts & Korst, 1989]. It is analogous to the physical annealing process used to produce a strong metal (with a regular crystalline structure). During annealing, a molten metal is cooled gradually in order to avoid imperfections in the crystalline structure of the metal due to freezing.

Architecture

We illustrate the architecture of a Boltzmann machine for units arranged in a two-dimensional array. The units within each row are fully interconnected, as shown in Figure 7.3. Similarly, the units within each column are also fully interconnected. The weights on each of the connections is $-p$ (where $p > 0$). In addition, each unit has a self-connection, with weight $b > 0$. The connections shown in the figure, with the proper choice of values for b and p as discussed in the next section, will form a portion of the Boltzmann machine to solve the traveling salesman problem. In keeping with the most common notation for neural network solutions of this problem, we label a typical unit $U_{i,j}$.

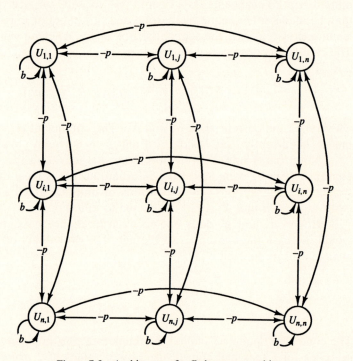

Figure 7.3   Architecture for Boltzmann machine.

Algorithm

Setting the Weights.   The weights for a Boltzmann machine are fixed so that the net will tend to make state transitions toward a maximum of the consensus function defined on page 339. If we wish the net illustrated in Figure 7.3 to have exactly one unit "on" in each row and in each column, we must choose the values of the weights p and b so that improving the configuration corresponds to increasing the consensus.

Each unit is connected to every other unit in the same row with weight $-p$ ($p > 0$); similarly, each unit is connected to every other unit in the same column with weight $-p$. These weights are penalties for violating the condition that at most one unit be "on" in each row and each column. In addition, each unit has a self-connection, of weight $b > 0$. The self-connection weight is an incentive (bonus) to encourage a unit to turn "on" if it can do so without causing more than one unit to be on in a row or column.

If $p > b$, the net will function as desired. The correct choice of weights to ensure that the net functions as desired can be deduced by considering the effect on the consensus of the net in the following two situations.

If unit $U_{i,j}$ is "off" ($u_{i,j} = 0$) and none of the units connected to $U_{i,j}$ is "on," changing the status of $U_{i,j}$ to "on" will increase the consensus of the net by the amount b. This is a desirable change; since it corresponds to an increase in consensus, the net will be more likely to accept it than to reject it.

On the other hand, if one of the units in row i or in column j (say, $U_{i,j+1}$ is already "on"), attempting to turn unit $U_{i,j}$ "on" would result in a change of consensus by the amount $b - p$. Thus, for $b - p < 0$ (i.e., $p > b$), the effect would be to decrease the consensus. The net will tend to reject this unfavorable change.

Bonus and penalty connections, with $p > b$, will be used in the net for the traveling salesman problem to represent the constraints for a valid tour.

Application Procedure.   The application algorithm given here is expressed in terms of units arranged in a two-dimensional array, as is needed for the traveling salesman problem with n cities. There are n^2 units. The weight between unit $U_{i,j}$ and unit $U_{I,J}$ is denoted $w(i, j; I, J)$. For the architecture shown in Figure 7.3,

$$w(i, j; I, J) = -p \quad \text{if } i = I \text{ or } j = J \text{ (but not both);}$$

$$w(i, j; i, j) = b.$$

The application procedure is as follows:

Step 0.   Initialize weights to represent the constraints of the problem.
Initialize the control parameter (temperature) T.
Initialize activations of units (random binary values).

Step 1.   While stopping condition is false, do Steps 2–8.

   Step 2.   Do Steps 3–6 n^2 times. (This constitutes an epoch.)

      Step 3.   Choose integers I and J at random between 1 and n.
(Unit U_{IJ} is the current candidate to change its state.)

      Step 4.   Compute the change in consensus that would result:

$$\Delta C = [1 - 2u_{I,J}][w(I, J; I, J)$$
$$+ \sum_{i,j \neq I,J} w(i, j; I, J)u_{i,j}].$$

Step 5.   Compute the probability of acceptance of the change:

$$A(T) = \frac{1}{1 + \exp\left(-\dfrac{\Delta C}{T}\right)}.$$

Step 6.   Determine whether or not to accept the change.
Let R be a random number between 0 and 1.
If $R < A$, accept the change:

$$u_{I,J} = 1 - u_{I,J}. \text{ (This changes the state of unit } U_{I,J}.)$$

If $R \geq A$, reject the proposed change.

Step 7.   Reduce the control parameter:

$$T(\text{new}) = 0.95T(\text{old}).$$

Step 8.   Test stopping condition:
If there has been no change of state for a specified number of epochs, or if the temperature has reached a specified value, stop; otherwise continue.

Initial Temperature.   The initial temperature should be taken large enough so that the probability of accepting a change of state is approximately 0.5, regardless of whether the change is beneficial or detrimental. However, since a high starting temperature increases the required computation time significantly, a lower initial temperature may be more practical in some applications.

Cooling Schedule.   Theoretical results [Geman & Geman, 1984] indicate that the temperature should be cooled slowly according to the logarithmic formula

$$T_B(k) = \frac{T_0}{\log(1 + k)},$$

where k is an epoch. An epoch is n^2 attempted unit updates, where n is the number of cities and n^2 is the number of units in the net. However, published results are often based on experimentation with both the starting temperature and the cooling schedule.

We have used the exponential cooling schedule $T(\text{new}) = \alpha T(\text{old})$, reduced after each epoch [Kirkpatrick, Gelatt, & Vecchi, 1983]. A larger α (such as $\alpha = 0.98$) allows for fewer epochs at each temperature; a smaller α (such as $\alpha = 0.9$) may require more epochs at each temperature.

Application: traveling salesman problem

Summary of Nomenclature

n number of cities in the tour (there are n^2 units in the net)
i index designating a city; $1 \leq i \leq n$.

j index designating position in tour, mod n; i.e., $j = n + 1 \rightarrow j = 1$, $j = 0 \rightarrow j = n$.

$U_{i,j}$ unit representing the hypothesis that the ith city is visited at the jth step of the tour.

$u_{i,j}$ activation of unit $U_{i,j}$; $u_{i,j} = 1$ if the hypothesis is true, 0 if it is false.

$d_{i,k}$ distance between city i and city k, $k \neq i$.

d maximum distance between any two cities.

Architecture. For this application, it is convenient to arrange the units of the neural net in a grid, as illustrated in Figure 7.2. The rows of the grid represent cities to be visited, the columns the position of a city in the tour.

The connection pattern for the neural net is as follows:

- $U_{i,j}$ has a self-connection of weight b; this represents the desirability of visiting city, i at stage j.

- $U_{i,j}$ is connected to all other units in row i with penalty weights $-p$; this represents the constraint that the same city is not to be visited twice.

- $U_{i,j}$ is connected to all other units in column j with penalty weights $-p$; this represents the constraint that two cities cannot be visited simultaneously.

- $U_{i,j}$ is connected to $U_{k,j+1}$ for $1 \leq k \leq n$, $k \neq i$, with weight $-d_{i,k}$; this represents the distance traveled in making the transition from city i at stage j to city k at stage $j + 1$.

- $U_{i,j}$ is connected to $U_{k,j-1}$ for $1 \leq k \leq n$, $k \neq i$, with weight $-d_{i,k}$; this represents the distance traveled in making the transition from city k at stage $j - 1$ to city i at stage j.

Setting the Weights. The desired neural net will be constructed in two steps. First, a neural net will be formed for which the maximum consensus occurs whenever the constraints of the problem are satisfied, i.e., when exactly one unit is "on" in each row and in each column. Second, we will add weighted connections to represent the distances between the cities. In order to treat the problem as a maximum consensus problem, the weights representing distances will be negative.

A Boltzmann machine with weights representing the constraints (but not the distances) for the traveling salesman problem is illustrated in Figure 7.3. If $p > b$, the net will function as desired (as explained earlier).

To complete the formulation of a Boltzmann neural net for the traveling salesman problem, weighted connections representing distances must be included. In addition to the weights described before and shown in Figure 7.3 (which represent the constraints), a typical unit $U_{i,j}$ is connected to the units $U_{k,j-1}$ and $U_{k,j+1}$ (for all $k \neq i$) by weights that represent the distances between city i and city k. The distance weights are shown in Figure 7.4 for the typical unit $U_{i,j}$. Note that units in the last column are connected to units in the first column by connections representing the appropriate distances also. However, units in a particular column are not connected to units in columns other than those immediately adjacent to the said column.

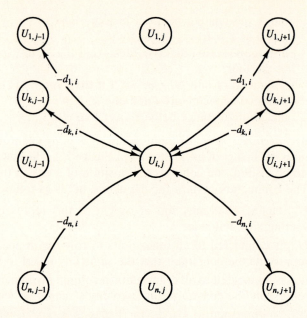

Figure 7.4 Boltzmann neural net for the "traveling salesman problem"; weights representing distances for unit $U_{i,j}$.

We now consider the relation between the constraint weight b and the distance weights. Let d denote the maximum distance between any two cities on the tour. Assume that no city is visited in the jth position of the tour and that no city is visited twice. In this case, some city, say, i, is not visited at all; i.e., no unit is "on" in column j or in row i. Since allowing $U_{i,j}$ to turn on should be encouraged, the weights should be set so that the consensus will be increased if it turns on. The change in consensus will be $b - d_{i,k1} - d_{i,k2}$, where $k1$ indicates the city visited at stage $j - 1$ of the tour and $k2$ denotes the city visited at stage $j + 1$ (and city i is visited at stage j). This change is greater than or equal to $b - 2d$; however, equality will occur only if the cities visited in positions $j - 1$ and $j + 1$ are both the maximum distance, d, away from city i. In general, requiring the change in consensus to be positive will suffice, so we take $b > 2d$.

Thus, we see that if $p > b$, the consensus function has a higher value for a feasible solution (one that satisfies the constraints) than for a nonfeasible solution, and if $b > 2d$ the consensus will be higher for a short feasible solution than for a longer tour.

Sample Results.

Example 7.1 A Boltzmann machine for the traveling salesman problem: large bonus and penalty weights

The Boltzmann machine described in the previous sections was used to solve the traveling salesman problem; 100 different starting configurations were employed, each with approximately half the units "on." With $T_0 = 20.0$, $b = 60$, and $p = 70$, valid tours were produced in 20 or fewer epochs for all 100 initial configurations. In these experiments, it was rare for the net to change its configuration once a valid tour was found. Typically, a valid tour was found in 10 or fewer epochs). An epoch consisted of each unit attempting to change its state. The cooling schedule was $T(\text{new}) = 0.9T(\text{old})$ after each epoch. Five tours of length less than 4 were found:

TOUR										LENGTH
G	F	D	E	A	C	B	J	I	H	3.036575
D	A	I	J	G	F	H	E	C	B	3.713347
B	J	H	A	F	G	I	E	D	C	3.802492
H	I	E	J	A	B	C	D	F	G	3.973623
J	A	F	H	D	E	C	B	G	I	3.975433

The best tour found is illustrated in Figure 7.5.

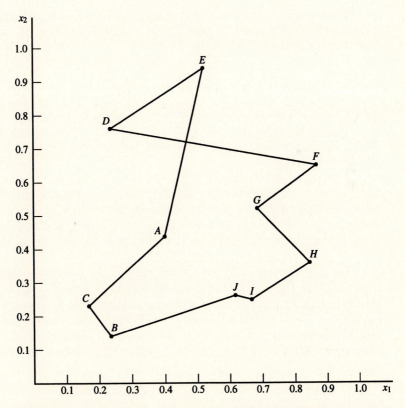

Figure 7.5 Best tour for traveling salesman problem from Boltzmann machine (100 initial configurations).

Example 7.2 A Boltzmann machine for the traveling salesman problem: smaller bonus and penalty weights

With somewhat smaller values of b and p ($b = 30$, $p = 35$), 100 valid tours were again found for each of the 100 initial configurations. The net changed from a valid tour to a second valid (and shorter) tour in approximately 25% of these trials. However, none of the tours was shorter than that found in Example 7.1. Using even smaller values ($b = 6$, $p = 7$), the net was unable to find valid tours (in 20 epochs) for any of the 100 initial configurations. More epochs would not be likely to help, as the temperature after 20 epochs is quite low.

Analysis

The traveling salesman problem is a nice model for a variety of constrained optimization problems. It is, however, a difficult problem for the Boltzmann machine, because in order to go from one valid tour to another, several invalid tours must be accepted. By contrast, the transition from valid solution to valid solution may not be as difficult in other constrained optimization problems.

Equilibrium.   The net is in *thermal equilibrium* (at a particular temperature T) when the probabilities, P_α and P_β, of two configurations of the net, α and β, obey the Boltzmann distribution

$$\frac{P_\alpha}{P_\beta} = \exp\left[\frac{E_\beta - E_\alpha}{T}\right],$$

where E_α is the energy of configuration α and E_β is the energy of configuration β. At higher temperatures, the probabilities of different configurations are more nearly equal. At lower temperatures, there is a stronger bias toward configurations with lower energy.

Starting at a sufficiently high temperature ensures that the net will have approximately equal probability of accepting or rejecting any proposed state transition. If the temperature is reduced slowly, the net will remain in equilibrium at lower temperatures. It is not practical to verify directly the equilibrium condition at each temperature, as there are too many possible configurations.

Energy Function.   As mentioned earlier, the constraint satisfaction problems to which the Boltzmann machine without learning is applied can be formulated as either maximization or minimization problems. Also, as described in Section 2.1.2 with regard to pattern classification, the use of a bias (self-connection, or connection to a unit that is always "on") and the use of a threshold are equivalent. Ackley, Hinton, and Sejnowski (1985) define the energy of a configuration as

$$E = -\sum_i \sum_{j<i} w_{ij} x_i x_j + \sum_i \theta_i x_i,$$

where θ_i is a threshold and self-connections (or biases) are not used. The difference in energy between a configuration with unit X_k "off" and one with X_k "on" (and the state of all other units remaining unchanged) is

$$\Delta E(k) = -\theta_k + \sum_i w_{ik}x_i.$$

If the units change their activations randomly and asynchronously, and the net always moves to a lower energy (rather than moving to a lower energy with a probability that is less than 1, as described in the preceding sections), the discrete Hopfield net results.

To simplify notation, one may include a unit in the net that is connected to every other unit and is always "on." This allows the threshold to be treated as any other weight, so that

$$E = -\sum_i \sum_{j<i} w_{ij}x_ix_j.$$

The energy gap between the configuration with unit X_k "off" and that with unit X_k "on" is

$$\Delta E(k) = \sum_i w_{ik}x_i.$$

Variations.   The acceptance condition in the algorithm on page 342 is closely related (but not identical) to the Metropolis condition, which is:

Set output of unit to 1 with probability

$$A(T) = \frac{1}{1 + \exp\left(-\dfrac{\Delta C}{T}\right)},$$

regardless of the current activity of the unit [Metropolis, Rosenbluth, Rosenbluth, Teller, & Teller, 1953].

The Boltzmann machine, as well as the Cauchy machine presented in Section 7.1.4, can be described in terms of a Markov chain process. For this purpose, the following notation is useful:

$x(k)$ Configuration of the net at stage k of the process (binary vector of activations of the units).

n Dimension of x (number of units in the net).

$\|\Delta x\|$ Number of units whose activations change in going from current configuration to new configuration.

$T(k)$ Temperature at stage k of the process.

At stage k of the process, the probability of transition from the current configuration of the net, $x(k)$, to any other configuration is a function of the distance $\|\Delta x\|$ between the two configurations and the change in consensus that would result if the transition occurs. The change in consensus depends on the current temperature. Each stage of the Markov process can be considered to consist of three steps: Generate a potential new configuration of the net, accept or reject the new configuration, and reduce the temperature according to the annealing schedule.

For the Boltzmann machine, the generating probability is given by the Gaussian distribution:

$$G(k) = T(k)^{-.5n} \exp\left(\frac{-\|\Delta x\|^2}{T(k)}\right).$$

Thus, all configurations that are at the same distance from the current configuration (i.e., all that involve the same number of units changing their state simultaneously) are equally likely to be generated as the candidate configuration. In the preceding discussion of the Boltzmann machine, all configurations in which exactly one unit changes its state are equally likely to be chosen as the candidate state at any time. Configurations in which more than one unit changes its state ($\Delta x > 1$) are generated with probability zero.

The probability of accepting the new configuration depends on the current temperature and the change in consensus ΔC that would result and is

$$A(k, T(k)) = \frac{1}{1 + \exp\left(-\dfrac{\Delta C}{T(k)}\right)}.$$

This form of analysis is useful for deriving the theoretical lower bounds on the cooling schedules for the Boltzmann and Cauchy machines; see Jeong and Park (1989) for a discussion of cooling schedules using these generating and acceptance probabilities.

We shall limit our considerations to formulations of the Boltzmann, Hopfield, and other related nets in which only one unit updates its output signal at any time. In this context, the motivation for the acceptance probability as the (approximate) integral of the Gaussian probability density function [Takefuji, 1992] will provide a useful framework for relating the Boltzmann machine to the Gaussian and Cauchy machines in Section 7.1.3 and 7.1.4, respectively.

7.1.2 Continuous Hopfield Net

A modification of the discrete Hopfield net (Section 3.4.4), with continuous-valued output functions, can be used either for associative memory problems (as with the discrete form) or constrained optimization problems such as the traveling salesman problem. As with the discrete Hopfield net, the connections between units are bidirectional, so that the weight matrix is symmetric; i.e., the connection from unit U_i to unit U_j (with weight w_{ij}) is the same as the connection from U_j to U_i (with weight w_{ji}). For the continuous Hopfield net, we denote the internal activity of a neuron as u_i; its output signal is $v_i = g(u_i)$.

If we define an energy function

$$E = 0.5 \sum_{i=1}^{n} \sum_{j=1}^{n} w_{ij} v_i v_j + \sum_{i=1}^{n} \theta_i v_i,$$

then the net will converge to a stable configuration that is a minimum of the energy function as long as

$$\frac{d}{dt} E \leq 0.$$

For this form of the energy function, the net will converge if the activity of each neuron changes with time according to the differential equation

$$\frac{d}{dt} u_i = -\frac{\partial E}{\partial v_i} = -\sum_{j=1}^{n} w_{ij} v_j - \theta_i,$$

as is shown later in this section.

In the original presentation of the continuous Hopfield net [Hopfield, 1984], the energy function is

$$E = -0.5 \sum_{i=1}^{n} \sum_{j=1}^{n} w_{ij} v_i v_j - \sum_{i=1}^{n} \theta_i v_i + \frac{1}{\tau} \sum_{i=1}^{n} \int_0^{v_i} g_i^{-1}(v) \, dv,$$

where τ is, of course, the time constant. If the activity of each neuron changes with time according to the differential equation

$$\frac{d}{dt} u_i = -\frac{u_i}{\tau} + \sum_{j=1}^{n} w_{ij} v_j + \theta_i,$$

the net will converge. The argument is essentially the same as that given in the proof of convergence of the Hopfield net later. Because the time scale is arbitrary, in applications, the time constant τ in the decay term is usually taken to be 1.

In the Hopfield-Tank solution of the traveling salesman problem [Hopfield & Tank, 1985], each unit has two indices. The first index—x, y, etc.—denotes the city, the second—i, j, etc.—the position in the tour.

The Hopfield-Tank energy function for the traveling salesman problem is

$$E = \frac{A}{2} \sum_x \sum_i \sum_{j \neq i} v_{x,i} v_{x,j}$$

$$+ \frac{B}{2} \sum_i \sum_x \sum_{y \neq x} v_{x,i} v_{y,i}$$

$$+ \frac{C}{2} [N - \sum_x \sum_i v_{x,i}]^2$$

$$+ \frac{D}{2} \sum_x \sum_{y \neq x} \sum_i d_{x,y} v_{x,i} (v_{y,i+1} + v_{y,i-1}).$$

The differential equation for the activity of unit $U_{X,I}$ is

$$\frac{d}{dt} u_{X,I} = -\frac{u_{X,I}}{\tau} - A \sum_{j \neq I} v_{X,j} - B \sum_{y \neq X} v_{y,I} - C[N - \sum_x \sum_i v_{x,i}]$$

$$-D \sum_{y \neq X} d_{X,y}(v_{y,I+1} + v_{y,I-1}).$$

The output signal is given by applying the sigmoid function (with range between 0 and 1), which Hopfield and Tank expressed as

$$v_i = g(u_i) = 0.5[1 + \tanh(\alpha u_i)].$$

Architecture for traveling salesman problem

The units used to solve the 10-city traveling salesman problem are arranged as shown in Figure 7.2. The connection weights are fixed and are usually not shown or even explicitly stated. The weights for interrow connections correspond to the parameter A in the energy equation; there is a contribution to the energy if two units in the same row are "on." Similarly, the intercolumnar connections have weights B; the distance connections appear in the fourth term of the energy equation. More explicitly, the weights between units U_{xi} and U_{yj} are

$$w(x,i; y,j) = -A\delta_{xy}(1 - \delta_{ij}) - B\delta_{ij}(1 - \delta_{xy}) - C$$

$$-Dd_{xy}(\delta_{i,j+1} + \delta_{i,j-1}),$$

where δ_{ij} is the so-called Dirac delta, which is 1 if $i = j$ and 0 otherwise. In addition, each unit receives an external input signal

$$I_{xi} = +C N$$

The parameter N is usually taken to be somewhat larger than the number of cities, n.

Algorithm

The basic procedure for solving the traveling salesman problem using a continuous Hopfield net is described in the algorithm that follows. For convenience, we may think of the computations in Step 2 as constituting an epoch, i.e., each unit has had, on average, one opportunity to update its activity level. It may be desirable to ensure that each unit does update its activity. The method of initializing the activations is discussed following the algorithm.

Step 0.   Initialize activations of all units.
                  Initialize Δt to a small value.
Step 1.   While the stopping condition is false, do Steps 2–6.
                  Step 2.   Perform Steps 3–5 n^2 times (n is the number of cities).
                        Step 3.   Choose a unit at random.
                        Step 4.   Change activity on selected unit:

$$u_{x,i}(\text{new}) = u_{x,i}(\text{old})$$
$$+ \Delta t[-u_{x,i}(\text{old}) - A\sum_{j \neq i} v_{x,j}$$
$$- B\sum_{y \neq x} v_{y,i} - C\{N - \sum_x \sum_j v_{x,j}\}$$
$$- D\sum_{y \neq x} d_{x,y}(v_{y,i+1} + v_{y,i-1})].$$

                        Step 5.   Apply output function:

$$v_{x,i} = 0.5[1 + \tanh(\alpha u_{x,i})].$$

                  Step 6.   Check stopping condition.

Hopfield and Tank used the following parameter values in their solution of the problem:

$$A = B = 500, C = 200, D = 500, N = 15, \alpha = 50.$$

The large value of α gives a very steep sigmoid function, which approximates a step function. Furthermore, the large coefficients and a correspondingly small Δt result in very little contribution from the decay term ($u_{x,i}(\text{old}) \Delta t$). The initial activity levels ($u_{x,i}$) were chosen so that $\sum_x \sum_i v_{x,i} = 10$ (the desired total activation for a valid tour). However, some noise was included so that not all units started with the same activity (or output signal).

Application

Hopfield and Tank [1985] claimed a high rate of success in finding valid tours; they found 16 from 20 starting configurations. Approximately half of the trials produced one of the two shortest paths. The best tour found was

$$A \quad D \quad E \quad F \quad G \quad H \quad I \quad J \quad B \quad C,$$

with length 2.71 (see Figure 7.6).

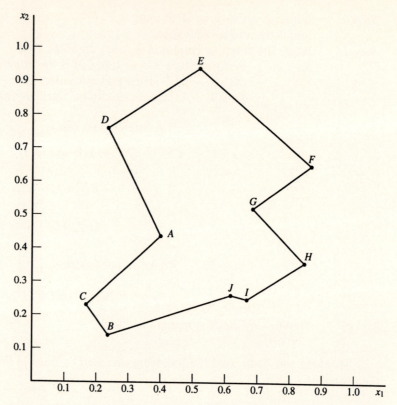

Figure 7.6 Best tour for traveling salesman problem found by Hopfield and Tank [1985].

Analysis

Variations.   Many other researchers have had difficulty achieving success comparable to that reported by Hopfield and Tank (1985). In this section, we describe two variations on the Hopfield-Tank model [Wilson & Pawley, 1988; Szu, 1988].

Wilson and Pawley.   Wilson and Pawley (1988) provide a somewhat more detailed statement of the Hopfield-Tank algorithm than do those authors themselves, together with an analysis of their attempts to duplicate the results published previously. They used the time step $\Delta t = 10^{-5}$ and stopped their simulations when they found that they had a valid tour, a frozen net, or a total of 1,000 epochs performed. In testing for a valid tour, a unit was considered to be "on" if its activation was greater than 0.9 and "off" if its activation was less than 0.1. A net was frozen if no activations changed by more than 10^{-35}.

In 100 attempts, allowing 1,000 epochs on each, Wilson and Pawley found 15 valid tours (45 froze and 40 failed to converge). Some of the better tours produced the following results:

A	D	E	F	G	H	J	I	B	C	(length 2.71)
A	D	E	G	F	H	J	I	C	B	(see Figure 7.7)
A	C	B	D	E	G	F	H	J	I	

Since the starting city and direction of each tour are not indicated in the authors' paper, the tours are represented here with the same starting city and direction of travel.

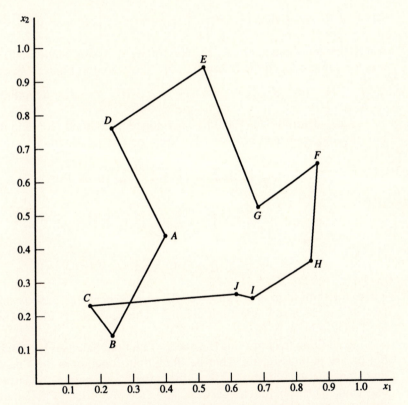

Figure 7.7 One of the better tours produced by Wilson and Pawley (1988). (They also found the best tour, illustrated in Figure 7.6.)

In attempting to obtain a success rate for valid tours that would approach the rate achieved by Hopfield and Tank, Wilson and Pawley tried a number of variations of the Hopfield and Tank algorithm. They experimented with different parameter values, different initial activity configurations (including starting with a larger total activity level and decreasing it over the first 1,000 iterations), and imposing a large distance penalty for visiting same city twice, none of which helped much. Fixing the starting city helped on the Hopfield-Tank cities, but not on other randomly generated sets of cities.

One variation that did improve the ability of the net to generate valid tours was a modification of the initialization procedure. The Willshaw initialization is based on the rationale that cities on opposite sides of the square probably should be on opposite sides of the tour [Durbin & Willshaw, 1987]. The starting activity of each unit is biased to reflect this fact. Cities far from the center of the square received a stronger bias than those near the middle. The formula, in terms of the ith city and jth position, is

$$\text{bias}(i, j) = \cos\left[\text{atan}\left(\frac{y_i - 0.5}{x_i - 0.5}\right) + \frac{2\pi(j - 1)}{n}\right]\sqrt{(x_i - 0.5)^2 + (y_i - 0.5)^2}.$$

where the coordinates of the ith city are x_i, y_i. Using this initialization, Wilson and Pawley produced the following tour, illustrated in Figure 7.8, in 166 epochs:

$$A \quad B \quad C \quad D \quad E \quad F \quad G \quad H \quad I \quad J \quad \text{(length 2.83)}$$

Szu.   Harold Szu (1988) has developed a modified Hopfield net for solving the traveling salesman problem. He uses the energy function

$$E = \frac{A}{2}\sum_x\sum_i\sum_{j\neq i} v_{x,i}v_{x,j} + \frac{B}{2}\sum_i\sum_x\sum_{y\neq x} v_{x,i}v_{y,i}$$

$$+ \frac{C}{2}\{\sum_x [1 - \sum_i v_{x,i}]^2 + \sum_i [1 - \sum_x v_{x,i}]^2\}$$

$$+ \frac{D}{2}\sum_x\sum_{y\neq x}\sum_i d_{x,y}v_{x,i}(v_{y,i+1} + v_{y,i-1}),$$

where the third term now expresses the requirement that exactly one unit should be "on" in each row and each column. The coefficients in the energy function are taken to be $A = B = C = D = 1$.

In addition to improving the energy function, Szu uses continuous activities, but binary output signals. That is, the output function is the "hard limiter," or unit step function, rather than the differentiable sigmoid function used by Hopfield and Tank. This step function is also called the McCulloch-Pitts input/output function [Takefuji, 1992].

The architecture Szu uses is the same as that for the Boltzmann or Hopfield-Tank models. However, Szu updates the activities of the units simultaneously, rather than sequentially. He performs n^2 such updates and then tests for a valid tour. The specific values in the following algorithm are based on a computer code to solve the traveling salesman problem using the first five of the cities in the Hopfield-Tank sample problem [Szu, 1989].

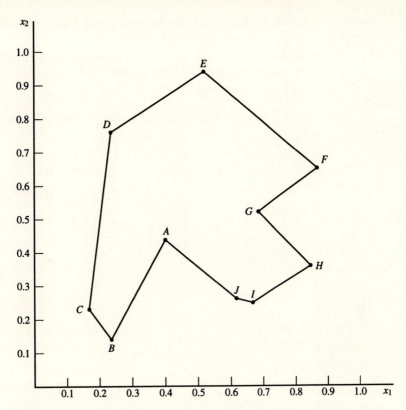

Figure 7.8 Tour for traveling salesman problem found using Willshaw initialization.

Fast Optimization Algorithm for TSP

Step 0.   Initialize distances between cities.

Set time step $\Delta t = 10^{-5}$.

Step 1.   Perform Steps 2–8 the specified number of times.

(Generate the specified number of tours.)

    Step 2.   Initialize activities of all units.

    Use random values between -0.0005 and $+0.0005$.

    Increase activity of unit U_{11} by 0.005.

        Step 3.   Do Steps 4–7 n^2 times

            Step 4.   Do Steps 5–6 for each unit.

                Step 5.   Calculate all terms for change in activity.

                Step 6.   Update activity.

            Step 7.   Apply binary output function to each unit.

        Step 8.   Test for valid tour.

Example 7.3 Sample results for traveling salesman problem using Szu's fast optimization algorithm

Published results for the fast optimization algorithm show 91 valid tours obtained from 1,000 trials (tours generated) [Szu, 1988]. The best tour found was

$$D \quad E \quad F \quad G \quad H \quad I \quad J \quad C \quad B \quad A \qquad \text{(length 2.7693)},$$

illustrated in Figure 7.9. The other tours found that were of length less than 3.5 are:

TOUR										LENGTH
J	H	G	F	E	D	B	A	C	I	3.3148
A	C	B	G	J	I	H	F	D	E	3.3306
J	I	G	H	F	A	B	C	D	E	3.3647
A	E	G	F	H	J	I	C	B	D	3.3679
C	B	E	D	F	H	G	I	J	A	3.3822
A	F	D	E	G	H	J	I	C	B	3.4345
C	B	E	G	F	H	I	J	D	A	3.4917

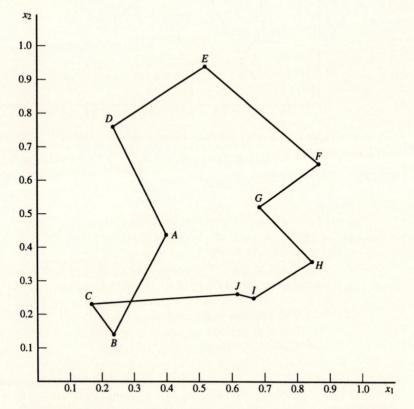

Figure 7.9 Best tour for traveling salesman problem from fast optimization algorithm [Szu, 1988].

Proof of Convergence of Hopfield Net.   For an energy function of the form

$$E = \sum_{i=1}^{n} \sum_{j=1}^{n} w_{ij} v_i v_j + \sum_{i=1}^{n} \theta_i v_i,$$

the Hopfield net will converge if the activations change according to the differential equation

$$\frac{du_i}{dt} = -\frac{\partial E}{\partial v_i},$$

as the simple calculations that follow show [Takefuji, 1992].

If $v_i = g(u_i)$ is monotonic and nondecreasing, then $dv_i/du_i \geq 0$. Since

$$\frac{dE}{dt} = \sum_i \frac{dv_i}{dt} \frac{\partial E}{\partial v_i}$$

$$= -\sum_i \frac{dv_i}{dt} \frac{du_i}{dt}$$

$$= -\sum_i \frac{dv_i}{du_i} \frac{du_i}{dt} \frac{du_i}{dt},$$

the energy is nonincreasing, as required.

In the original presentation of the continuous Hopfield net [Hopfield, 1984], the energy function includes an integral term:

$$E = -0.5 \sum_{i=1}^{n} \sum_{j=1}^{n} w_{ij} v_i v_j - \sum_{i=1}^{n} \theta_i v_i + \frac{1}{\tau} \sum_{i=1}^{n} \int_0^{v_i} g_i^{-1}(v)\, dv.$$

If the weight matrix is symmetric and the activity of each neuron changes with time according to the differential equation

$$\frac{d}{dt} u_i = -\frac{u_i}{\tau} + \sum_{j=1}^{n} w_{ij} v_j + \theta_i,$$

the net will converge. The argument is essentially the same as for the energy function without the integral term.

Takefuji (1992) has shown that convergence is not guaranteed for the energy function without the integral term if the neuron activations are updated using the differential equation with the decay term, $-\dfrac{u_i}{\tau}$.

7.1.3 Gaussian Machine

A general framework that includes the Boltzmann machine, Hopfield net, and other neural networks is known as the Gaussian machine [Akiyama, Yamashita, Kajiura, & Aiso, 1989]. An obvious minor extension of the description of the Gaussian machine allows it to include the Cauchy machine also (see Section 7.1.4).

A Gaussian machine is described by the following three parameters:

α slope parameter of the sigmoid function,
T temperature,
Δt time step.

The operation of the net consists of the following steps:

1. Calculating the net input to unit U_i:

$$\text{net}_i = \sum_{j=1}^{N} w_{ij} v_j + \theta_i + \epsilon,$$

where ϵ is random noise, which depends on the temperature T.

2. Changing the activity level of unit U_i:

$$\frac{\Delta u_i}{\Delta t} = -\frac{u_i}{\tau} + \text{net}_i.$$

3. Applying the output function

$$v_i = f(u_i) = 0.5[1 + \tanh(\alpha u_i)],$$

where the binary step function corresponds to $\alpha = \infty$.

The Hopfield machine corresponds to a Gaussian machine with $T = 0$ (no noise). The Boltzmann machine is obtained by setting $\Delta t = \tau = 1$, to obtain $\Delta u_i = -u_i + \text{net}_i$, or

$$u_i(\text{new}) = \text{net}_i = \sum_{j=1}^{N} w_{ij} v_j + \theta_i + \epsilon.$$

If the noise obeys a Gaussian distribution with mean of zero and standard deviation $\sigma = T \sqrt{8/\pi}$, then the distribution of outputs has the same behavior as a Boltzmann machine with probabilistic acceptance of state transitions.

Integrating the Gaussian noise distribution (approximately), we find the approximate Boltzmann acceptance function:

$$\int_0^{\infty} \frac{1}{\sqrt{2\pi\sigma^2}} \exp \frac{(x - u_i)^2}{2\sigma^2} \, dx \approx A(i, T) = \frac{1}{1 + \exp\left(-\dfrac{u_i}{T}\right)}.$$

Note that $u_i = \Delta C(i)$. Noise obeying a logistic rather than a Gaussian distribution will give a Gaussian machine that is identical to the Boltzmann machine with the Metropolis acceptance function, i.e., setting the output to 1 with probability

$$A(i, T) = \cfrac{1}{1 + \exp\left(-\cfrac{u_i}{T}\right)},$$

regardless of the unit's original state.

The equivalence of adding noise to the net input of a unit and using a probabilistic state transition provides a simple framework for extending the Gaussian machine (especially the Boltzmann machine form) to include the Cauchy machine, as described in the next section. (See also Takefuji, 1992, for further discussion of these ideas.)

7.1.4 Cauchy Machine

A modification of the Boltzmann machine, known as the Cauchy machine, or *fast simulated annealing,* is based on adding more noise to the net input to increase the likelihood of escaping from a neighborhood of a local minimum [Szu & Hartley, 1987]. The unbounded variance of the Cauchy distribution allows for occasional larger changes in the configuration of the system than does the bounded variance of the Gaussian distribution. Noise based on the Cauchy distribution is called "colored noise," in contrast to the "white noise" of the Gaussian distribution. The addition of Cauchy noise is equivalent to using an acceptance probability that makes it more likely for the net to accept a bad move.

The Cauchy machine can be included in the Gaussian machine framework by setting

$$\Delta t = \tau = 1,$$

to obtain

$$\Delta u_i = -u_i + \text{net}_i$$

or

$$u_i(\text{new}) = \text{net}_i = \sum_{j=1}^{N} w_{ij} v_j + \theta_i + \epsilon,$$

and taking the noise to obey a Cauchy distribution with mean zero and standard deviation $\sigma = T \sqrt{8/\pi}$, rather than a Gaussian or logistic distribution. Integrating the Cauchy noise distribution, we find the Cauchy acceptance function:

$$\int_0^\infty \frac{1}{\pi} \frac{T \, dx}{T^2 + (x - u_i)^2} = \frac{1}{2} + \frac{1}{\pi} \arctan\left(\frac{u_i}{T}\right) = A(i, T)$$

Note that $u_i = \Delta C(i)$.

One of the potential advantages of the Cauchy machine is the possibility of using a faster cooling schedule. The Cauchy cooling schedule is $T = T_0/k$, where k is the number of iterations (epochs) that have been performed. (See Jeong & Park, 1989, for a proof of this.) As has been observed [Szu, 1990], and as the examples in the next section illustrate, the faster cooling schedule is needed to help the net stabilize.

Application

Example 7.4 A hybrid Boltzmann-Cauchy machine for the traveling salesman problem

The traveling salesman problem was solved using the same architecture, algorithm, and parameters as for the Boltzmann machine in the previous section (including the cooling schedule mentioned there); only the acceptance probability was changed.

Tours of length less than 4 were found for nine starting configurations, but no tours were as short as the best found with the Boltzmann acceptance probability. However, five of the tours found were of length less than 3.5, whereas only one of the tours generated by the Boltzmann machine was that good.

TOUR										LENGTH
J	G	E	D	A	B	C	I	H	F	3.3341
I	J	H	E	D	A	C	B	F	G	3.3968
I	J	D	A	B	C	E	G	F	H	3.4649
I	E	D	F	G	H	A	C	B	J	3.4761
J	F	E	A	C	B	I	H	G	D	3.8840
F	C	B	A	D	G	E	I	J	H	3.8944
H	J	A	I	B	C	F	D	E	G	3.9045
C	J	H	I	G	D	E	F	B	A	3.9513
F	H	D	J	I	A	C	B	G	E	3.9592

Example 7.5 A Cauchy machine for the traveling salesman problem

The Cauchy machine solution to the traveling salesman problem was repeated using the faster Cauchy cooling schedule. In this case, the Cauchy net found tours of length less than 4 for 11 of the 100 random starting configurations. The shortest was

| G | I | C | B | D | A | E | F | H | J | (length 3.63). |

Analysis

Extremely good results have been reported [Szu, 1990] for the Cauchy machine solution of the traveling salesman problem by using a clever mapping of the problem onto a one-dimensional space. Taking as fixed the first city to be visited, we find that there are $(n - 1)!$ permutations of the remaining cities, i.e., there are $(n - 1)!$ distinct tours. The relation between a tour and an integer between 0 and $(n - 1)! - 1$ is found by representing the integer in terms of the numbers $(n - 1)!, \ldots, 2!, 1!, 0!$. For example, with $n = 5$, the integer

$$15 = 0 \times 4! + 2 \times 3! + 1 \times 2! + 1 \times 1! + 0 \times 0!,$$

which gives a representation index of (0, 2, 1, 1, 0). This corresponds to the tour that visits the cities *A, D, C, E,* and *B,* in that order [Szu & Maren, 1990].

For convenience in describing the process of obtaining the new tour from the base tour A, B, C, D, E, and the index, we denote the index as $(b_4, b_3, b_2, b_1, b_0)$. We then have the following:

1. Starting with the most significant bit (the coefficient of 4!), we take the city in position $1 + b_4$ and move it to the left b_4 places to get the new tour. Since $b_4 = 0$, the first city (A) does not move.

2. Next, we take the city in position $2 + b_3$ and move it to the left b_3 places. Since $b_3 = 2$, the city currently in position $2 + 2$ (i.e., city D) is moved 2 places to the left, giving the new tour $A\ D\ B\ C\ E$.

3. The city in position $3 + b_2$ is moved b_2 places to the left; i.e., the city in position $3 + 1 = 4$ (city C) is moved left 1 place to give $A\ D\ C\ B\ E$.

4. The city in position $4 + b_1$ is moved b_1 places to the left; i.e., the city in position $4 + 1 = 5$ (city E) is moved 1 place to the left, giving $A\ D\ C\ E\ B$.

5. Finally, the city in position $5 + b_0$ moves b_0 spaces.

Note that b_0 and b_4 are always 0. Thus, at the first step, the city in position 1 does not move. The value of b_3 can be 0, 1, 2, or 3; hence, at the second step, the city in position 2, 3, 4, or 5 may move. However, if the city in position 2 moves, it moves 0 steps; if the city in position 3 moves, it moves 1 step, etc. Thus, whichever city moves, it will move to the second position. Similarly, b_2 can take on values of 0, 1, or 2. So the city that is currently in position 3, 4, or 5 will move (depending on the value of b_2); it will move left to the third position. At the fourth step, b_1 has the value of either 0 or 1; thus, the city in position 4 or 5 will move (to the fourth position). Since b_0 is always 0, at the last step the city in the fifth position moves 0 steps to the left [Szu, 1990].

With this one-dimensional representation of the traveling salesman problem, new states can be generated by adding noise according to the Cauchy distribution. The new state is then accepted, based on the Cauchy acceptance function presented earlier. Szu does not discuss implementing this form of the Cauchy machine in the framework of a strictly local neural network.

The Cauchy machine can be viewed as a Gaussian machine with Cauchy distribution noise, as discussed earlier in this section. This is equivalent to treating the Cauchy machine as a Boltzmann machine, with uniform probability of generating each state, but with the Cauchy acceptance function rather than the Boltzmann acceptance function. It is also possible to view the differences between the Cauchy machine and the Boltzmann machine as a result of using a different distribution for the generation of states (i.e., the Cauchy distribution rather than the Gaussian distribution), but the same acceptance function. For example, analysis of the cooling schedule for the Cauchy machine can be based on such an approach [Jeong & Park, 1989]. The probability distribution for generating a new configuration of the net is the Cauchy distribution

$$G(k) = \frac{T(k)}{[T(k)^2 + \|\Delta x\|^2]^{.5(n+1)}},$$

rather than the Gaussian distribution used for the Boltzmann machine. The probability of accepting a new configuration of the net is given by the same function for the Cauchy machine as for the Boltzmann machine; i.e., the probability of accepting a change of state is

$$A(k, T(k)) = \frac{1}{1 + \exp\left(-\dfrac{\Delta C}{T(k)}\right)}.$$

However, a faster annealing schedule can be used; that is, the temperature parameter can be reduced more quickly than in the Boltzmann machine. The annealing schedule for the Cauchy machine is

$$T_C(k) = \frac{T_0}{k}.$$

Further analysis [Jeong & Park, 1989] has shown that the annealing schedule for the Cauchy machine can be taken to be

$$T_C(k) = \frac{T_0}{k^\alpha},$$

where $1 \leq \alpha < 2$.

7.2 A FEW MORE NETS THAT LEARN

There are numerous extensions to the learning algorithms and network architectures we have discussed. In this section, we consider two feedforward self-organizing nets that use modified Hebbian learning rather than competition and four types of learning for recurrent nets. The learning procedure for the Boltzmann machine (Section 7.2.2) is one way of incorporating the advantages of probabilistic changes in activations in a net that learns specified input-output relations. We then discuss several methods of training recurrent nets using backpropagation.

7.2.1 Modified Hebbian Learning

The self-organizing nets in Chapters 4 and 5 use a competitive layer as part of the process of clustering similar input patterns. Other types of information can also be obtained from self-organizing nets. In this section we consider two types of nets that extract information about the training patterns using unsupervised training without competition [Oja, 1992]. The first type learns the principal components (eigenvectors corresponding to the largest eigenvalues) of the correlation matrix of the training vectors. The second type finds the parameters for an optimal curve or surface fit to the training patterns. These nets use a single layer of linear units. (The output function is the identity function.)

Principal components

One Output Unit.   The simplest net uses a single output unit with a modified Hebbian learning rule that causes the weight vector to approach the eigenvector of unit length corresponding to the largest eigenvalue of the correlation matrix of the input vectors [Oja, 1982]. The correlation matrix is the average of the weight matrices to store the training vectors, as described in Chapter 2:

$$C_{ij} = \frac{1}{P} \sum_{p=1}^{P} x_i x_j.$$

If the input vectors have a mean of zero, this is the covariance matrix, which, in general, is written

$$C_{ij} = \frac{1}{P} \sum_{p=1}^{P} (x_i - m_i)(x_j - m_j),$$

where m_i is the average of the ith components of the input vectors.

If patterns are presented repeatedly, the weights found by the "plain Hebb" rule will continue to grow. Although the weights can be renormalized to prevent this, the modified Hebb rule suggested by Oja causes the weight vector to approach unit length automatically. The Oja learning rule is

$$\Delta w_i = \alpha \, y(x_i - y \, w_i),$$

where $y = \sum_i x_i w_i$ is the output of the net. Hertz, Krogh, and Palmer (1991) prove that the weights have the desired properties and point out that the weight updates depend on the difference between the input x_i and the backpropagated output $y \, w_i$. This modification is the standard Hebb rule, $\Delta w_i = \alpha \, y \, x_i$, with a decay term that is proportional to the output squared. Oja's rule maximizes the average squared output (y^2).

Example 7.6 Using a modified Hebb rule to find principal components

The graph in Figure 7.10 is a simplified illustration of an example presented by Hertz, Krogh and Palmer (1991). They used input from a two-dimensional Gaussian distribution and found that the average weight vector pointed toward the center of the distribution. The average weight vector was of approximately unit length.

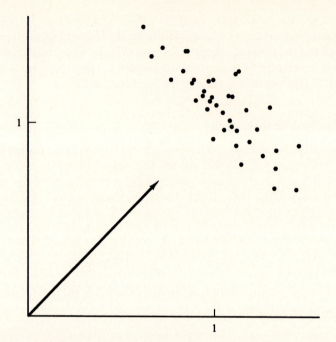

Figure 7.10 Weight vector for modified Hebbian learning; adapted from Hertz, Krogh, and Palmer (1991).

***M* Output Units**   The preceding ideas have been extended to several output units [Sanger, 1989; Oja, 1989]. The learning rules developed by Sanger and Oja are similar, and each reduces to the Oja learning rule for one output unit when $M = 1$. The Sanger learning rule causes the net to find the first M principal components, in order, as the weight vectors for the M output units. This rule can be extended to nonlinear output functions also. For the linear units we have been considering in this section, the Sanger rule is

$$\Delta w_{ij} = \alpha y_j \left(x_i - \sum_{k=1}^{j} y_k w_{ik} \right),$$

where $y_j = \sum_i x_i w_{ij}$ is the output of the jth output unit.

The Oja rule finds weight vectors that span the same subspace, but are not necessarily the individual eigenvectors. The Oja M-unit rule is

$$\Delta w_{ij} = \alpha y_j \left(x_i - \sum_{k=1}^{M} y_k w_{ik} \right).$$

(See Hertz, Krogh, and Palmer, 1991, and Oja, 1992, for further discussion of these rules.)

The use of backpropagation for data compression is often described as *self-supervised backpropagation*. The hidden units project onto the subspace of the first M principal components, with results and dynamics that are similar to those produced by the nets discussed in this section [Sanger, 1989; Hertz, Krogh & Palmer, 1991].

Minor components

A modified anti-Hebbian learning rule can be used to find the parameters for optimal curve or surface fitting [Xu, Oja, & Suen, 1992]. The standard least squares (LS) approach to the common problem of fitting a line to a set of data consists of minimizing the vertical distance from the data points to the line. In many applications, the optimal line is the line found by minimizing the sum of the squares of the distances from the points to the line, where the distances are measured in a direction perpendicular to the estimated line. This is the TLS (total least squares) method. (See Golub & Van Loan, 1989.) Unfortunately, the computations for TLS are more involved than for LS. However, the desired solution can be obtained from the minimum eigenvalue, and its corresponding normalized eigenvector, for the covariance matrix of the data; i.e., the problem reduces to finding the minor component of the data set.

As in the case of principal components discussed before, we consider a single linear output unit with response

$$y = \sum_{i=1}^{n} x_i w_i.$$

The learning rule is now an anti-Hebbian rule in which

$$\Delta \mathbf{w} = -\alpha y (\mathbf{x} - y\, \mathbf{w}).$$

The normalized form of the learning rule scales the weight vector to unit length before multiplying by the output of the net, to give

$$\Delta \mathbf{w} = -\alpha\, y \left(\mathbf{x} - y\, \frac{\mathbf{w}}{\|\mathbf{w}\|} \right).$$

The straight line that best fits the P data points, $(d_1(p), d_2(p))$, $p = 1, \ldots, P$, is given by

$$w_1(x_1 + m_1) + w_2(x_2 + m_2) = 0$$

where (m_1, m_2) is the vector mean of the data. The coefficients w_1 and w_2 are the weights found by the following algorithm [Xu, Oja, & Suen 1992].

Step 0.   Compute vector mean for the data:

$$\mathbf{m} = \frac{1}{P} \sum_{p=1}^{P} \mathbf{d}(p).$$

Form training input points (for $p = 1, \ldots, P$):

$$\mathbf{x}(p) = \mathbf{d}(p) - \mathbf{m}.$$

Initialize weight vector $\mathbf{w}$:

Use uniform random distribution on $[0, 1]$ for each component.

Step 1.   For each input vector (presented in random order), do Steps 2–5:

    Step 2.   Compute output:

$$y = x_1 w_1 + x_2 w_2.$$

    Step 3.   Update weights:

$$\mathbf{w}(\text{new}) = \mathbf{w}(\text{old}) - \alpha\, y(\mathbf{x} - y\, \mathbf{w}).$$

    Step 4.   Reduce learning rate.

    Step 5.   If weights have stopped changing, or if maximum number of vector presentations have been made, then stop; otherwise continue.

    The performance of this approach has been found for 500 data points (d_1, d_2), generated by adding Gaussian noise to points on a straight line [Xu, Oja, & Suen, 1992]. In the simplest case these authors report, the initial value of $\alpha = 0.01$ was reduced linearly over the first 500 time steps (vector presentations) to $\alpha = 0.0025$ and held constant for the remainder of the 3,000 learning steps. There was very little variation in the solution values over the last 500 time steps.

    Figure 7.11 shows a result similar to that found by Xu, Oja, and Suen (1992), illustrated, however, for only a few data points.

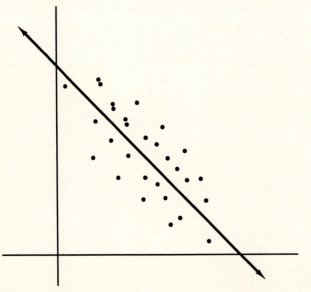

Figure 7.11 Line of best fit.

7.2.2 Boltzmann Machine with Learning

Following the development in Ackley, Hinton, and Sejnowski (1985), we now consider a learning rule for the Boltzmann machine. The simple relationship between the energy of a state in a Boltzmann machine and the weights of the network leads to the following expression relating the weights to the probability of a particular state α of the net:

$$\frac{\partial \ln P_\alpha}{\partial w_{ij}} = \frac{1}{T} [x_i x_j - PF_{ij}].$$

Here, $x_i x_j$ is 1 if units X_i and X_j are both "on" in state α (and is 0 otherwise), and PF_{ij} is the probability of finding units X_i and X_j both "on" when the system is in equilibrium (with no units clamped, i.e., held fixed at the desired values).

The most interesting situations for which a learning algorithm is needed are the cases in which only partial information about the global states of the system is available. Thus, we assume that the net consists of visible units (input and output units) and hidden units. During training, the activations of the visible units are clamped. After training, some of the visible units (the input units) are clamped, and the net is allowed to find the correct values for the other visible units (the output units). Hidden units are never clamped. For a net with v visible units, there are 2^v possible states for which probability distributions might be known. In general, unless the number of hidden units is extremely large, a perfect model of all possible states is not possible.

Based on information-theoretic arguments, the agreement between the desired probabilities for the visible units and the probabilities of the visible units when the net is at equilibrium can be increased by changing the weights. Furthermore, the weight changes can be made on the basis of local information. The weight change rule is

$$\Delta w_{ij} = \mu [PC_{ij} - PF_{ij}],$$

where PC_{ij} is the probability that units X_i and X_j are both "on" when the visible units are clamped and the net is in equilibrium and PF_{ij} is the probability that units X_i and X_j are both "on" when the system is "running free," i.e., no units are clamped, and the net is in equilibrium.

As originally presented, the algorithm uses a fixed weight-step increment if $PC_{ij} > PF_{ij}$ and the same-sized decrement for the weights if $PC_{ij} < PF_{ij}$. Difficulties can occur when only a few of the 2^v possible states for the visible units are specified. Rather than trying to demand that other (nonspecified) states never occur (which would require infinitely large energy for those states, which in turn would require infinitely large weights), it is recommended that one use noisy inputs with low, but nonzero, probabilities.

In the next several sections, we consider the performance of the Boltzmann machine in learning a simple "encoder problem," as described by Ackley, Hinton, and Sejnowski (1985). The problem is to train the net to reproduce the input pattern on the output units after passing the signals through a hidden layer that has fewer units than the input and output layers. This is a simple form of the data compression application of backpropagation discussed in Chapter 6.

Architecture

We consider a Boltzmann machine consisting of four input units, two hidden units, and four output units, often called a *4–2–4 net*. The difference between this architecture and the typical architecture for a backpropagation net (with four input units, two hidden units, and four output units) is the presence of interconnections among the input units, among the hidden units, and among the output units in the Boltzmann machine. To simplify the diagram, the weights are not shown on the connections in Figure 7.12. A bias is also used for each unit, but is not shown.

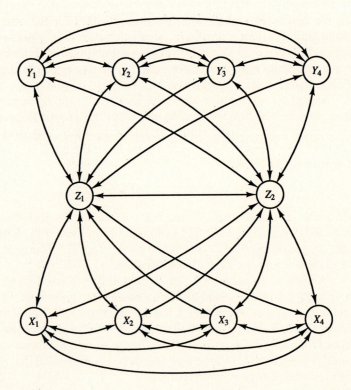

Figure 7.12   Architecture for Boltzmann machine solution of 4–2–4 encoder problem.

Algorithm

The learning process for the Boltzmann machine is influenced by choices of the learning rate (or the use of a fixed-size weight adjustment) and the length of time over which the estimates of *PC* and *PF* are gathered. The algorithm presented here follows a published description of the process [Ackley, Hinton, & Sejnowski, 1985].

An *epoch* is the same number of attempted activation updates as there are unclamped units. This allows each unit, on average, one chance to update its activation on each epoch.

The training process is as follows:

Step 0.   Initialize all weights to zero.

Do Steps 1–38 until weights stabilize or differences between *PC* and *PF* are small.

Step 1.   For each training vector, do Steps 2–17 the specified number of times.

    Step 2.   Clamp values for visible units.

        (Fix the values of the input and output units.)

    Step 3.   Allow net to reach equilibrium: Do Steps 4–10.

        Step 4.   Initialize activations of hidden units:

            Set to "on" or "off" at random.

            Initialize temperature.

        Step 5.   For specified annealing schedule, do Steps 6–10.

            Step 6.   For specified number of epochs, do Steps 7–9.

                Step 7.   Choose unit at random.

                Step 8.   Unit determines ΔE = net input.

                Step 9.   Set unit to "on," with probability

$$P = \cfrac{1}{1 + \exp\left(-\cfrac{\Delta E}{T}\right)},$$

                regardless of previous activation.

            Step 10.   Reduce temperature.

    Step 11.   Gather statistics for PC_{ij}; Do Steps 12–17.

        Step 12.   For specified number of epochs, do Steps 13–16.

            Step 13.   Choose unit at random.

            Step 14.   Unit determines ΔE = net input.

            Step 15.   Set unit to "on," with probability

$$P = \cfrac{1}{1 + \exp\left(-\cfrac{\Delta E}{T}\right)},$$

            regardless of previous activation.

Step 16.   Record activation data for unit.

Step 17.   For each pair of units, determine fraction of the time that the units are both "on."

(This gives probabilities for PC_{ij} for current training run.)

Step 18. Find average value of PC_{ij} for each pair of units i, j, over all training runs.

Step 19. Find average value of PC_{ij} over all training patterns.

Step 20. Gather statistics for free-running net:

Do Steps 21–35 for specified number of cycles.

Step 21.   Allow net to reach equilibrium: Do Steps 22–28.

Step 22.   Initialize activations of all units:

Set to "on" or "off" at random.

Initialize temperature.

Step 23.   For specified annealing schedule, do Steps 24–28.

Step 24.   For specified number of epochs, do Steps 25–27.

Step 25.   Choose unit at random.

Step 26.   Unit determines ΔE = net input.

Step 27.   Set unit to "on," with probability

$$P = \frac{1}{1 + \exp\left(-\dfrac{\Delta E}{T}\right)},$$

regardless of previous activation.

Step 28.   Reduce temperature.

Step 29.   Gather statistics for PF_{ij}; Do Steps 30–35.

Step 30.   For specified number of epochs, do Steps 31–34.

Step 31.   Choose unit at random.

Step 32.   Unit determines ΔE = net input.

Step 33.   Set unit to "on," with probability

$$P = \frac{1}{1 + \exp\left(-\dfrac{\Delta E}{T}\right)},$$

regardless of previous activation.

Step 34.   Record activation data for unit.

Step 35.   For each pair of hidden units, determine fraction of the time that they are both "on."

(This gives probabilities for PF_{ij} for current cycle.)

Step 36. Find average value of PF_{ij} over all cycles.

Step 37. Update weights:

Increment w_{ij} by 2 if $PC_{ij} > PF_{ij}$.

Decrement w_{ij} by 2 if $PC_{ij} < PF_{ij}$.

Step 38. Test whether training is complete.

Application.   Ackley, Hinton, and Sejnowski (1985) have illustrated the use of a Boltzmann machine to learn a variety of encoder problems. We have used the architecture geared for the simplest problem, the 4–2–4 encoder in Figure 7.12. This net consists of four input units, two hidden units, and four output units. The units are fully interconnected, with the exception of connections directly between input units and output units. In general, for an encoder problem, there are two groups of visible units with V units in each group and H hidden units with

$$\ln 2 \le H < V.$$

A bias is also used for each unit.

The training vectors (environmental vectors) are the following vectors:

INPUT				OUTPUT			
(1	0	0	0)	(1	0	0	0)
(0	1	0	0)	(0	1	0	0)
(0	0	1	0)	(0	0	1	0)
(0	0	0	1)	(0	0	0	1)

That is, each set of visible units can have only one unit "on" at any time, and we desire the pattern of activations in the two sets of visible units to match, even though there is no direct communication between the two groups.

Because weights can become very large if the net does not receive training information for many possible configurations, noisy versions of these training vectors were used. On each presentation of a training pattern, the component that is 1 in the true training vector is set to 0, with probability 0.15. The components that are 0 are set to 1, with probability 0.05.

The annealing schedule was

- Two epochs at $T = 20$
- Two epochs at $T = 15$
- Two epochs at $T = 12$, and
- Four epochs at $T = 10$.

In other words, following the previous algorithm, the initial temperature was 20. Now, there are two unclamped units during the first phase of the training cycle (when the statistics to find PC_{ij} are determined). Thus, Steps 7–9 are performed four times (two epochs, each of which consists of two attempts to update units) at $T = 20$; the temperature is reduced to $T = 15$, and Steps 7–9 are performed four more times. Then they are performed four times with $T = 12$ and eight times with $T = 10$.

Statistics were gathered for 10 epochs at $T = 10$; that is, Steps 13–16 were performed 20 times. The average fraction of the time that units i and j are both "on" is determined.

This process is repeated for each of the four training vectors, and the results for all of the training vectors are averaged.

The process of determining PF_{ij} uses the same annealing schedule and gathers statistics for the same number of epochs at the final temperature ($T = 10$). However, since no units are clamped during this second phase, each epoch consists of 10 attempts to update units.

Once the values of PC_{ij} and PF_{ij} have been found, the weights are updated and the entire weight update cycle (Steps 1–38) is repeated, until the weights have stabilized or the differences between PC_{ij} and P_{ij} are sufficiently small.

In 250 tests of the 4–2–4 encoder problem, the net always found one of the global minima and remained at that solution. As many as 1,810 weight update cycles were required, but the median number was 110 [Ackley, Hinton, & Sejnowski, 1985].

After training, the net can be applied by clamping the input units and allowing the net to reach equilibrium (for example, following the same annealing schedule as given for the training algorithm). The activations of the output units then give the response of the net.

7.2.3 Simple Recurrent Net

Several neural networks have been developed to learn sequential or time-varying patterns. Unlike the recurrent nets with symmetric weights or the feedforward nets, these nets do not necessarily produce a steady-state output. In this section, we consider a simple recurrent net [Elman, 1990; Hertz, Krogh, & Palmer, 1991] that can be used to learn strings of characters [Servan-Schreiber, Cleeremans, & McClelland, 1989]. This net can be considered a "partially recurrent" net, in that most of the connections are feedforward only. A specific group of units receives feedback signals from the previous time step. These units are known as *context units*. The weights on the feedback connections to the context units are fixed, and information processing is sequential in time, so training is essentially no more difficult than for a standard backpropagation net.

Architecture

The architecture for a simple recurrent net is as shown in Figure 7.13.

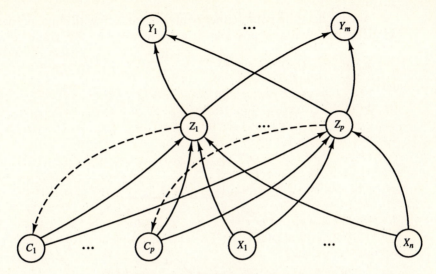

Figure 7.13   Architecture for simple recurrent net.

Algorithm

At time t, the activations of the context units are the activations (output signals) of the hidden units at the previous time step. The weights from the context units to the hidden units are trained in exactly the same manner as the weights from the input units to the hidden units. Thus, at any time step, the training algorithm is the same as for standard backpropagation.

Application

One example of the use of a simple recurrent net demonstrates the net's ability to learn an unlimited number of sequences of varying length [Servan-Schreiber, Cleeremans, & McClelland, 1989]. The net was trained to predict the next letter in a string of characters. The strings were generated by a small finite-state grammar in which each letter appears twice, followed by a different character. A diagram of the grammar is given in Figure 7.14. The string begins with the symbol B and ends with the symbol E.

At each decision point, either path can be taken with equal probability. Two
examples of the shortest possible strings generated by this grammar are

   B   P   V   V   E

and

   B   T   X   S   E

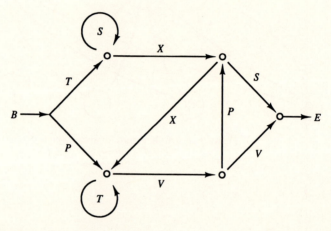

Figure 7.14 One grammar for simple recurrent net (Reber, 1967; Servan-Schreiber et al., 1989).

The training patterns for the neural net consisted of 60,000 randomly generated strings ranging in length from 3 to 30 letters (not including the Begin and
End symbols).

The neural net architecture for this example had six input units (one for each
of the five characters, plus one for the Begin symbol) and six output units (one
for each of the five characters, plus one for the End symbol). There were three
hidden units (and therefore, three context units). In a specific case, the net might
be displayed as in Figure 7.15. With the architecture as illustrated, the input
pattern for the letter *B* (the Begin symbol) would correspond to the vector
(1, 0, 0, 0, 0, 0).

Training the net for a particular string involves several steps, the number
depending on the length of the string. At the beginning of training, the activations
of the context units are set to 0.5. The first symbol (the Begin symbol) is presented
to the input units, and the net predicts the successor. The error (the difference
between the predicted and the actual successor specified by the string) is determined and backpropagated, and the weights are adjusted. The context units receive a copy of the hidden units' activations, and the next symbol in the string

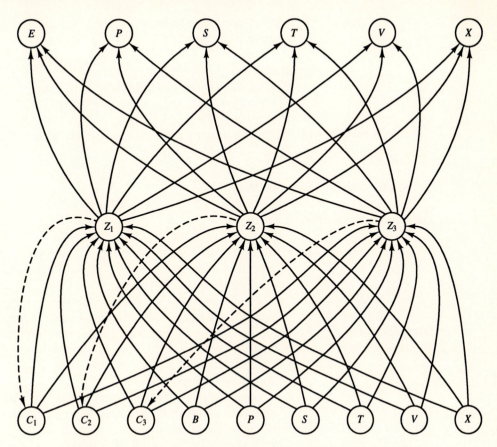

Figure 7.15 Simple recurrent net to learn context-sensitive grammar. (Elman, 1990; Servan-Schreiber et al., 1989).

(which was the target value for the output units on the first step of training) is presented to the input units. Training continues in this manner until the End symbol is reached.

The training algorithm for a context-sensitive grammar in the example given is as follows:

For each training string, do Steps 1–7.

Step 1.   Set activations of context units to 0.5.
Step 2.   Do Steps 3–7 until end of string.
   Step 3.   Present input symbol.
   Step 4.   Present successor to output units as target response.
   Step 5.   Calculate predicted successor.
   Step 6.   Determine error, backpropagate, update weights.

Step 7.   Test for stopping condition:
          If target $= E$, then
                    stop;
          otherwise,
                    Copy activations of hidden units to context units;
                    continue.

As a specific example of the training process, suppose the string

    B T X S E

is used for training. Then we have:

Step 2.   Begin training for this string.
        Step 3.   Input symbol B, i.e., $(1, 0, 0, 0, 0, 0)$.
        Step 4.   Target response is T, i.e., $(0, 0, 0, 1, 0, 0)$.
        Step 5.   Compute predicted response, a real-valued vector with
                             components between 0 and 1.
        Step 6.   Determine error, backpropagate, update weights.
        Step 7.   Copy activations of hidden units to context units.
Step 2.   Training for second character of the string.
        Step 3.   Input symbol T, i.e., $(0, 0, 0, 1, 0, 0)$.
        Step 4.   Target response is X, i.e., $(0, 0, 0, 0, 0, 1)$.
        Step 5.   Compute predicted response,
        Step 6.   Determine error, backpropagate, update weights.
        Step 7.   Copy activations of hidden units to context units.
Step 2.   Training for third character of the string.
        Step 3.   Input symbol X, i.e., $(0, 0, 0, 0, 0, 1)$.
        Step 4.   Target response is S, i.e., $(0, 1, 0, 0, 0, 0)$.
        Step 5–7.   Train net and update activations of context units.
Step 2.   Training for fourth character of the string.
        Step 3.   Input symbol S, i.e., $(0, 1, 0, 0, 0, 0)$.
        Step 4.   Target response is E, i.e., $(1, 0, 0, 0, 0, 0)$.
        Steps 5–6.   Train net.
        Step 7.   Target response is the End symbol;
                             training for this string is complete.

After training, the net can be used to determine whether a string is a valid string, according to the grammar. As each symbol is presented, the net predicts the possible valid successors of that symbol. Any output unit with an activation of 0.3 or greater indicates that the letter it represents is a valid successor to the current input. To determine whether a string is valid, the letters are presented to the net sequentially, as long as the net predicts valid successors in the string. If the net fails to predict a successor, the string is rejected. If all successors are predicted, the string is accepted as valid.

The reported results for 70,000 random strings, 0.3% of which were valid according to the grammar, are that the net correctly rejected all of the 99.7% of the strings that were invalid and accepted all of the valid strings. The net also performed perfectly on 20,000 strings from the grammar and on a set of extremely long strings (100 or more characters in length).

7.2.4 Backpropagation in Time

We now consider a network in which the outputs from the net at one time step become the inputs at the next time step. These nets can be trained for several time steps by making copies of the net (with the same weights), training each copy, and then averaging the weight updates. This process, originally introduced by Rumelhart, Hinton, and Williams (1986), is called "backpropagation in time" [Hertz, Krogh, & Palmer, 1991] or sometimes "recurrent backpropagation" [Hecht-Nielsen, 1990].

Architecture

A simple example of a backpropagation in time net is shown in Figure 7.16. An expanded form of a backpropagation in time net is illustrated in Figure 7.17. A generalization of this allows for both external inputs and recurrent signals from the previous time step, as shown in Figure 7.18. As in the simple recurrent net discussed in the previous section, the recurrent connections have weights fixed at 1; the adjustable weights are shown.

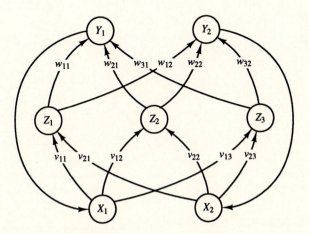

Figure 7.16 A recurrent multilayer net in which the outputs at one time step become inputs at the next step.

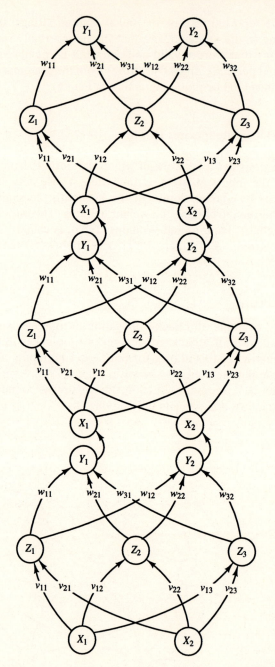

Figure 7.17 The recurrent multilayer net of Figure 7.16 expanded for three time steps.

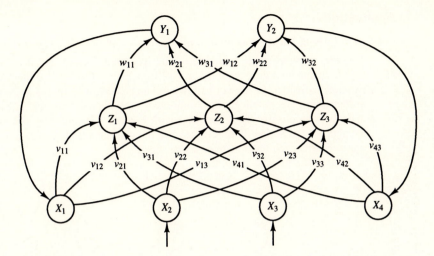

Figure 7.18   A recurrent multilayer net with external and recurrent inputs at each step.

Algorithm

The training algorithm using backpropagation in time for a recurrent net of the form illustrated in Figure 7.16 or 7.18 is based on the observation that the performance of such a net for a fixed number of time steps N is identical to the results obtained from a feedforward net with $2N$ layers of adjustable weights. For example, the results produced by the net in Figure 7.16 after three time steps could also be obtained from the net shown in Figure 7.17.

The training process consists of a feedforward pass through the entire expanded network (for the desired number of time steps). The error is then computed for each output layer (i.e., for each time step). The weight adjustments for each copy of the net are determined individually (i.e., computed) and totaled (or averaged) over the number of time steps used in the training. Finally, all copies of each weight are updated. Training continues in this way for each training pattern, to complete an epoch. As with standard backpropagation, typically, many epochs are required.

Note that it is not necessary actually to simulate the expanded form of the net for training. The net can run for several time steps, determining the information on errors and the weight updates at each step and then totaling the weight corrections and applying them after the specified number of steps.

In addition, information on errors does not need to be available for all output units at all time steps. Weight corrections are computed using whatever information is available and then are averaged over the appropriate number of time steps. In the example in the next section, information on errors is supplied only at the second time step; no responses are specified after the first time step.

Application

Example 7.7 Using backpropagation in time to form a simple shift register

A neural network with no hidden units has been trained to act as a simple shift register using backpropagation in time [Rumelhart, Hinton, & Williams, 1986a]. For example, consider the network shown in Figure 7.19, with three input units and three output units. (In practice, these units can be the same, but we will treat them as distinct to emphasize the similarities with Figures 7.16 and 7.18. For simplicity, the weights are not shown in the figure or in the diagrams that follow. In addition to the units shown, each unit receives a bias signal.

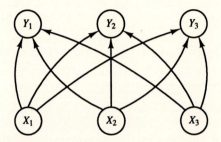

Figure 7.19   Recurrent net used as shift register.

The training patterns consist of all binary vectors with three components; the target associated with each vector is the pattern shifted two positions to the left (with wraparound). This is the desired response of the net after two time steps of processing. The expanded form of the net is shown in Figure 7.20.

This example illustrates the fact that it is not required to have information on errors at the intermediate time steps. If the net were told the desired response after one time step, the solution would be very simple. Instead, the weights in both copies of the net are adjusted on the basis of errors after two time steps. In general, a combination of information on errors at the final level and at any or all intermediate levels may be used.

Rumelhart, Hinton, and Williams (1986a, 1986b) found that the net consistently learned the weights required for a shift register in 200 or fewer epochs of training, with a learning rate of 0.25, as long as the bias weights were constrained negative. The same conclusions apply to the net with five input (and output) units. In either of these cases, if the biases are not restricted to be negative, other solutions to the training can also result. These give the desired results after an even number of time steps, but not after an odd number of time steps.

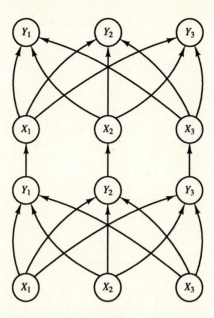

Figure 7.20   Expanded diagram of recurrent net used as shift register.

Example 7.8 Using backpropagation in time to produce a damped sinusoid function

Backpropagation in time can also be used to train a neural net to produce a damped sinusoid function, as illustrated in Figure 7.21. The input units represent function values at several time steps, and the output unit gives the function value at the next time step. In a simple example, shown in Figure 7.22, there are four input units and five hidden units. The number of input units required depends on the frequency ω of the sinusoidal oscillation in the target function

$$f(t) = \frac{\sin(\omega t)}{\omega t}.$$

For $\omega = \pi$, seven input units are sufficient. The results shown in Figure 7.21 are based on $\omega = 0.5$; the network has 10 input units and 10 hidden units.

At time step t, X_1 receives the net's computed function value, $f(t - 1)$, from Y. X_2 receives the previous function value, $f(t - 2)$, from X_1, X_3 receives $f(t - 3)$ from X_2, and X_4 receives $f(t - 4)$ from X_3.

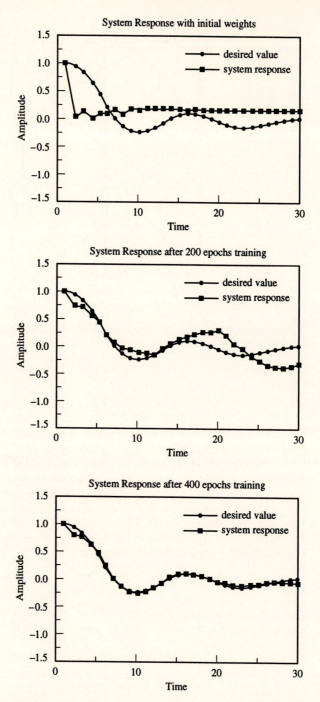

Figure 7.21   Target function and computed response during training.

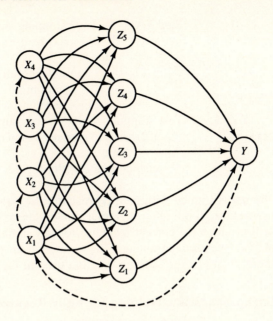

Figure 7.22   Recurrent neural net.

Here again, we can think of the training process for the net as consisting of many copies of the net, but it is not necessary actually to program each copy separately. The training process is as follows:

Step 0.   Initialize weights (small random values).
Step 1.   Until stopping condition for training, do Steps 2–9.
  Step 2.   Initialize activations. (Set to small random values.)
  Step 3.   Present initial function value, $f(0)$, to input unit X_1.
  Step 4.   Until stopping condition for epochs, do Steps 5–8.
    Step 5.   Calculate the response of the net:

$$y = f(1).$$

  Step 6.   Calculate error for current time step.
        Find weight updates by backpropagation.
        (Do not change the weights.)
    Step 7.   Update the activations:

$$x_4 = x_3,$$

$$x_3 = x_2,$$

$$x_2 = x_1,$$

$$x_1 = y.$$

Step 8.   Test for stopping condition for epoch:
              If $y > $ max, or if number of time steps > 30,
              then
                    apply weight updates and
                          continue with Step 9;
                    else continue with Step 4.
Step 9.   Test stopping condition for training:
              If error $<$ tolerance or total number of epochs $>$ limit,
                    stop;
              else continue with Step 1.

If the net is trained with data representing several damped sine functions, with the same periods but different initial amplitudes (at $t = 0$), then after training, the net will produce the correct response for any initial amplitude that is within the range of values used in training. This shows that the net has learned something more general than just the specific function used in the training.

7.2.5 Backpropagation Training for Fully Recurrent Nets

Backpropagation can be applied to a recurrent net with an arbitrary pattern of connections between the units. The process described here is the recurrent backpropagation presented by Hertz, Krogh, and Palmer (1991), based on work by Pineda [1987, 1988, 1989] and Almeida [1987, 1988].

The activations of our general recurrent net are assumed to obey the evolution equations of Cohen and Grossberg (1983) and Hopfield (1984), namely,

$$\tau \frac{dv_i}{dt} = -v_i + g(x_i + \sum_j v_j w_{ij}),$$

where x_i is the external input to unit V_i and τ is a time constant. We assume that at least one stable attractor exists, i.e.,

$$v_i = g(x_i + \sum_j v_j w_{ij}).$$

To train a recurrent net using backpropagation, we assume that target values are specified for some units, which we call output units. The error is defined in the usual manner for these units, i.e.,

$$E = \frac{1}{2} \sum_k (t_k - v_k)^2,$$

where the summation ranges over all the output units.

Gradient descent applied to this net gives a weight update rule that requires a matrix inversion at each step. However, if we write the weight updates as

$$\Delta w_{pq} = \alpha \delta_q v_p$$

where

$$\delta_q = g'(x_q + \sum_j v_j w_{qj}) y_q$$

(in which the matrix inversion is included in the term y_q), we find [Hertz, Krogh, & Palmer, 1991] that the y_q terms obey a differential equation of exactly the same form as the evolution equations for the original network. Thus, the training of the network can be described by the following algorithm:

Step 1.   Allow the net to relax to find the activations v_i; i.e., solve the equation

$$\tau \frac{dv_i}{dt} = -v_i + g(x_i + \sum_j v_j w_{ij}).$$

Define the equilibrium net input value for unit q:

$$h_q = (x_q + \sum_j v_j w_{qj}).$$

Step 2.   Determine the errors for the output units, E_q.

Step 3.   Allow the net to relax to find the y_q; i.e., solve the equation

$$\tau \frac{dy_q}{dt} = -y_q + E_q + \sum_k g'(h_k) w_{qk} y_k.$$

The weight connections of the original net have been replaced by $g'(h_k)w_{qk}$, and the activation function is now the identity function. The error term, E_q, plays the role of the external input.

Step 4.   Update the weights:

$$\Delta w_{pq} = \alpha v_p g'(h_q) y_q,$$

where v_p is the equilibrium value of unit p,
y_q is the equilibrium value of the "matrix inverse unit," and
h_q is the equilibrium net input to unit q.

Applications of recurrent backpropagation have included pattern completion [Almeida, 1987], vision [Qian & Sejnowski, 1988], and control of robot manipulators [Barhen, Gulati, & Zak, 1989]. (See Hertz, Krogh, & Palmer, 1991, for more details of the derivation of the algorithm.)

7.3 ADAPTIVE ARCHITECTURES

7.3.1 Probabilistic Neural Net

The probabilistic neural net [Specht, 1988, 1990] is constructed using ideas from classical probability theory, such as Bayesian classification and classical estimators for probability density functions, to form a neural network for pattern classification. The description here gives only the simplest form of the net. (See the Gaussian potential-function network [Lee, 1992] for a more extensive discussion.) Note that the term "probabilistic" in Specht's net does not mean that the operation of the net is stochastic, as is the case for the Boltzmann machine.

The problem we consider is to classify input vectors into one of two classes in a Bayesian-optimal manner. Bayesian decision theory allows for a cost function to represent the fact that it may be worse to misclassify a vector that is actually a member of Class A (by mistakenly assigning it to Class B) than it is to misclassify a vector that truly belongs to Class B. (Or, of course, the worse situation may be misclassification in the other direction.) The Bayes decision rule states that an input vector should be classified as belonging to Class A if

$$h_A c_A f_A(\mathbf{x}) > h_B c_B f_B(\mathbf{x}),$$

where h_A is the *a priori* probability of occurrence of patterns in Class A, c_A is the cost associated with classifying a vector as belonging to Class B when it actually belongs to Class A, and $f_A(\mathbf{x})$ is the probability density function for Class A; corresponding definitions apply to quantities with the subscript B.

The boundary between the region where the input vector is assigned to Class A and the region where it is classified as belonging to Class B is given by

$$f_A(\mathbf{x}) = \frac{h_B c_B}{h_A c_A} f_B(\mathbf{x}).$$

Usually, the *a priori* probabilities h_A and h_B are known or can be estimated accurately; for example, they can be taken to be the proportion of input patterns that belong to each of these classes.

The costs associated with misclassification, c_A and c_B, are application dependent; if no other information is available, they can be taken to be equal to each other.

Thus, the main question in applying the Bayes decision rule is how to estimate the probability density functions $f_A(\mathbf{x})$ and $f_B(\mathbf{x})$ from the training patterns. In general, a probability density function must be nonnegative everywhere, must be integrable, and the integral over all $\mathbf{x}$ must equal 1.

The probabilistic neural net uses the following estimator for the probability density function:

$$f_A(\mathbf{x}) = \frac{1}{(2\pi)^{n/2}\sigma^n} \frac{1}{m_A} \sum_{i=1}^{m_A} \exp\left[-\frac{(\mathbf{x} - \mathbf{x}_{Ai})^T(\mathbf{x} - \mathbf{x}_{Ai})}{2\sigma^2}\right].$$

In this equation, $\mathbf{x}_{Ai}$ is the ith training pattern from Class A, n is the dimension of the input vectors, m_A is the number of training patterns in Class A, and σ is a smoothing parameter corresponding to the standard deviation of the Gaussian distribution. (The role of σ will be discussed shortly.)

The operation of the net for Bayes-optimal classification is based on the fact that $f_A(\mathbf{x})$ serves as an estimator as long as the parent density is smooth and continuous. This means that $f_A(\mathbf{x})$ aymptotically approaches the parent density function as the number of data points used for the estimation increases. The function $f_A(\mathbf{x})$ is a sum of Gaussian distributions, but the result is not limited to being a Gaussian distribution.

The use of this parent density function estimator, together with the Bayes decision rule

$$h_A c_A f_A(\mathbf{x}) > h_B c_B f_B(\mathbf{x}),$$

gives good results, but suffers from the disadvantages that the entire training set must be stored and the computation needed to classify an unknown vector is proportional to the size of the training set.

Nonetheless, a neural network can be constructed using these ideas. Training is instantaneous, and the net can perform classifications with as few as one training vector from each class. Of course, the net's ability to generalize improves as more training patterns are presented.

After training, application of the net is very rapid, since the net can compute each term in the estimate for $f_A(\mathbf{x})$ in parallel.

Architecture

The probabilistic neural net for classifying input vectors into one of two classes (say, Class A and Class B) consists of four types of units: input units, pattern units, summation units, and an output unit. Pattern units are of two types, those for Class A and those for Class B. The architecture of this net is shown in Figure 7.23. The weights from the summation units to the output unit are $v_A = 1$,

$$v_B = -\frac{h_B c_B}{h_A c_A} \frac{m_A}{m_B}.$$

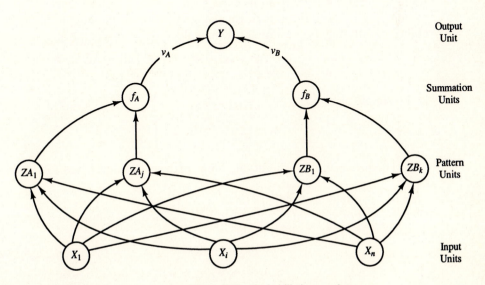

Figure 7.23 Probabilistic neural net.

Algorithm

The probabilistic neural net is constructed as training progresses. Each pattern unit represents one of the two classes to which training patterns belong; there is one pattern unit (of the appropriate type) that corresponds to each training pattern. Training patterns are first normalized to unit length. The weight vector for the pattern unit ZA_j is simply the jth training vector that belongs to Class A. As each training pattern is presented, a new pattern unit corresponding to the correct class is added to the net, its weights are set, and the unit is also connected to the correct summation unit.

The algorithm for constructing the net is as follows:

Step 1. For each training input pattern $\mathbf{x}(p)$, $p = 1, \ldots, P$, do Steps 2–3.

  Step 2. Create pattern unit Z_p:
             Weight vector for unit Z_p:

$$\mathbf{w}_p = \mathbf{x}(p).$$

             (Unit Z_p is either a ZA unit or a ZB unit.)

  Step 3. Connect pattern unit to summation unit:
             If $\mathbf{x}(p)$ belongs to Class A, then connect pattern unit Z_p (a ZA unit) to summation unit S_A.
             Otherwise, connect pattern unit Z_p (a ZB unit) to summation unit S_B.

Application

Input patterns for classification must be normalized to unit length. For vectors of unit length, the term needed for the summation in the definition of the probability density function estimator f, namely,

$$\exp\left[-\frac{(\mathbf{x} - \mathbf{w}_j)^T(\mathbf{x} - \mathbf{w}_j)}{2\sigma^2} \right],$$

is, by simple algebra,

$$\exp\left[\frac{z\_in_j - 1}{\sigma^2} \right].$$

The procedure for classifying input patterns (of unit length) is as follows:

Step 0. Initialize weights.

Step 1. For each input pattern to be classified, do Steps 2–4.

  Step 2. Pattern units:
             Compute net input:

$$z\_in_j = \mathbf{x} \cdot \mathbf{w}_j = \mathbf{x}^T \mathbf{w}_j.$$

Compute output as

$$z = \exp\left[\frac{z\_in_j - 1}{\sigma^2}\right].$$

Step 3.   Summation units:

Sum the inputs from the pattern units to which they are connected. The summation unit for Class B multiples its total input by

$$v_B = -\frac{h_B c_B}{h_A c_A}\frac{m_A}{m_B}.$$

Step 4.   Output (decision) unit:

The output unit sums the signals from f_A and f_B.
The input vector is classified as Class A if the total input to the decision unit is positive.

The net can be used for classification as soon as an example of a pattern from each of the two classes has been presented to it. However, the ability of the net to generalize improves as it is trained on more examples.

Typically, the *a priori* probabilities of Class A and Class B will be the ratio of the number of training patterns in Class A to the number of training patterns in Class B. If that is the case, then

$$\frac{h_B}{h_A}\frac{m_A}{m_B} = 1,$$

and the expression for v_B simplifies to

$$v_B = -\frac{c_B}{c_A}.$$

This ratio depends on the significance of a decision, not on the statistics of the situation. If there is no reason to bias the decision, we take $K = -1$.

Analysis

Varying σ gives control over the degree of nonlinearity of the decision boundaries for the net. A decision boundary approaches a hyperplane for large values of σ and approximates the highly nonlinear decision surface of the nearest neighbor classifier for values of σ that are close to zero.

Reported results [Specht, 1988, 1990] indicate that the net is relatively insensitive to the choice of σ. This type of net has been used to classify electrocardiograms as normal or abnormal. One application used input patterns with 46 components (before normalization). With 249 training patterns and 63 testing patterns, peak performance was obtained for σ between 4 and 6, with results almost as good for σ ranging from 3 to 10 [Specht, 1967].

7.3.2 Cascade Correlation

In addition to the probabilistic neural net, cascade correlation [Fahlman & Le-biere, 1990] is another network that builds its own architecture as training pro-gresses. It is based on the premise that the most significant difficulty with current learning algorithms (such as backpropagation) for neural networks is their slow rate of convergence. This is due, at least in part, to the fact that all of the weights are being adjusted at each stage of training. A further complication is the fixity of the network architecture throughout training.

Cascade correlation addresses both of these issues by dynamically adding hidden units to the architecture—but only the minimum number necessary to achieve the specified error tolerance for the training set. Furthermore, a two-step weight-training process ensures that only one layer of weights is being trained at any time. This allows the use of simpler training rules (the delta rule, perceptron, etc.) than for multilayer training. In practice, a modification of backpropagation known as *QuickProp* [Fahlman, 1988] is usually used. QuickProp is described later in this section.

A cascade correlation net consists of input units, hidden units, and output units. Input units are connected directly to output units with adjustable weighted connections. Connections from inputs to a hidden unit are trained when the hidden unit is added to the net and are then frozen. Connections from the hidden units to the output units are adjustable.

Cascade correlation starts with a minimal network, consisting only of the required input and output units (and a bias input that is always equal to 1). This net is trained until no further improvement is obtained; the error for each output unit is then computed (summed over all training patterns).

Next, one hidden unit is added to the net in a two-step process. During the first step, a candidate unit is connected to each of the input units, but is not connected to the output units. The weights on the connections from the input units to the candidate unit are adjusted to maximize the correlation between the candidate's output and the residual error at the output units. The residual error is the difference between the target and the computed output, multiplied by the derivative of the output unit's activation function, i.e., the quantity that would be propagated back from the output units in the backpropagation algorithm. When this training is completed, the weights are frozen and the candidate unit becomes a hidden unit in the net.

The second step in which the new unit is added to the net now commences. The new hidden unit is connected to the output units, the weights on the con-nections being adjustable. Now all connections to the output units are trained. (The connections from the input units are trained again, and the new connections from the hidden unit are trained for the first time.)

A second hidden unit is then added using the same process. However, this unit receives an input signal both from the input units and from the previous hidden unit. All weights on these connections are adjusted and then frozen. The connections to the output units are then trained. The process of adding a new unit, training its weights from the input units and previously added hidden units, and then freezing the weights, followed by training all connections to the output units, is continued until the error reaches an acceptable level or the maximum number of epochs (or hidden units) is reached.

In Figures 7.24 through 7.29, the weights to the hidden units (either from the input units or from the previously added hidden units), which are frozen before the hidden unit being added is connected to the output units, are denoted u (from input units to hidden units) or t (from the previous hidden unit to the new hidden unit). The weights from the input units directly to the output units are denoted w, and the weights from the hidden units to the output units are denoted v. The weights w and v are trained during the second step of each stage of the algorithm. Figure 7.29 shows the diagram as originally given [Fahlman & Lebiere, 1990].

Architecture

A cascade correlation net with three input units and two output units is shown in Figures 7.24 through 7.29 during the first stages of construction and learning. The bias input unit is shown by the symbol 1, its signal.

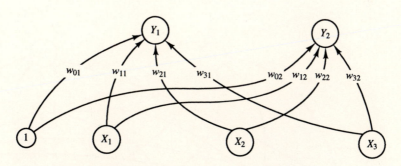

Figure 7.24   Cascade correlation net, Stage 0: No hidden units.

Figure 7.25 shows the net at the first step of Stage 1. There is one candidate unit, Z_1, which is not connected to the output units. The weights shown are trained and then frozen. Figure 7.26 shows the second step of Stage 1. The hidden unit, Z_1, has been connected to output units. The weights shown in Figure 7.25 are now frozen. Weights to the output units are trained, and the error is computed for each output unit.

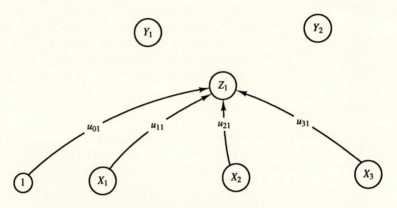

Figure 7.25   Cascade correlation net, Stage 1: One candidate unit, Z_1.

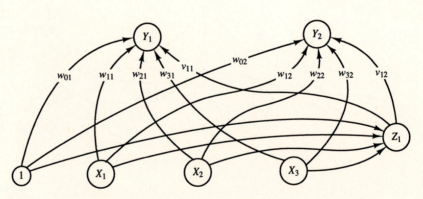

Figure 7.26   Cascade correlation net, Stage 1: One hidden unit, Z_1.

Figure 7.27 shows Stage 2, in which a new candidate unit, Z_2, receives signals from the input units and the previous hidden unit Z_1. Z_2 is not connected to output units during the first step of training. The weights shown are trained and then frozen. Weights on connections from X's to Z_1 are also frozen. In Figure 7.28, the new hidden unit, Z_2, has been connected to output units. The weights shown in Figure 7.27 are now frozen. Weights to output units are trained, and the error is computed.

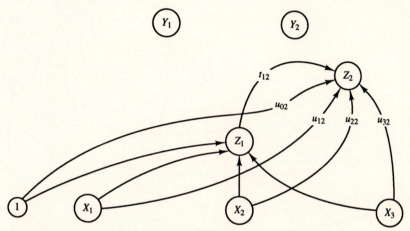

Figure 7.27 Cascade correlation net, Stage 2: New candidate unit, Z_2. (Only the weights being adjusted are shown.)

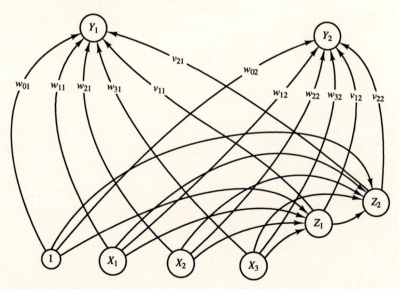

Figure 7.28 Cascade correlation net, Stage 2: Hidden unit Z_2 connected to output units.

Figure 7.29 shows the same net as in Figures 7.24 through 7.28, but in the style of the original paper [Fahlman & Lebiere, 1990] and with all weights (to this stage of network development) shown. Weights u and t are frozen, and weights w and v are retrained at each stage.

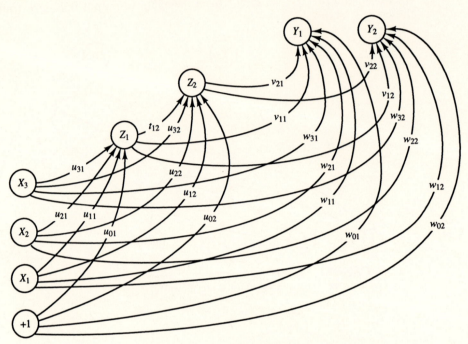

Figure 7.29   Cascade correlation network.

Algorithm

The training process for cascade correlation involves both adjusting the weights and modifying the architecture of the net. We use the same notation as for nets in previous chapters:

n	dimension of input vector (number of input units)
m	dimension of output vector (number of output units)
P	total number of training patterns
X_i	input units, $i = 1, \ldots, n$
Y_j	output units, $j = 1, \ldots, m$
$\mathbf{x}(p)$	training input vector, $p = 1, \ldots, P$
$\mathbf{t}(p)$	target vector for input vector $\mathbf{x}(p)$
$\mathbf{y}(p)$	computed output for input vector $\mathbf{x}(p)$

$E_j(p)$ residual error for output unit Y_j for pattern p:

$$E_j(p) = (y_j(p) - t_j(p))y_j(p)'$$

E_av_j average residual error for output unit Y_j:

$$E\_av_j = \frac{1}{P} \sum_{p=1}^{P} E_j(p)$$

$z(p)$ computed activation of candidate unit for input vector $\mathbf{x}(p)$

z_av average activation, over all patterns $p = 1, \ldots, P$, of candidate unit:

$$z\_av = \frac{1}{P} \sum_{p=1}^{P} z(p)$$

In the algorithm that follows, the weights on all new connections are initialized to small random values. Biases are included on all hidden units and on all output units. This is usually indicated by including an additional input unit whose activation is always 1.

The "correlation" is defined as

$$C = \sum_{j=1}^{m} \left| \sum_{p=1}^{P} (z(p) - z\_av)(E_j(p) - E\_av_j) \right|.$$

This is actually the covariance between the output of the candidate, z, and the "residual output error" [Fahlman & Lebiere, 1990]. The residual output error is the product of the true error and the derivative of the activation function for the output unit.

In a manner similar to that for the derivation of backpropagation we find that

$$\frac{\partial C}{\partial u_i} = \sum_{j=1}^{m} \sigma_j \sum_{p=1}^{P} z(p)' \, x_i(p)(E_j(p) - E\_av_j),$$

where σ_j is the sign of

$$\sum_{p=1}^{P} (z(p) - z\_av)(E_j(p) - E\_av_j),$$

z' is the derivative of the activation function for the candidate unit, and x_i is the input signal received by the candidate unit from input unit X_i.

Training can use any standard method, such as the delta rule (Section 3.1.2), for training a single-layer net. QuickProp (see shortly) is often used. The weights are adjusted to minimize the error in Steps 1 and 3 of the algorithm and to maximize C in Steps 2 and 4. The activation functions may be linear, sigmoid, Gaussian, or any other differentiable functions, depending on the application.

The algorithm for cascade correlation is as follows:

Step 1.   Start with required input and output units.
Train the net until the error reaches a minimum:
   If the error is acceptably small, stop;
   if not, compute $E_j(p)$ for each training pattern p, E_av_j, and proceed to Step 2.

Step 2.   Add first hidden unit.

  Step 3.   A candidate unit Z is connected to each input unit.
   Initialize weights from input units to Z
      (small random values).
   Train these weights to maximize C.
   When the weights stop changing, they are frozen.

  Step 4.   Train all weights v to the output units
   (from the input units and the hidden unit or units).
   If acceptable error or maximum number of units has been reached, stop.
   Else proceed to Step 5.

Step 5.   While stopping condition is false, do Steps 6 and 7.
(Add another hidden unit.)

  Step 6.   A candidate unit Z is connected to each input unit
   and each previously added hidden unit.
   Train these weights to maximize C.
   (Weights from the input units to the previously added hidden units have already been frozen.)
   When these weights stop changing, they are frozen.

  Step 7.   Train all weights v to the output units
   (from the input units and the hidden unit or units).
   If acceptable error or maximum number of units has been reached, stop.
   Else continue.

Step 5 is only a slight modification of Step 2. Simply note that after Step 2, each time a new unit is added to the net, it receives input from all input units and from all previously added hidden units. However, only one layer of weights is being trained during the unit's candidate phase. The weights from the input units to the previously added hidden units have already been frozen; only the weights to the candidate from the input units and the other hidden units are being trained. When this phase of learning stops, those weights are frozen permanently.

Variations of this technique include using several candidate units (a pool of candidates) at Step 3 or Step 6 and then choosing the best candidate to add to the net after training. This is especially beneficial in a parallel computing environment, where the training can be done simultaneously. Starting with different initial random weights for each candidate reduces the risk of the candidates getting stuck during training and being added to the net with its weights frozen at undesired values.

Cascade correlation is especially suitable for classification problems. A modified version has been developed for problems involving approximations to functions.

QuickProp.   QuickProp [Fahlman, 1988] is a heuristic modification to the backpropagation algorithm based on the assumptions that the curve of error versus weight can be approximated by a parabola which is concave up and that the change in the slope of the error curve which a particular weight "sees" is not affected by other weights that are also changing. Although these assumptions are described as risky [Fahlman, 1988], very significant speedups (compared with training with backpropagation) are reported for many problems. The slope referred to is the sum of the partial derivatives of the error with respect to the given weight, summed over all training patterns.

QuickProp uses information about the previous weight change and the value of the slope, defined as

$$S(t) = \sum_{p=1}^{P} \frac{\partial E(p)}{\partial w},$$

where the partial derivatives are summed over all patterns in the epoch.

In terms of the notation we used for standard backpropagation, the slope for a weight from a hidden unit to an output unit is

$$S_{jk}(t) = -\sum_{p=1}^{P} \delta_k(p)z_j(p),$$

and similarly, the slope for the weight from an input unit to a hidden unit is

$$S_{ij}(t) = -\sum_{p=1}^{P} \delta_j(p)x_i(p).$$

The new weight change is defined to be

$$\Delta w(t) = \frac{S(t)}{S(t-1) - S(t)} \Delta w(t-1).$$

The initial weight change can be taken to be

$$\Delta w(0) = -\alpha S(0),$$

where α is the learning rate. Thus, the first step in QuickProp is simply batch updating for backpropagation.

There are three cases that we must consider in analyzing the behavior of this algorithm. If the current slope is in the same direction as the previous slope, but is smaller in magnitude than the previous slope, then the weight change will be in the same direction as that carried out in the previous step. If the current slope is in the opposite direction from the previous slope, then the weight change will be in the opposite direction to the weight change carried out in the previous step. If the current slope is in the same direction as the previous slope, but is the same size or larger in magnitude than the previous slope, then the weight change would be infinite, or the weights would be moved away from the minimum and toward a maximum of the error.

To prevent the difficulties that occur in the third case, weight changes are limited so that if they would be too large, or if they would be uphill, a factor times the previous step is used instead of the change given by the formula for $\Delta w(t)$.

A further refinement is used whenever the current slope is of the same sign as the previous slope. In that case, a small multiple of the current slope is added to the weight change computed in the preceding formula. This prevents weight changes from being frozen (which would occur when a nonzero slope was encountered after a previous zero slope if no provision was made to correct the weight update rule).

The QuickProp weight update rule achieves impressive speedups in a number of instances, although it may also fail to converge in situations where backpropagation would eventually reach an acceptable answer. QuickProp's convergence difficulties can be avoided by letting the algorithm run for a fairly small number of epochs and then restarting it if it has not converged [Fahlman, 1988].

7.4 NEOCOGNITRON

The neocognitron [Fukushima, Miyake, & Ito, 1983; Fukushima, 1988] is an example of a hierarchical net in which there are many layers, with a very sparse and localized pattern of connectivity between the layers. It is an extension of an earlier net known as the cognitron [Fukushima, 1975]. The cognitron is a self-organizing net; the neocognitron is trained using supervised learning. The training will be described following a discussion of the motivation for the net and its architecture.

The neocognitron was designed to recognize handwritten characters—specifically, the Arabic numerals 0, 1, . . . , 9. The purpose of the network is to make its response insensitive to variations in the position and style in which the digit is written. The structure of the net is based on a physiological model of the visual system [Hubel & Wiesel, 1962]. The details of the description of the neocognitron given here are based on Fukushima, Miyake, and Ito (1983), which has been included in two compilations of important articles on neural networks [Anderson & Rosenfeld, 1988; Vemuri, 1988]. Later variations differ only in a few details of the architecture and training patterns.

The architecture of the neocognitron consists of several layers of units. The units within each layer are arranged in a number of square arrays. A unit in one layer receives signals from a very limited number of units in the previous layer; similarly, it sends signals to only a few units in the next layer. The input units are arranged in a single 19×19 square array. The first layer above the input layer has 12 arrays, each consisting of 19×19 units. In general, the size of the arrays decreases as we progress from the input layer to the output layer of the net. The details of the architecture are described in Section 7.4.1.

The layers are arranged in pairs, an S-layer followed by a C-layer. The S arrays are trained to respond to a particular pattern or group of patterns. The C arrays then combine the results from related S arrays and simultaneously thin out the number of units in each array.

The motivation for the multiple copies of the arrays in each layer will become clearer when we consider the training of the net. For now, we simply note that each array (within a layer) is trained to respond to a different pattern of signals (or feature of the original input). Each unit in a particular array "looks for" that feature in a small portion of the previous layer.

Training progresses layer by layer. The weights from the input units to the first layer are trained and then frozen. Then the next trainable weights are adjusted, and so forth. The weights between some layers are fixed, as are the connection patterns, when the net is designed. Examples of training patterns for the neocognitron are given in Section 7.4.2, along with a more detailed description of the training process.

7.4.1 Architecture

The architecture of the neocognitron consists of nine layers. After the input layer, there are four pairs of layers. The first layer in each pair consists of S cells, the second of C cells. We shall denote the layers as Input, $S1$, $C1$, $S2$, $C2$, $S3$, $C3$, $S4$, and $C4$. The $C4$ layer is the output layer.

The units in each layer are arranged in several square arrays (or cells), according to the following table:

LAYER	NUMBER OF ARRAYS	SIZE OF EACH ARRAY
Input	1	19×19
$S1$	12	19×19
$C1$	8	11×11
$S2$	38	11×11
$C2$	22	7×7
$S3$	32	7×7
$C3$	30	7×7
$S4$	16	3×3
$C4$	10	1×1

Figure 7.30 shows the architecture of the neocognitron. We denote the array (or cell) within a layer by a superscript; i.e., the first array in the first S-layer is $S1^1$, the second array in the first S-layer is $S1^2$, etc.

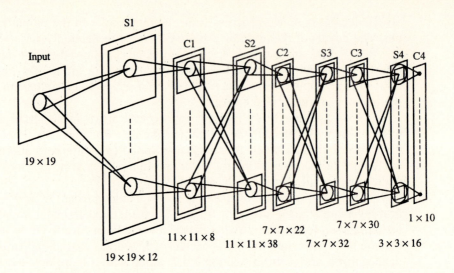

Figure 7.30 Architecture of neocognitron. Superscripts denote the cells, or arrays of units, within each layer. Adapted from [Fukushima, et al., 1983]. © 1983 IEEE

Each unit in one of the arrays receives signals from a small group of units in the previous layer. The units within a specific array (in a particular layer) are designated by subscripts; a typical unit in the first array (in the first S-layer) is $S1_{i,j}^1$. Depending on whether the unit is in a C-layer or an S-layer, it will receive signals from the designated units in one or more of the arrays in the previous layer.

To make this idea more concrete, consider the units in the input layer, which are arranged in a 19×19 array, and the units of the first S-layer, $S1$. The $S1$ layer consists of 12 19×19 arrays of units, $S1^1, \ldots, S1^{12}$. A unit in one of the $S1$ arrays receives signals from a 3×3 array of units in the input layer. For example, unit $S1_{i,j}^1$ receives signals from input units $U_{i-1,j-1}$, $U_{i-1,j}$, $U_{i-1,j+1}$, $U_{i,j-1}$, $U_{i,j}$, $U_{i,j+1}$, $U_{i+1,j-1}$, $U_{i+1,j}$, and $U_{i+1,j+1}$. Furthermore, unit $S1_{i,j}^2$ receives signals from exactly the same input units, and so on for all of the 12 arrays in the first S-layer. (If i or j is equal to 1 or 19, that $S1$ unit receives signals from only four input units, rather than nine.) A unit in an S-layer array receives signals from the designated units in all of the arrays in the previous C-layer.

The second of each pair of layers is called a C-layer. The C-layers serve to "thin out" the number of units in each array (by receiving input from a somewhat larger field of view). An array in a C-layer will receive input from one, two, or three of the arrays in the preceding S-layer. When the array receives signals from more than one S array, the C array combines similar patterns from the S-layer. The first C-layer, $C1$, consists of eight 11×11 square arrays of units. The $C1^1$ array receives signals from the $S1^1$ array. More specifically, each unit in the $C1^1$ array receives signals from a 5×5 field of units in the $S1^1$ array. The $C1^2$ array not only serves to condense the signals from a region of S units, but also combines the signals corresponding to similar patterns for which the $S1^2$ array or the $S1^3$ array was trained. Thus, a $C1^2$ unit receives signals from a region in the $S1^2$ array and from the same region in the $S1^3$ array. In what follows, we first consider the pattern of connections between arrays at different levels and then describe the "field of view" for units within each array.

Each $S2$ array receives signals from all of the $C1$ arrays; that is, each unit in an $S2$ array receives signals from the same portion of each of the $C1$ arrays in the preceding layer. Similarly, each $S3$ array receives signals from all $C2$ arrays, and each $S4$ array receives signals from all $C3$ arrays. However, as has been mentioned, the arrays in a $C1$-layer receive signals from only one, or at most a few, of the $S1$ arrays in that same level. Specifically, the connection pattern is as follows:

CONNECTIONS FROM $S1$ TO $C1$

$$S1^1 \qquad\quad \rightarrow C1^1$$
$$S1^2, S1^3 \;\; \rightarrow C1^2$$
$$S1^4 \qquad\quad \rightarrow C1^3$$
$$S1^5, S1^6 \;\; \rightarrow C1^4$$
$$S1^7 \qquad\quad \rightarrow C1^5$$
$$S1^8, S1^9 \;\; \rightarrow C1^6$$
$$S1^{10} \qquad\; \rightarrow C1^7$$
$$S1^{11}, S1^{12} \rightarrow C1^8$$

The motivation for these connection patterns becomes clearer if we look ahead to the patterns used for training the weights from the input layer to the $S1$-layer. The $S1^1$ array is trained to respond to a small horizontal line segment as shown in Figure 7.31. The $S1^2$ and $S1^3$ arrays both respond to line segments at

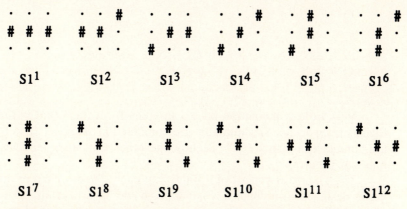

Figure 7.31 Training patterns for $S1$-layer of neocognitron. Adapted from [Fukushima et al., 1983] © 1983 IEEE

approximately a 22-degree angle from the horizontal. The $C1^2$ array serves to combine the signals from these two arrays. In a similar manner, $S1^5$ and $S1^6$ respond to different forms of segments between diagonal and vertical; their signals are then combined into a single $C1$ array.

The connection patterns from the $S2$ arrays to the $C2$ arrays are based on the same ideas and are as follows:

CONNECTIONS FROM $S2$ TO $C2$

$S2^1, S2^2, S2^3$	$\rightarrow$	$C2^1$
$S2^2, S2^3, S2^4$	$\rightarrow$	$C2^2$
$S2^5$	$\rightarrow$	$C2^3$
$S2^6, S2^7, S2^8$	$\rightarrow$	$C2^4$
$S2^7, S2^8, S2^9$	$\rightarrow$	$C2^5$
$S2^{10}$	$\rightarrow$	$C2^6$
$S2^{11}, S2^{12}$	$\rightarrow$	$C2^7$
$S2^{13}, S2^{14}$	$\rightarrow$	$C2^8$
$S2^{15}, S2^{16}$	$\rightarrow$	$C2^9$
$S2^{17}, S2^{18}$	$\rightarrow$	$C2^{10}$
$S2^{19}$	$\rightarrow$	$C2^{11}$
$S2^{20}, S2^{21}$	$\rightarrow$	$C2^{12}$
$S2^{22}, S2^{23}, S2^{24}$	$\rightarrow$	$C2^{13}$
$S2^{25}$	$\rightarrow$	$C2^{14}$
$S2^{26}$	$\rightarrow$	$C2^{15}$
$S2^{27}, S2^{28}, S2^{29}$	$\rightarrow$	$C2^{16}$
$S2^{30}, S2^{31}$	$\rightarrow$	$C2^{17}$
$S2^{32}$	$\rightarrow$	$C2^{18}$
$S2^{33}$	$\rightarrow$	$C2^{19}$
$S2^{34}$	$\rightarrow$	$C2^{20}$
$S2^{35}, S2^{36}$	$\rightarrow$	$C2^{21}$
$S2^{37}, S2^{38}$	$\rightarrow$	$C2^{22}$

Very little combination occurs in going from the $S3$-layer to the $C3$-layer. Signals from arrays $S3^{23}$ and $S3^{24}$ are combined in array $C3^{23}$, and $S3^{30}$ and $S3^{31}$ are combined in array $C3^{29}$. Each of the other $C3$ arrays receives signals from only one $S3$ array.

The $C4$ arrays consist of a single unit each, one for each of the 10 digits the net is designed to recognize. Signals from the $S4$ arrays are combined to form the net's final response. The connection patterns from the $S4$ arrays to the $C4$ arrays are as follows:

CONNECTIONS FROM *S*4 TO *C*4

$$S4^1, S4^2 \quad \rightarrow \quad C4^1$$
$$S4^3, S4^4 \quad \rightarrow \quad C4^2$$
$$S4^5 \quad \rightarrow \quad C4^3$$
$$S4^6, S4^7 \quad \rightarrow \quad C4^4$$
$$S4^8, S4^9 \quad \rightarrow \quad C4^5$$
$$S4^{10} \quad \rightarrow \quad C4^6$$
$$S4^{11}, S4^{12} \quad \rightarrow \quad C4^7$$
$$S4^{13} \quad \rightarrow \quad C4^8$$
$$S4^{14} \quad \rightarrow \quad C4^9$$
$$S4^{15}, S4^{16} \quad \rightarrow \quad C4^{10}$$

We now consider the receptive field for a unit in one of the arrays (at each of the different levels of the net). The unit in question will receive signals from the designated region in whichever of the arrays its array is connected to, as has been described. A unit in any of the $S1$ arrays "sees" a 3×3 portion of the input pattern; i.e., unit $S1^1_{i,j}$ receives signals from input units $U_{i,j}$, $U_{i,j-1}$, $U_{i,j+1}$, $U_{i-1,j-1}, \ldots, U_{i+1,j+1}$, and Unit $S1^2_{i,j}$ receives signals from the same nine input units.

A unit in a $C1$ array "sees" a 5×5 portion of one or two $S1$ arrays. Units in the corner of an array "see" only a part of the region they would "see" if they were situated in the center of the array, because part of their "field of view" falls outside of the array(s) from which they receive their signals. "Thinning" occurs because the size of each $C1$ array is smaller than the $S1$ array. The "field of view" of a $C1$ unit is shown in Figure 7.32; the x's simply indicate the position of the units. It is convenient to view the $C1$ array as being positioned on top of the

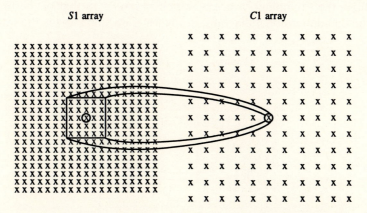

Figure 7.32 Connections from one $S1$-layer array to one $C1$-layer unit.

corresponding $S1$ array. The $C1$ array then extends beyond the $S1$ array, so the corner units of the $C1$ array receive signals from four units in the $S1$ array.

It is convenient to abbreviate the information in Figure 7.32 by looking at a one-dimensional slice of the two-dimensional pattern; this is shown in Figure 7.33.

Figure 7.33   Cross section of connections from $S1$ array to $C1$ array.

At the second level, each $S2$ unit "sees" a 3×3 region of each of the eight $C1$ arrays. Since each of the $S2$ arrays is the same size (11×11) as each of the $C1$ arrays, no "thinning" occurs at this level. The only $S2$ units that do not receive signals from nine $C1$ units (in each of the eight $C1$ arrays) are the corner units in the $S2$ arrays; they receive signals from four $C1$ units (in each of the eight $C1$ arrays). The "one-dimensional slice" diagram is shown in Figure 7.34.

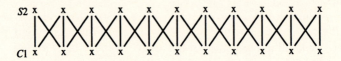

Figure 7.34   Cross section of connections from $C1$ array to $S2$ array.

The $C2$ units see a 5×5 region of the $S2$ array (or arrays) from which they receive signals. The "field of view" of a $C2$ array is indicated in Figure 7.35, to show the "thinning" process, which is similar to that of the first layer. Again, the one-dimensional diagram, in Figure 7.36, summarizes the information.

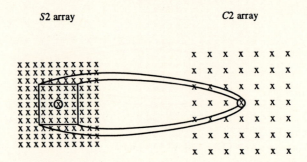

Figure 7.35   Connections from $S2$ array to $C2$ array.

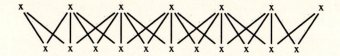

Figure 7.36 Cross section of connections from $S2$ array to $C2$ array.

Each $S3$ unit ''sees'' a 3×3 region of each of the 22 $C2$ arrays; no ''thinning'' occurs. Each of the $C3$ arrays ''sees'' a 3×3 region of the $S3$ array(s) to which it is connected. No ''thinning'' occurs in the third level either, because the $C3$ arrays are the same size as the $S3$ arrays, namely, 7×7. Since the $C3$ unit with coordinates i,j ''sees'' the region of the $S3$ array centered at i,j, no diagram is necessary.

Each of the $S4$ arrays ''sees'' a 5×5 region of each of the 30 $C3$ arrays. The ''field of view'' of the $S4$ arrays is shown in Figure 7.37. Note that the reduction of the number of units takes place between Levels 3 and 4, rather than within a level, as has been the case previously. Also, observe that, instead of skipping units, the units at the corners are now treated differently (neglected, if you will).

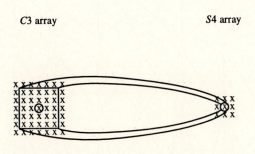

Figure 7.37 Connections from $C3$ array to $S4$ array.

Each $C4$ array is actually a single unit that ''sees'' the entire 3×3 $S4$ array (or arrays).

We can summarize all of the information concerning the connection patterns between units in the various layers in a single cross-sectional diagram, as shown in Figure 7.38.

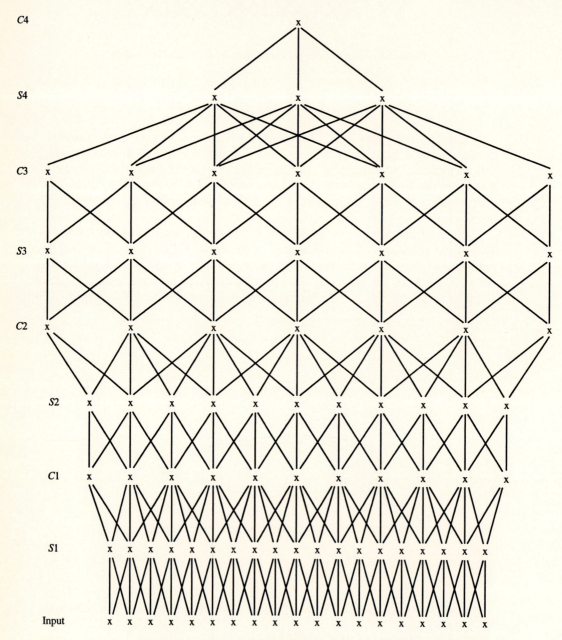

Figure 7.38   Cross section of connection patterns for neocognitron. Adapted from
[Fukushima et al., 1983] © 1983 IEEE

7.4.2 Algorithm

The output signal of a unit in an S-type cell (a cell in any of the S-layers) is a function of the excitatory signals it receives from units in the previous layer and the inhibitory signals it receives from those same units. The mechanism is described in terms of an intermediate, or auxiliary, unit (denoted here as a V unit) whose signal to the S unit is proportional to the (weighted) Euclidean norm of the signal sent by the input units. We adopt the following notation:

c_i	output from C unit
s_i	output from S unit
v	output from V unit
w_i	adjustable weight from C unit to S unit
w_0	adjustable weight from V unit to S unit
t_i	fixed weight from C unit to V unit
u_i	fixed weight from S unit to C unit

The signal sent by the inhibitory unit V is

$$v = \sqrt{\sum \sum t_i c_i^2},$$

where the summations are over all units that are connected to V in any array and over all arrays. The input layer is treated as the $C0$ level.

Thus, a typical S unit forms its scaled input,

$$x = \frac{1 + e}{1 + v\,w_0} - 1,$$

where

$$e = \sum_i c_i w_i$$

is the net excitatory input from C units, and $v\,w_0$ is the net input from the V unit. The output signal is

$$S = \begin{cases} x & \text{if } x \geq 0 \\ 0 & \text{if } x < 0. \end{cases}$$

The inhibitory signal serves to normalize the response of the S unit in a manner somewhat similar to that used in ART2.

The output of a C layer unit is a function of the net input it receives from all of the units, in all of the S arrays, that feed into it. As was shown in the description of the architecture, that input is typically from 9 or 25 units in each of one, two, or three S arrays. The net input is

$$c\_in = \sum_i s_i u_i.$$

The output is

$$
c = \begin{cases} \dfrac{c\_in}{a \,+\, c\_in} & \text{if } c\_in > 0 \\ 0 & \text{otherwise} \end{cases}
$$

The parameter a depends on the level and is 0.25 for Levels 1, 2, and 3 and 1.0 for Level 4.

Training Process

The neocognitron is trained layer by layer. The weights from the C units to the S unit are adaptable, as is the weight from the V unit to the S unit. The weights from the C units to the V unit are fixed.

The weights from an S-layer array to the corresponding C-layer array are fixed. They are stronger for units that are closer, but no particular metric is specified. As an example of the type of weight pattern that might be used, consider the "taxicab metric," in which the distance from the S-layer unit $S_{i-k,j-h}$ to the C-layer unit $C_{i,j}$ is $|k| + |h|$. A possible array of weights could be based on the formula for the weight from $S_{i-k,j-h}$ to $C_{i,j}$, i.e.,

$$
u(S_{i-k,j-h}; C_{i,j}) = \frac{1}{1 + |k| + |h|}.
$$

For a 5×5 connection region, which is what we have for the connections from the $S2$-layer to the $C2$-layer, the weights would be:

1/5	1/4	1/3	1/4	1/5
1/4	1/3	1/2	1/3	1/4
1/3	1/2	1	1/2	1/3
1/4	1/3	1/2	1/3	1/4
1/5	1/4	1/3	1/4	1/5

The pattern of weights is the same for every $C2$ unit.

The fixed weights from the C units to the inhibitory V units are also set to decrease monotonically as a function of distance.

The weights to the S-layer units (from the input units or from the C-layer units in the previous level) are trained sequentially. The weights from the input units to the $S1$ units are trained and then frozen. Next, the weights from the $C1$ units to the $S2$ units are trained and fixed. The process continues, level by level, until the output layer is reached. We describe the process in detail next.

Training the Weights from the Input Units to the S1 Units.   Each of the 12 arrays in the $S1$-layer is trained to respond to a different 3×3 input pattern. The training feature patterns for each of the arrays of the $S1$ layer are shown in Figure 7.31.

Each unit in array $S1^1$ responds to the pattern (a horizontal segment) when it appears in the portion of the input array from which that particular unit receives signals. The pattern of weights to all units in $S1^1$ is the same. In order to train all of the units in the $S1^1$ array, we only have to train one unit (namely, the center unit). The training pattern for the $S1^1$ array is presented to the center of the input array (and a target signal is sent to the $S1$ arrays designating that the center unit of the $S1^1$ array is the unit to be trained). The weight from input unit $i + k$, $j + h$ to $S1^1$ unit i, j is adjusted as follows:

$$\Delta w(I_{i+k,j+h}; S1^1_{i,j}) = \alpha t(I_{i+k,j+h}; S1^1_{i,j})c_{i+k,j+h}.$$

For the first S-layer, the signal $c_{i+k,j+h}$ is simply the input signal. The weight $t(I_{i+k,j+h}; S1^1_{i,j})$ is the fixed weight to the inhibitory unit. Thus, the weight adjustment is proportional to the signal received by the inhibitory unit. The weight from the inhibitory unit to the S unit is adjusted by an amount

$$\Delta w_0 = \alpha c_{i,j}.$$

The initial values for the adjustable weights are 0, and the learning rate α is usually taken to be relatively large, so the S unit being trained learns its desired response after only a few presentations of the pattern. When the weights for the center unit are determined, each of the other units in the $S1^1$ array is given exactly the same set of weights. In this manner, the center unit is trained to respond when the input pattern is presented in the center of the input field, but the other units in the $S1^1$ array respond to the same input pattern (in this case, a small horizontal line segment) when it appears in other parts of the input field.

In a similar manner, the center unit of the $S1^2$ array is trained to respond to the input pattern given for it in Figure 7.31. Once the weights are determined, all other units in this array are assigned the same weights. Training continues in the same way for each of the 12 arrays in the $S1$ layer.

A diagram of the cross section of the receptive field for an $S1$ unit (see Figure 7.39) illustrates the reason that the training patterns for this level are only 3×3: That is all an $S1$ unit can "see."

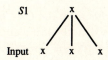

Figure 7.39   Cross section of receptive field for $S1$ unit.

Training the Weights from the C1 Units to the S2 Units. The center unit in each array in the S2-layer receives signals from nine units in each of the C1 arrays. Each S2 array is trained to respond to a small number of patterns. For example, the training patterns for the $S2^4$ array might be several variations of the pattern shown in Figure 7.40.

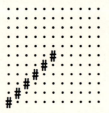

Figure 7.40   Sample training pattern for $S2^4$ array.

As described before, the training pattern is presented to the center of the input field, and the center unit in the $S2^4$ array is designated to learn the pattern. The weights for the nine connections from each of the C1 arrays are adjusted using the same learning rule as for the first layer. Note that in general, very few of the C1 arrays will respond to the input signal, so the actual connection pattern (with nonzero weights) from the C1 level to the S2 level is not as extensive as the general description would indicate. Although this training pattern is a pure diagonal line segment, training patterns for other S2 arrays involve combinations of the simpler patterns to which the S1 and C1 arrays have already been trained to respond.

As for the first layer, once the center unit has learned its training patterns (typically, four variations of essentially the same pattern), the other units in that array have their weights fixed at the same values as the center unit. Training of each array in the S2-layer proceeds in the same manner. When all the arrays are trained, the weights are fixed and we proceed to the next level of adjustable weights.

The cross-sectional diagram for the receptive fields, shown in Figure 7.41, illustrates the reason that the training patterns for this level are 11 × 11. If we trace back the connections from the center unit at the S2 level to the input level, we see that an 11 × 11 region in the input plane influences the S2-level array.

Training the Weights from the C2 Units to the S3 Units. The training of the S3-level arrays follows exactly the same procedure as that for the lower levels (whose weights are now fixed). The receptive field of the center unit is now the entire input array, so the training patterns are 19 × 19.

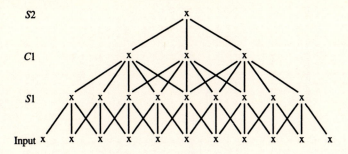

Figure 7.41   Receptive field for center unit in an S2-level array.

Training the Weights from the C3 Units to the S4 Units.   The final training of the weights, for the 16 S4 units, is based on various sample patterns discussed in the next section.

Sample Training Patterns

Sample training patterns for the S2 arrays are shown in Figure 7.43. Figure 7.44 illustrates one sample training pattern for each of the arrays at the S3 level and S4 level; typically, two or three variations of the given pattern would be used for training each array.

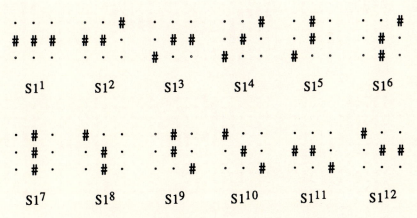

Figure 7.42   Training patterns for S1-layer of neocognitron.

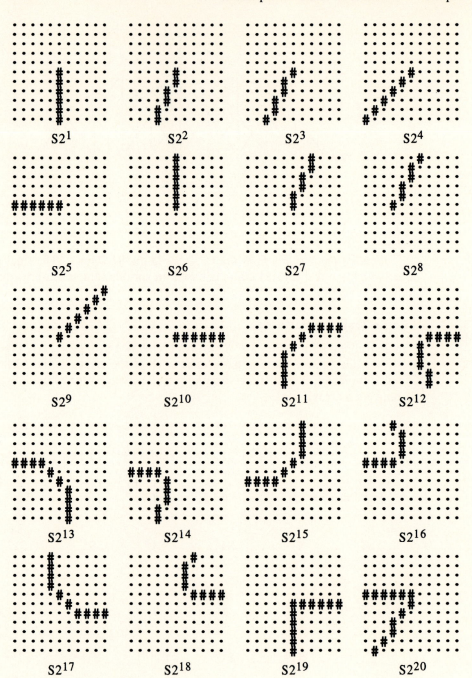

Figure 7.43 Sample training patterns for $S2$-level arrays. Adapted from [Fuku-shima et al., 1983] © 1983 IEEE

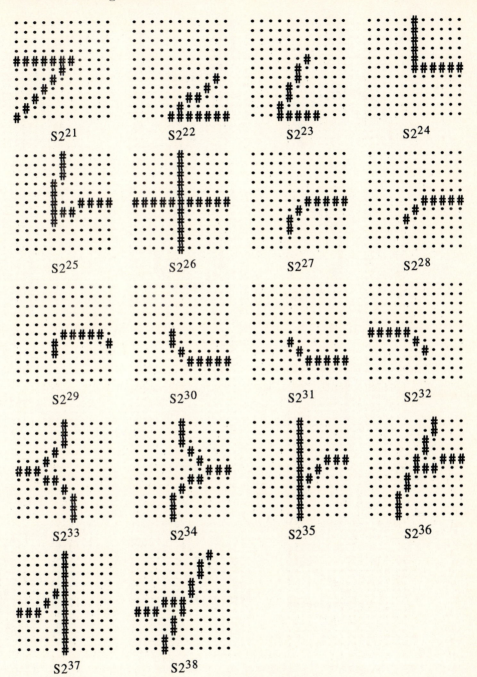

Figure 7.43   *(Continued)*

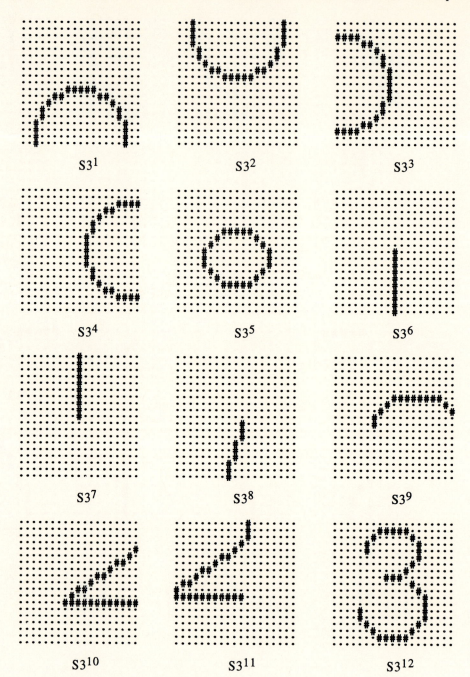

Figure 7.44 Sample training patterns for $S3$-level and $S4$-level arrays. Adapted from [Fukushima et al., 1983] © 1983 IEEE

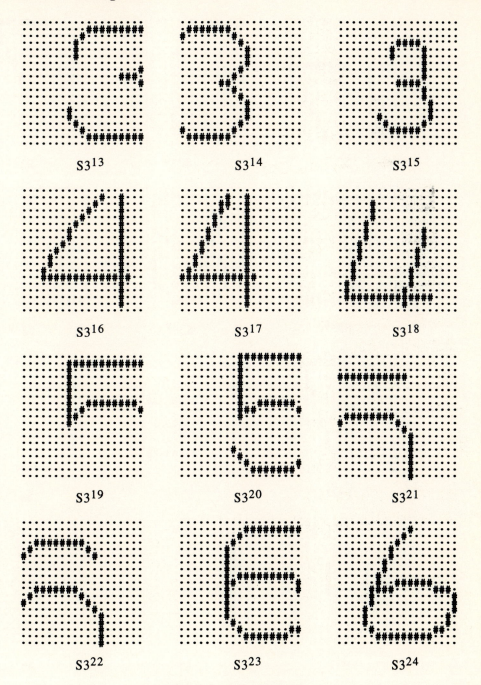

Figure 7.44   *(Continued)*

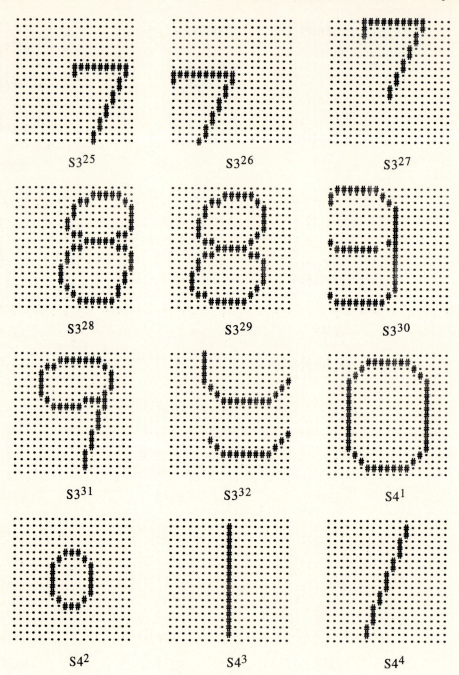

Figure 7.44 *(Continued)*

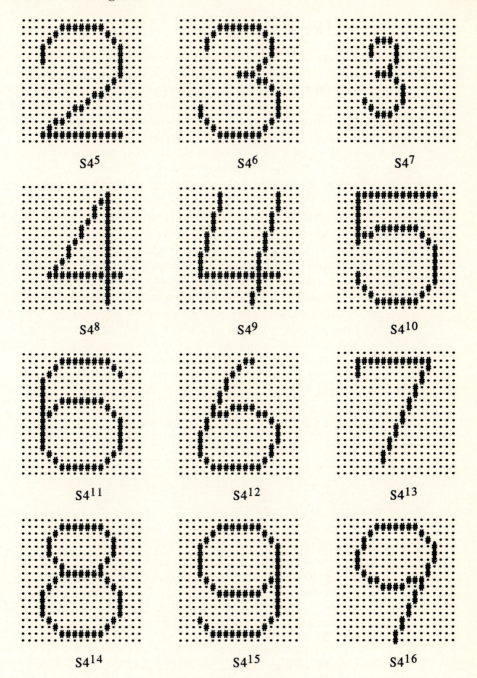

Figure 7.44   *(Continued)*

7.5 SUGGESTIONS FOR FURTHER STUDY

7.5.1 Readings

For further information on the neural networks surveyed in this chapter, the reader is encouraged to consult the specific references included in the text for each network. Also journals and conference proceedings are excellent sources for recent developments in this rapidly evolving field.

Neural Networks, the journal of the INNS (International Neural Network Society), *IEEE Transactions on Neural Networks,* and *Neural Computation* are among the more established journals devoted to the theory and application of neural networks. Proceedings from conferences sponsored by INNS and IEEE (often jointly) and the Advances in Neural Information Processing Systems series are invaluable for staying abreast of new results. The conference proceedings include papers on the latest findings in learning theory (associative memory, self organization, supervised learning, reinforcement learning), applications (including image analysis, vision, robotics and control, speech, signal processing, and pattern recognition) and neural network implementations (electronic and optical neurocomputers). Other journals and conferences devoted in whole or in part to neural networks also provide excellent articles. It is hoped that the reader will find continued exploration of neural networks to be interesting and rewarding.

7.5.2 Exercises

7.1 Consider the following Boltzmann machine (without learning):

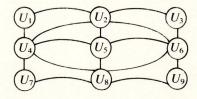

The connections shown each have a weight of -2. In addition, each unit has a selfconnection of weight 1.

Assume that unit U_2 is "on" and all the other units are "off." Describe what happens in each of the following cases:

a. $T = 10$, U_9 attempts to turn "on."
b. $T = 1$, U_9 attempts to turn "on."
c. $T = 10$, U_6 attempts to turn "on."
d. $T = 1$, U_6 attempts to turn "on."

7.2 The Boltzmann machine is being used to solve the traveling salesman problem. The distances between the four cities making up the tour are as follows:

	A	*B*	*C*	*D*
A	0	4	3	2
B	4	0	5	6
C	3	5	0	1
D	2	6	1	0

Use the penalty $p = 20$ and the bonus $b = 15$ for the constraints. Use the temperature $T = 100$. For the tour *CDBAC:*

a. What are the activations of the units, i.e., which are "on" and which are "off"?

b. Draw the network diagram for the tour, with connections shown only between active units. (Indicate all constraint and distance connections between active units.)

c. Compute the change in consensus ΔC for each active unit (if it attempts to turn "off").

d. For each active unit, compute the probability of changing its state.

7.3 The Boltzmann machine is being used to solve the traveling salesman problem for the following cities in the following positions:

CITY	POSITION			
	1	**2**	**3**	**4**
A	U_{A1}	U_{A2}	U_{A3}	U_{A4}
B	U_{B1}	U_{B2}	U_{B3}	U_{B4}
C	U_{C1}	U_{C2}	U_{C3}	U_{C4}
D	U_{D1}	U_{D2}	U_{D3}	U_{D4}

The distances between the four cities are as follows:

	A	*B*	*C*	*D*
A	0	6	8	5
B	6	0	10	5
C	8	10	0	5
D	5	5	5	0

Use the penalty $p = 20$ and the bonus $b = 10$ for the constraints.

a. Show all weighted connections for unit U_{C3}.

b. Determine the value of the consensus for the network if the units have the following activations:

CITY		POSITION		
	1	**2**	**3**	**4**
A	1	0	0	0
B	0	1	0	0
C	0	0	1	1
D	0	0	1	0

c. Determine the value of the consensus for the network if the units have the following activations:

	POSITION			
CITY	1	2	3	4
A	1	0	0	0
B	0	1	0	0
C	0	0	1	0
D	0	0	0	1

d. Which of configurations (in parts b and c) satisfies all of the constraints of the traveling salesman problem?

e. What is the effect on the consensus (i.e., find ΔC) if the activation of unit U_{C3} is reversed, with the net as shown in part b? With the net as shown in part c?

f. For each of the cases considered in part e, find the probability of accepting the change if $T = 10$ and if $T = 1$.

7.4 In the traveling salesman problem, the distances between the cities are given in the following matrix:

	A	*B*	*C*	*D*	*E*
A	0	8	10	20	5
B	8	0	26	20	9
C	10	26	0	10	5
D	20	20	20	0	5
E	5	9	5	5	0

Use $p = 70$, $b = 60$, and $T = 100$.

a. Draw the network with the connections and weights to represent the constraint (but not the distances) for this problem.

b. Draw the network with the connections and weights to represent the distances (but not the constraints) for this problem.

c. For the tour *BACEDB*, what are the activations of the units? That is, which units are "on" and which units are "off" in each column?

d. What is the value of the consensus function C for this configuration?

e. Compute ΔC for each unit that is currently turned "on."

f. For each unit that is currently turned "on," compute the probability of changing its state from "on" to "off."

g. Discuss the sequence of events that must occur in order for the activations to change to an improved configuration.

7.5.3 Project

7.1 This project generated the sample results in Example 7.1; however, your results will not necessarily be identical, because of the random nature of certain aspects of the process.

Write a computer program to implement the Boltzmann machine without learning to solve the five-city traveling salesman problem with distances given as follows:

	A	B	C	D	E
A	0	8	10	20	5
B	8	0	26	20	9
C	10	26	0	10	5
D	20	20	10	0	5
E	5	9	5	5	0

Besides the distances, the weights that are needed are $b > 2d$ and $p > b$, where d is. Try $b = 60$ and $p = 70$.

Using a random number generator that returns a value between 0 and 1, you can choose the random unit that will attempt to change its consensus by taking $i = 5r_1$ and $j = 5r_2$, where r_1 and r_2 are two different random numbers and i and j are truncated to integers. The wraparound aspect of the distances can be implemented by defining jplus $= j + 1$, followed by if jplus $= 6$, then jplus $= 1$ (and similarly for jminus).

After the probability of acceptance A is computed, generate a random number (again between 0 and 1, with a uniform distribution). If the number generated is less than A, the activation of unit (i, j) is changed. Reversal of the activation can be accomplished by

$$x(i, j) = x(i, j) + 1 \bmod 2.$$

Start with $T = 10$ and reduce T linearly to 0. Run your program several times, starting with different initial units "on" each time. Try different starting temperatures, etc., also.

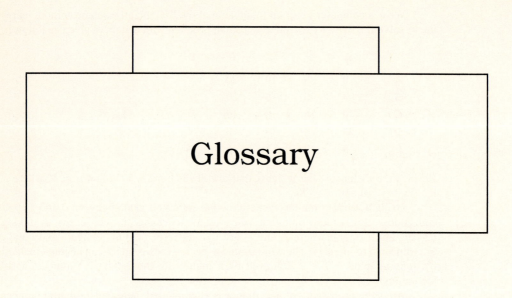

Glossary

Accretion: Approximation formed by combining information from several training patterns (as in counterpropagation), as opposed to interpolating between training patterns.

Activation: A node's level of activity; the result of applying the activation function to the net input to the node. Typically this is also the value the node transmits.

Activation function: A function that transforms the net input to a neuron into its activation. Also known as a transfer, or output, function.

ADALINE (ADAptive LInear NEuron): Developed by Bernard Widrow, an ADALINE's output is $+1$ if the weighted sum of its inputs is greater than a threshold, -1 otherwise. The weights are calculated by the delta rule, which is also known as the Widrow Hoff rule [Widrow & Hoff, 1960].

Adaptive resonance theory (ART): Adaptive resonance theory is a quantitative explanation of learning and memory developed by Gail Carpenter, Stephen Grossberg and others. ART1 and ART2 are neural net architectures based on adaptive resonance theory. Each of these neural nets self-organizes the input data into categories with the variation allowed within a category depending on a user selected vigilance parameter. ART1 is used for binary input, ART2 for continuous input [Carpenter & Grossberg, 1987a, 1987b].

Algorithm: A computational procedure; a neural net training algorithm is a step by step procedure for setting the weights of the net. Training algorithms are also known as learning rules.

Annealing schedule: Plan for systematic reduction of temperature parameter in a neural network that uses simulated annealing.

Architecture: Arrangement of nodes and pattern of connection links between them in a neural network.

Associative memory: A neural net in which stored information (patterns, or pattern pairs) can be accessed by presenting an input pattern that is similar to a stored pattern. The input pattern may be an inexact or incomplete version of a stored pattern.

Asynchronous: Process in which weights or activations are updated one at a time, rather than all being updated simultaneously. The discrete Hopfield net uses asynchronous updates of the activations. BAM may use either synchronous or asynchronous updates.

Autoassociator: A neural net used to store patterns for future retrieval [Mc-Clelland & Rumelhart, 1988]. The net consists of a single slab of completely interconnected units, trained using the Hebb rule. The activations in this net may become very large, very quickly because a unit's connection to itself acts as a self-reinforcing feedback. See also Associative Memory, Brain-State-in-a-Box, and Hopfield net.

Autoassociative memory: An associative memory in which the desired response is the stored pattern.

Axon: Long fiber over which a biological neuron transmits its output signal to other neurons.

Backpropagation: A learning algorithm for multilayer neural nets based on minimizing the mean, or total, squared error.

Bias(j): the weight on the connection between node j and a mythical unit whose output is always 1; i.e. a term which is included in the net input for node j along with the weighted inputs from all nodes connected to node j.

Bidirectional associative memory (BAM): A recurrent heteroassociative neural net developed by Bart Kosko [Kosko, 1988, 1992a].

Binary: 0 or 1.

Binary sigmoid: Continuous, differentiable S-shaped activation function whose values range between 0 and 1. See sigmoid.

Bipolar: -1 or 1.

Bipolar sigmoid: Continuous, differentiable S-shaped activation function whose values range between -1 and 1. See sigmoid.

Bivalent: Either binary, bipolar, or any other pair of values.

Boltzmann machine (without learning): A class of neural networks used for solving constrained optimization problems. In a typical Boltzmann machine, the weights are fixed to represent the constraints of the problem and the function to be optimized. The net seeks the solution by changing the activations (either 1 or 0) of the units based on a probability distribution and the effect that the change would have on the energy function or consensus function for the net [Aarts & Korst, 1989]. See also simulated annealing.

Boltzmann machine (with learning): A net that adjusts its weights so that the equilibrium configuration of the net will solve a given problem, such as an encoder problem [Ackley, Hinton, & Sejnowski, 1985].

Bottom-up weights: Weights from the F1 layer to the F2 layer in an adaptive resonance theory neural net.

Boundary contour system (BCS): Neural network developed by Stephen Grossberg and Ennio Mingolla for image segmentation problems [Grossberg & Mingolla, 1985a, 1985b]. See also discussion by Maren (1990).

Brain-state-in-a-box (BSB): Neural net developed by James Anderson to overcome the difficulty encountered when an auto-associator neural net iterates, namely the activations of the units may grow without bound. In the BSB neural net the activations are constrained to stay between fixed upper and lower bounds (usually -1 and $+1$) [Anderson, 1972]. See also autoassociator.

Capacity: The capacity of a neural net is the number of patterns that can be stored in the net.

Cascade correlation: Neural net designed by Scott Fahlman that adds only as many hidden units as are required to achieve a stated error tolerance [Fahlman & Lebiere, 1990].

Cauchy machine: A modification, developed by Harold Szu, to the Boltzmann machine; the Cauchy machine uses a faster annealing schedule than the Boltzmann machine [Szu & Hartley, 1987].

Clamped: Held equal to, as in 'input pattern clamped on input units'.

Classification: Problem in which patterns are to be assigned to one of several classes. In this text the term "classification" is used only for supervised learning problems in which examples of the desired group assignments are known.

Cluster unit: A unit in a competitive layer of a neural net; an output unit in a self-organizing net such as ART or SOM or a hidden unit in a counterpropagation net.

Clustering: Grouping of similar patterns together. In this text the term "clustering" is used only for unsupervised learning problems in which the desired groupings are not known in advance.

Competitive learning: Unsupervised learning in which a competitive neural net (or subnet) adjusts its weights after the winning node has been chosen. See competitive neural net.

Competitive neural net: A neural net (or subnet) in which a group of neurons compete for the right to become active (have a non-zero activation). In the most extreme (and most common) example, the activation of the node with the largest net input is set equal to 1 and the activations of all other nodes are set equal to 0; this is often called "winner-take-all". MAXNET is an example of a competitive neural net which can be used as a subnet in the Hamming and other neural nets [Lippmann, 1987].

Conscience: Mechanism to prevent any one cluster formed in a counterpropagation neural net from claiming an unfair proportion of the input vectors; the first clusters are most likely to have an advantage.

Consensus function: Function to be maximized by a net (such as the Boltzmann machine without learning) for solving a constrained optimization problem.

Constrained optimization problem: A problem in which the desired solution gives the maximum or minimum value of a quantity, subject to satisfying certain constraints. See the traveling salesman problem for a classic example.

Content addressable memory: A method of storing information in which the information can be addressed by a partial pattern (the content) rather than by the location of the information (as is the case in traditional computing); see also associative memory.

Context unit: An input layer unit in a simple recurrent net that receives information from the hidden units at the previous time step [Elman, 1990].

Convergence: Recurrent nets converge if the configuration (pattern of activations of the units) eventually stops changing; iterative training processes converge if the weight updates reach equilibrium (stop changing).

Correlation encoding: Storage of information based on the correlation between input and output patterns, as in the Hebb rule.

Coulomb energy nets: Neural nets based on a Coulomb energy function with minima at specified points, allowing construction of nets with arbitrarily large storage capacity [Bachmann, Cooper, Dembo, & Zeitouni, 1987]. Coulomb energy nets can also learn [Scofield, 1988].

Counterpropagation: A neural network developed by Robert Hecht-Nielson based on a two-stage training process. During the first phase, clusters are formed from the input patterns; during the second phase weights from the clusters to the output units are adjusted to produce the desired response [Hecht-Nielsen, 1987a, 1987b].

Crosstalk: Interference that occurs when the patterns stored in an associative memory are not mutually orthogonal. If the net is given one of the stored patterns as input, the response will be a combination of the desired output and the target pattern(s) for the other stored pattern(s) that are not orthogonal to the input pattern.

Decision boundary: Boundary between regions where the input vector will produce a positive response and regions where the response will be negative.

Delta-bar-delta: Modification to the backpropagation learning rule. Each unit has its own learning rate; these learning rates are increased when several weight changes are made in the same direction, decreased when weight changes on successive steps are in opposite directions [Jacobs, 1988].

Delta rule: Learning rule based on minimization of squared error for each training pattern; used for single layer perceptron. Also called Least Mean Square (LMS) or Widrow-Hoff learning.

Dendrites: The portion of a biological neuron that receives incoming signals from other neurons.

Dot product: A vector-vector product that produces a scalar result. The net input to a neuron is the dot product of the input pattern and the weight vector. The dot product also can be used to measure similarity of vectors (that are of the same length); the more similar the vectors, the larger their dot product.

Echo cancellation: An early application of Adalines to the area of telephone communication [Widrow & Stearns, 1985].

Encoder problem: A problem in which the target output is the same as the input pattern, but the input signal is required to pass through a constriction before the output signal is produced, i.e. there are fewer hidden units than there are input or output units in the neural net.

Energy function: Function (of the weights and activations of a neural network) that is monotone non-increasing and bounded below. If such a function can be found for an iterative process, such as a Hopfield net or BAM, the convergence of the process in guaranteed.

Epoch: One presentation of each training pattern.

Euclidean distance: The Euclidean distance, D, between vectors $(x_1, x_2, \ldots, x_n)$ and $(y_1, y_2, \ldots, y_n)$ is defined by:

$$D^2 = \sum_{i=1}^{n} (x_i - y_i)^2.$$

Excitatory connection: Connection link between two neurons with a positive weight; it serves to increase the response of the unit that receives the signal. In contrast, see inhibitory connection.

Exemplar: A vector that represents the patterns placed on a cluster; this may be formed by the neural net during training, as in SOM, or specified in advance, as in the Hamming net.

Extended delta rule: Learning rule based on minimizing the error of a single layer net in which the output units may have any differentiable function for their activation function. (The standard delta rule assumes that the output units have the identity function for their activation function during the training process.)

Fast learning: Learning mode for ART in which it is assumed that all weight updates reach equilibrium on each learning trial.

Fault tolerance: A neural net is fault tolerant if removing some nodes from the net makes little difference in the computed output. Also, neural nets are in general tolerant of noise in the input patterns.

Feedforward: A neural net in which the signals pass from the input units to the output units (possibly through intermediate layers of hidden units) without any connections back to previous layers. In contrast, recurrent nets have feedback connections.

Fixed weight nets: Neural nets in which the weights do not change. Examples include Hopfield nets (discrete and continuous), Boltzmann machine without learning, and MAXNET.

Gain: See weight; also called strength or synapse.

Gaussian machine: A three parameter description of a class of neural nets that includes Boltzmann machines, Hopfield nets, and others [Akiyama, Yamashita, Kajiura, & Aiso, 1989].

Gaussian potential function: An example of a radial basis function.

$$g(x) = \exp(-x^2).$$

Generalization: The ability of a neural net to produce reasonable responses to input patterns that are similar, but not identical, to training patterns. A balance between memorization and generalization is usually desired.

Generalized delta rule: The delta rule (with arbitrary differentiable activation functions) for multilayer neural nets. See backpropagation.

Grandmother cells: Processing elements which store a single pattern, for example a single neuron that fires only when the input pattern is an image of your grandmother.

Grossberg learning: Learning rule for the output units in counterpropagation nets. This is a special case of (Grossberg) outstar learning.

Hamming distance: The number of differing bits in two binary or bipolar vectors.

Hamming network: A fixed-weight neural network which places an input pattern into the appropriate category or group based on the Hamming distance between the input pattern and the (prespecified) exemplar vector for each category [Lippmann, 1987].

Handwritten character recognition: One example of pattern recognition problems to which many types of neural networks are being applied.

Hebb net: A simple net trained using the Hebb rule. When used for pattern association problems (the most typical use of the Hebb rule) the nets are usually known as autoassociative or heterassociative nets.

Hebb rule: A learning algorithm based on the premise that the strength of the connection between two neurons should be increased if both neurons are behaving in the same manner (both have positive activations or both have negative activations). Also known as correlation encoding.

Heteroassociative net: A neural net designed to associate input pattern—output pattern pairs, where the input pattern and the output pattern are not identical. The weights for such nets are usually found by the Hebb rule.

Hidden units: Units that are neither input units nor output units.

Hopfield net: Fully interconnected (except no self-connections of a unit to itself) single layer net used as an autoassociative net or for constraint satisfaction problems [Hopfield, 1984].

Inhibitory connection: Connection link between two neurons such that a signal sent over this link will reduce the activation of the neuron that receives the signal. This may result from the connection having a negative weight, or from the signal received being used to reduce the activation of a neuron by scaling the net input the neuron receives from other neurons.

Input units: Units that receive signals from outside the neural net; typically they transmit the input signal to all neurons to which they are connected, without modification. Their activation function is the identity function.

Iteration: One performance of a calculation (or group of calculations) that must, in general, be repeated several times. In the neural network literature this term may be used to mean an "epoch" or a "learning trial". In this text, the term "iteration" is usually reserved for processes that occur within a single learning trial, as in weight update iterations in ART (fast learning mode).

Kohonen learning rule: Weight update rule in which the new weight is a convex combination of the old weight and the current input pattern. The coefficient that multiplies the input pattern, the learning rate, is gradually reduced during the learning process.

Kohonen self-organizing map: A clustering neural net, with topological structure among cluster units.

Layer: Pattern of weighted connections between two slabs of neurons; in neural net literature the term layer is also used frequently for a group of neurons that function in the same way (a slab).

Learning algorithms: Procedures for modifying the weights on the connection links in a neural net (also known as training algorithms, learning rules).

Learning rate: A parameter that controls the amount by which weights are changed during training. In some nets the learning rate may be constant (as in standard backpropagation); in others it is reduced as training progresses to achieve stability (for example, in Kohonen learning).

Learning trial: One presentation of one training pattern (especially in ART nets).

Learning vector quantization (LVQ): A neural net for pattern classification; trained using one of several variations on Kohonen learning. The input space is divided into regions that are represented by one or more output units (each of which represents an output class). The weight vector for an output unit is also known as a codebook vector [Kohonen, 1989a].

Linear autoassociator: A simple recurrent autoassociative neural net.

Linear separability: Training patterns belonging to one output class can be separated from training patterns belonging to another class by a straight line, plane, or hyperplane. Linearly separable patterns can be learned by a single layer neural net.

Linear threshold units: Neurons that form a linear combination of their weighted input signals (their net input) and send an output signal (equal to 1) if the net input is greater than the threshold (otherwise the output is 0); see Perceptron.

Logic functions: Functions with bivalent inputs (true or false, 1 or 0, 1 or -1) and a single bivalent output. There are 16 different logic functions with two inputs.

Long term memory: The weights in a neural network represent the long term memory of the information the net has learned.

Lyapunov function: See energy function.

McCulloch-Pitts neuron: Generally regarded as the first artificial neuron; a McCulloch-Pitts neuron has fixed weights, a threshold activation function, and a fixed discrete (non-zero) time step for the transmission of a signal from one neuron to the next. Neurons, and networks of neurons, can be constructed to represent any problem that can be modeled by logic functions with unit time steps [McCulloch & Pitts, 1943]. A neuron that uses a threshold activation function (but does not satisfy the other requirements of the original model) is sometimes also called a McCulloch-Pitts neuron [Takefuji, 1992].

Madaline: Multi-Adaline; a neural net composed of many Adaline units [Widrow & Lehr, 1990].

Mean squared error: Sometimes used in place of 'squared error' or 'total squared error' in the derivation of delta rule and backpropagation training algorithms or in stopping conditions. Mean squared error may be the squared error divided by the number of output components, or the total squared error divided by the number of training patterns.

Mexican hat: A contrast enhancing competitive neural network (or pattern of connections within a layer of a neural network) [Kohonen, 1989a].

Memorization: The ability to recall perfectly a pattern that has been learned. In general, the objective for a neural net is a balance between memorization and generalization.

Missing data: Noise in a bivalent testing input pattern in which one or more components have been changed from the correct value to a value midway between the correct and the incorrect value, i.e. a $+1$, or a -1, has been changed to a 0.

Mistakes in the data: Noise in a bivalent testing input pattern in which one or more component has been changed from the correct value to the incorrect value, i.e. a $+1$ has been changed to a -1, or vice versa.

Momentum: A common modification to standard backpropagation training; at each step, weight adjustments are based on a combination of the current weight adjustment (as found in standard backpropagation) and the weight change from the previous step.

Multilayer perceptron: A neural net composed of three or more slabs (and therefore two or more layers of weighted connection paths); such nets are capable of solving more difficult problems than are single layer nets. They are often trained by backpropagation.

Neocognitron: Multi-stage pattern recognizer and feature extractor developed by Kunihiko Fukushima to recognize patterns (the alphabet or digits) even when the input image is distorted or shifted [Fukushima, Miyaka, & Ito, 1983].

Net input: The sum of the input signals that a neuron receives, each multiplied by the weight on the connection link, possibly along with a bias term.

Neural networks: Information processing systems, inspired by biological neural systems but not limited to modeling such systems. Neural networks consist of many simple processing elements joined by weighted connection paths. A neural net produces an output signal in response to an input pattern; the output is determined by the values of the weights.

Neural nets: Neural networks, also known as artificial neural nets (ANNs), connectionist models, parallel distributed processing models, massively parallel models, artificial information processing models.

Neurocomputing: The use of neural networks emphasizing their computational power, rather than their ability to model biological neural systems.

Neuron: See processing element; also called node or unit.

Node: See processing element; also called neuron or unit.

Noise: Small changes to the components of a training or testing input vector. Noise may be introduced into training patterns to create additional patterns or to improve the ability of the net to generalize. Noise may also be present as a result of inaccuracies in measurements, etc. Neural nets are relatively robust to noisy testing patterns; this is often called generalization.

Orthogonal vectors: Two vectors are orthogonal if their dot product is 0; an associative memory can store more orthogonal patterns than non-orthogonal patterns.

Outer product: Matrix product of a column vector with a row vector, result is a matrix.

Output: The value a node transmits.

Output unit: A unit whose activation can be observed and interpreted as giving the response of the net.

Outstar: A neural network structure developed by Stephen Grossberg in which an output unit receives both signals from other units and a training input. Differential equations control both the change in the activation of the unit, and the change in the weights [Grossberg, 1969]. See also discussion in [Caudill, 1989].

Pattern: Information processed by a neural network; a pattern is represented by a vector with discrete or continuous valued components.

Pattern associator: Neural net consisting of a set of input units connected to a set of output units by a single layer of adjustable weights, trained by the Hebb or delta learning rules. [McClelland & Rumelhart, 1988].

Pattern association: Problems in which the desired mapping is from an input pattern to an output pattern (which may be the same, or similar to the input pattern).

Pattern classification: Problems in which the desired mapping is from an input pattern to one (or more) of several classes to which the pattern does or does not belong. In this text, the term classification is reserved for problems in which the correct class memberships are known for the training patterns. In general pattern classification is also used for problems where similar patterns are grouped together; the groupings are then defined to be the classes.

Pattern classifier: A neural net to determine whether an input pattern is or is not a member of a particular class. Training data consists of input patterns and the class to which each belongs, but does not require a description of each class; the net forms exemplar vectors for each class as it learns the training patterns.

Perceptrons: Neural nets studied by Rosenblatt, Block, Minsky and Papert, and others; the term is often used to refer to a single layer pattern classification network with linear threshold units [Rosenblatt, 1962; Minsky & Papert, 1988].

Perceptron learning rule: Iterative training rule, guaranteed to find weights that will correctly classify all training patterns, if such weight exist, i.e. if the patterns are linearly separable.

Phonetic typewriter: An example of the application of neural networks, in this case SOM and LVQ, to problems in speech recognition [Kohonen, 1988].

Plasticity: The ability of a net to learn a new input pattern whenever it is presented. ART nets are designed to balance stability with plasticity. (In some other neural nets stability in learning is achieved by reducing the learning rate during training, which has the effect of reducing the net's ability to learn a new pattern presented late in the training cycle.)

Principal components: Eigenvectors corresponding to the largest eigenvalues of a matrix.

Probabilistic neural net: A net, developed by Donald Specht, to perform pattern classification using Gaussian potential functions and Bayes decision theory [Specht, 1988, 1990].

Processing element (PE): The computational unit in a neural network. Each processing element receives input signals from one or more other PEs, typically multiplied by the weight on the connection between the sending PE and the receiving PE. This weighted sum of the inputs is transformed into the PEs activation by the activation function. The PEs output signal (its activation) is then sent on to other PEs or used as output from the net. Processing elements are also called (artificial) neurons, nodes, or units.

QuickProp: A learning algorithm for multilayer neural nets, developed by Scott Fahlman, based on approximating the error surface by a quadratic surface [Fahlman & Lebiere, 1990].

Radial basis function: An activation function that responds to a local "field of view"; f(x) is largest for x = c and decreases to 0 as $|x - c| \to \infty$.

Recurrent net: A neural net with feedback connections, such as a BAM, Hopfield net, Boltzmann machine, or recurrent backpropagation net. In contrast, the signal in a feedforward neural net passes from the input units (through any hidden units) to the output units.

Restricted coulomb energy net: The neural network portion of the Nestor Learning System; used in many applications by the Nestor Corporation [Collins, Ghosh, & Scofield, 1988b]. See also Coulomb Energy Net.

Relaxation: A term used in neural networks, especially constraint satisfaction nets such as the Boltzmann Machine, to refer to the iterative process of gradually reaching a solution.

Resonance: The learning phase in ART, after an acceptable cluster unit has been selected; top-down and bottom-up signals "resonate" as the weight changes occur.

Saturate: An activation function that approaches a constant value for large magnitudes of the input variable is said to saturate for those values; since the derivative of the function is approximately zero in the saturation region, it is important to avoid such regions during early stages of the training process.

Self-organization: The process by which a neural net clusters input patterns into groups of similar patterns.

Self-organizing map (SOM): See Kohonen's self-organizing map.

Short term memory: The activations of the neurons in a neural net are sometimes considered to model the short term memory of a biological system.

Sigmoid function: An S-shaped curve; several common sigmoid functions are:

binary (logistic): $f(x) = \dfrac{1}{1 + \exp(-x)}$;

arctan (range from −1 to 1); $h(x) = \dfrac{2}{\pi} \arctan(x)$;

bipolar: $g(x) = \dfrac{2}{1 + \exp(-x)} - 1 = \dfrac{1 - \exp(-x)}{1 + \exp(-x)}$;

tanh: $\tanh(x) = \dfrac{\exp(x) - \exp(-x)}{\exp(x) + \exp(-x)} = \dfrac{1 - \exp(-2x)}{1 + \exp(-2x)}$.

Signals: Information received by the input units of a neural net, transmitted within the net, or produced by the output units.

Simulated annealing: The process of gradually decreasing the control parameter (usually called temperature) in the Boltzmann machine; this is used to reduce the likelihood of the net becoming trapped in a local minimum which is not the global minimum.

Single-layer perceptron: One of many neural nets developed by Rosenblatt in the 1950's, used in pattern classification, trained with supervision [Rosenblatt, 1958, 1959, 1962].

Single-layer neural net: A neural net with no hidden units; or equivalently, a neural net with only one layer of weighted connections.

Slab: A group of neurons with the same activation function and the same pattern of connections to other neurons; see layer.

Slope parameter: A parameter that controls the steepness of a sigmoid function by multiplying the net input.

Soma: The main cell body of a biological neuron.

Spanning tree data: A set of training patterns developed by Kohonen. The relationship between the patterns can be shown on a two-dimensional spanning tree diagram, in which the patterns that are most similar are closest together [Kohonen, 1989a].

Squared error: Sum over all output components of the square of the difference between the target and the computed output, for a particular training pattern. This quantity (or sometimes, for convenience, one half of this sum) is used in deriving the delta rule and backpropagation training algorithms. See also 'mean squared error' and 'total squared error'.

Stable state: A distribution of activations on neurons from which an iterative neural net will not move. A stable state may be a stored pattern; if it is not a stored pattern, the state is called a ''spurious stable state''.

Stability: The property of a dynamical process reaching equilibrium. In neural nets, stability may refer to the weight changes reaching equilibrium during training, or the activations reaching equilibrium for a recurrent net.

Step function: A function that is piecewise constant. Also called a Heaviside function, or threshold function.

Strength: See weight; also called gain or synapse.

Strictly local backpropagation: An alternative structure for backpropagation in which (the same) computations are spread over more units. This addresses questions of biological plausibility and also allows more customizing of the activation functions which can improve the performance of the net [D. Fausett, 1990].

Supervised training: Process of adjusting the weights in a neural net using a learning algorithm; the desired output for each of a set of training input vectors is presented to the net. Many iterations through the training data may be required.

Synapse: See weight; also called gain or strength. In a biological neural system, the synapse is the connection between different neurons, where their membranes almost touch and signals are transmitted from one to the other by chemical neurotransmitters.

Synchronous processing: All activations are changed at the same time. See also Asynchronous.

Synchronous updates: All weights are adjusted at the same time.

Target: Desired response of a neural net; used during supervised training.

Threshold: A value used in some activation functions to determine the unit's output. Mathematically the effect of changing the threshold is to shift the graph of the activation function to the right or left; the same effect can be accomplished by including a bias.

Threshold function: See step function.

Tolerance: User supplied parameter used in stopping conditions (as in "total squared error less than specified tolerance"), or in evaluating performance (as in "the response of the net is considered correct if the output signal is within a specified tolerance of the target values").

Top-down weights: Weights from the cluster (F2 layer) units to the input (F1 layer) units in an ART net.

Topological neighborhood: Used in a Kohonen self-organizing map to determine which cluster nodes will have their weights modified for the current presentation of a particular input pattern. The neighborhood is specified as all cluster nodes within a given radius of the input pattern; the radius may be decreased as clustering progresses (on subsequent cycles through the input patterns).

Total squared error: Used in stopping conditions for backpropagation training. The square of the error is summed over all output components and over all training patterns. See also, Squared Error and Mean Squared Error.

Training algorithm: A step by step procedure for adjusting the weights in a neural net. See also Learning Rule.

Training epoch: One cycle through the set of training patterns.

Transfer function: See activation function.

Traveling salesman problem: A classic constrained optimization problem in which a salesman is required to visit each of a group of cities exactly once, before returning to the starting city. It is desired to find the tour with the shortest length.

Truck-Backer-Upper: An example of a neural net solution to a problem from control theory, that of backing a truck and trailer up to a loading dock without jack-knifing the rig [Nguyen & Widrow, 1989].

Underrepresented classes: An output class for which significantly fewer training patterns are available than are present for other classes. The usual solution to the difficulties in learning such a class is to duplicate or create noisy versions of the training patterns for the underrepresented class.

Unit: See processing element; also called neuron or node.

Unsupervised learning: A means of modifying the weights of a neural net without specifying the desired output for any input patterns. Used in self-organizing neural nets for clustering data, extracting principal components, or curve fitting.

Vector: An ordered set of numbers, an n-tuple. An input pattern is an example of a vector.

Vector quantization: The task of forming clusters of input vectors in order to compress the amount of data without losing important information.

Vigilance parameter: A user specified parameter in ART clustering neural networks which determines the maximum difference between 2 patterns in the same cluster; the higher the vigilance, the smaller the difference that is permitted to occur between patterns on a cluster.

Weight: A value associated with a connection path between two processing elements in a neural network. It is used to modify the strength of a transmitted signal in many networks. The weights contain fundamental information concerning the problem being solved by the net. In many nets the weights are modified during training using a learning algorithm. The terms strength, synapse, and gain are also used for this value.

Widrow-Hoff learning rule: See the delta rule; also called Least Mean Squares (LMS).

Winner-Take-All: The most extreme form of competition in a neural net, in which only the winning unit (typically the unit with the largest input signal, or the unit whose weight vector is closest to the input pattern) remains active.

References

AARTS, E., & J. KORST. (1989). *Simulated Annealing and Boltzmann Machines*, New York: John Wiley & SONS.

ABU-MOSTAFA, Y. S., & J.-M. ST JACQUES. (1985). "Information Capacity of the Hopfield Model." *IEEE Transactions on Information Theory*, IT-31:461–464.

ACKLEY, D. H., G. E. HINTON, & T. J. SEJNOWSKI. (1985) "A Learning Algorithm for Boltzmann Machines." *Cognitive Science*, 9:147–169. Reprinted in Anderson & Rosenfeld [1988], pp. 638–649.

AHMAD, S., & G. TESAURO. (1989). "Scaling and Generalization in Neural Networks." In D. S. Touretzky, ed., *Advances in Neural Information Processing Systems 1*. San Mateo, CA: Morgan Kaufmann, pp. 160–168.

AKIYAMA, Y., A. YAMASHITA, M. KAJIURA, & H. AISO. (1989). "Combinatorial Optimization with Gaussian Machines." *International Joint Conference on Neural Networks, Washington, DC*, I:533–540.

ALLMAN, W. F. (1989). *Apprentices of Wonder: Inside the Neural Network Revolution*, New York: Bantam Books.

ALMEIDA, L. B. (1987). "A Learning Rule for Asynchronous Perceptrons with Feedback in a Combinatorial Environment." *IEEE First International Conference on Neural Networks, San Diego, Ca*, II:609–618.

ALMEIDA, L. B. (1988). "Backpropagation in Perceptrons with Feedback." In R. Eckmiller, & Ch. von der Malsburg, eds., *Neural Computers*. Berlin: Springer-Verlag, pp. 199–208.

ANDERSON, J. A. (1968). "A Memory Storage Model Utilizing Spatial Correlation Functions." *Kybernetics*, 5:113–119. Reprinted in Anderson, Pellionisz, & Rosenfeld [1990], pp. 79–86.

ANDERSON, J. A. (1972). "A Simple Neural Network Generating an Interactive Memory." *Mathematical Biosciences,* 14:197–220. Reprinted in Anderson & Rosenfeld [1988], pp. 181–192.

ANDERSON, J. A. (1986). "Cognitive Capabilities of a Parallel System." In E. Bienenstock, F. Fogelman-Souli, & G. Weisbuch, eds., *Disordered Systems and Biological Organization.* NATO ASI Series, F20, Berlin: Springer-Verlag.

ANDERSON, J. A., R. M. GOLDEN, & G. L. MURPHY. (1986). "Concepts in Distributed Systems." In H. H. Szu, ed., *Optical and Hybrid Computing,* 634:260–272, Bellington, WA: Society of Photo-Optical Instrumentation Engineers.

ANDERSON, J. A., A. PELLIONISZ, & E. ROSENFELD, eds. (1990). *Neurocomputing 2*: Directions for Research. Cambridge, MA: MIT Press.

ANDERSON, J. A., & E. ROSENFELD, eds. (1988). *Neurocomputing: Foundations of Research.* Cambridge, MA: MIT Press.

ANDERSON, J. A., J. W. SILVERSTEIN, S. A. RITZ, & R. S. JONES. (1977). "Distinctive Features, Categorical Perception, and Probability Learning: Some Applications of a Neural Model." *Psychological Review,* 84:413–451. Reprinted in Anderson & Rosenfeld [1988], pp. 287–326.

ANGENIOL, B., G. VAUBOIS, & J-Y. LE TEXIER. (1988). "Self-organizing Feature Maps and the Travelling Salesman Problem." *Neural Networks,* 1(4):289–293.

ARBIB, M. A. (1987). *Brains, Machines, and Mathematics* (2d ed.), New York: Springer-Verlag.

AROZULLAH, M., & A. NAMPHOL. (1990). "A Data Compression System Using Neural Network Based Architecture." *International Joint Conference on Neural Networks, San Diego, CA,* I:531–536.

BACHMANN, C. M., L. N. COOPER, A. DEMBO, & O. ZEITOUNI. (1987). "A Relaxation Model for Memory with High Storage Density." Proceedings of the National Academy of Sciences, 84:7529–7531. Reprinted in Anderson, Pellionisz, & Rosenfeld, [1990], pp. 509–511.

BARHEN, J., S. GULATI, & M. ZAK. (1989). "Neural Learning of Constrained Nonlinear Transformation." *Computer,* 22(6):67–76.

BARTO, A. G., & R. S. SUTTON. (1981). *Goal Seeking Components for Adaptive Intelligence: An Initial Assessment.* Air Force Wright Aeronautical Laboratories/Avionics Laboratory Tech Rep. AFWAL-TR-81-1070. Dayton, OH: Wright-Patterson AFB.

BAUM, E. B., & D. HAUSSLER. (1989). "What Size Net Gives Valid Generalization?" *Neural Computation,* 1(1):151–160.

BLOCK, H. D. (1962). "The Perceptron: A Model for Brain Functioning, I." *Reviews of Modern Physics,* 34:123–135. Reprinted in Anderson & Rosenfeld [1988], pp. 138–150.

BRYSON, A. E., & Y-C. HO. (1969). *Applied Optimal Control.* New York: Blaisdell.

CARPENTER, G. A., & S. GROSSBERG. (1985). "Category Learning and Adaptive Pattern Recognition, a Neural Network Model." *Proceedings of the Third Army Conference on Applied Mathematics and Computation,* ARO Report 86-1, pp. 37–56.

CARPENTER, G. A., & S. GROSSBERG. (1987a). "A Massively Parallel Architecture for a Self-Organizing Neural Pattern Recognition Machine." *Computer Vision, Graphics, and Image Processing,* 37:54–115.

CARPENTER, G. A., & S. GROSSBERG. (1987b). "ART2: Self-organization of Stable Category Recognition Codes for Analog Input Patterns." *Applied Optics,* 26:4919–4930. Reprinted in Anderson, Pellionisz, & Rosenfeld [1990], pp. 151–162.

CARPENTER, G. A., & S. GROSSBERG. (1990). "ART3: Hierarchical Search Using Chemical Transmitters in Self-organizing Pattern Recognition Architectures." *Neural Networks,* 3(4):129–152.

CATER, J. P. (1987). "Successfully Using Peak Learning Rates of 10 (and greater) in Back-propagation Networks with the Heuristic Learning Algorithm." *IEEE First International Conference on Neural Networks, San Diego, CA,* II:645–652.

CAUDILL, M. (1989). *Neural Networks Primer.* San Francisco, Miller Freeman.

CAUDILL, M., & C. BUTLER. (1990). *Naturally Intelligent Systems.* Cambridge, MA: MIT Press.

CHEN, S., C. F. N. COWAN, & P. M. GRANT. (1991). "Orthogonal Least Squares Learning Algorithm for Radial Basis Function Networks." *IEEE Transactions on Neural Networks,* 2:302–309.

COHEN, M. A., & S. GROSSBERG. (1983). "Absolute Stability of Global Pattern Formation and Parallel Memory Storage by Competitive Neural Networks." *IEEE Transactions on Systems, Man, and Cybernetics,* SMC-13:815–826.

COLLINS, E., S. GHOSH, & C. L. SCOFIELD. (1988a). "An Application of a Multiple Neural Network Learning System to Emulation of Mortgage Underwriting Judgements." *IEEE International Conference on Neural Networks, San Diego, CA,* II:459–466.

COLLINS, E., S. GHOSH, & C. L. SCOFIELD. (1988b). "A Neural Network Decision Learning System Applied to Risk Analysis: Mortgage Underwriting and Delinquency Risk Assessment." In M. Holtz, ed., *DARPA Neural Network Study: Part IV System Application,* pp. 65–79.

COTTRELL, G. W., P. MUNRO, & D. ZIPSER. (1989). "Image Compression by Back Propagation: An Example of Extensional Programming." In N. E. Sharkey, ed., *Models of Cognition: A Review of Cognitive Science.* Norwood, NJ: Ablex Publishing Corp, pp. 208–240.

DARPA. (1988). *DARPA Neural Network Study, Final Report,* Cambridge, MA: Massachusetts Institute of Technology, Lincoln Laboratory.

DAYHOFF, J. E. (1990). *Neural Network Architectures.* New York: VanNostrand Reinholt.

DEROUIN, E., J. BROWN, H. BECK, L. FAUSETT, & M. SCHNEIDER. (1991). "Neural Network Training on Unequally Represented Classes." In C. H. Dagli, S. R. T. Kumara, & Y. C. Shin, eds., *Intelligent Engineering Systems Through Artificial Neural Networks.* New York: ASME Press, pp. 135–141.

DURBIN, R., & D. J. WILLSHAW. (1987). "An Analog Approach to the Traveling Salesman Problem Using an Elastic Net Method." *Nature,* 326:689–691.

ELMAN, J. L. (1990). "Finding Structure in Time." *Cognitive Science,* 14:179–211.

FAHLMAN, S. E. (1988). "Faster-Learning Variations on Back-Propagation: An Empirical Study." In D. Touretsky, G. Hinton & T. Sejnowski, eds., *Proceedings of the 1988 Connectionist Models Summer School.* San Mateo, CA: Morgan Kaufmann, pp. 38–51.

FAHLMAN, S. E., & C. LEBIERE. (1990). "The Cascade-Correlation Learning Architecture." In D. S. Touretsky, ed., *Advances in Neural Information Processing Systems 2.* San Mateo, CA: Morgan Kaufmann, pp. 524–532.

FARHAT, N. H., D. PSALTIS, A. PRATA, & E. PAEK. (1985). "Optical Implementation of the Hopfield Model." *Applied Optics*, 24:1469–1475. Reprinted in Anderson & Rosenfeld [1988], pp. 653–660.

FAUSETT, D. W. (1990). "Strictly Local Backpropagation." *International Joint Conference on Neural Networks, San Diego, CA,* III:125–130.

FAUSETT, L. V. (1990). "An Analysis of the Capacity of Associative Memory Neural Nets." *IEEE Southcon/90 Conference Record, Orlando, FL,* pp. 228–233.

FUKUSHIMA, K. (1975). "Cognitron: A Self-organizing Multi-layered Neural Network." *Biological Cybernetics,* 20(3/4):121–136.

FUKUSHIMA, K. (1988). "Neocognitron: A Hierarchical Neural Network Model Capable of Visual Pattern Recognition." *Neural Networks,* 1(2):119–130.

FUKUSHIMA, K., S. MIYAKE, & T. ITO. (1983). "Neocognitron: A Neural Network Model for a Mechanism of Visual Pattern Recognition." *IEEE Transactions on Systems, Man, and Cybernetics,* 13:826–834. Reprinted in Anderson & Rosenfeld [1988], pp. 526–534.

FUNAHASHI, K. (1989). "On the Approximate Realization of Continuous Mappings by Neural Networks." *Neural Networks,* 2(3):183–192.

GEMAN, S., & D. GEMAN. (1984). "Stochastic Relaxation, Gibbs Distributions, and the Bayesian Restoration of Images," *IEEE Transactions on Pattern Analysis and Machine Intelligence,* PAMI-6:721–741. Reprinted in Anderson & Rosenfeld [1988], pp. 614–634.

GEVA, S., & J. SITTE. (1992). "A Constructive Method for Multivariate Function Approximation by Multilayer Perceptrons." *IEEE Transactions on Neural Networks,* 3(4): 621–624.

GLUCK, M. A., & D. E. RUMELHART, eds. (1990). *Neuroscience and Connectionist Theory.* Hillsdale, NJ: Lawrence Erlbaum Associates.

GOLUB, G. H., & C. F. VAN LOAN. (1989). *Matrix Computations* (2nd ed.), Baltimore, MD: Johns Hopkins University Press.

GROSSBERG, S. (1969). "Embedding Fields: A Theory of Learning with Physiological Implications." *Journal of Mathematical Psychology,* 6:209–239.

GROSSBERG, S. (1976). "Adaptive Pattern Classification and Universal Recoding, I: Parallel Development and Coding of Neural Feature Detectors." *Biological Cybernetics,* 23: 121–134. Reprinted in Anderson & Rosenfeld [1988], pp. 245–258.

GROSSBERG, S. (1980). "How Does a Brain Build a Cognitive Code?" *Psychological Review,* 87:1–51. Reprinted in Anderson & Rosenfeld [1988], pp. 349–400.

GROSSBERG, S. (1982). *Studies of Mind and Brain.* Boston: Reidel.

GROSSBERG, S., ed. (1987 and 1988). *The Adaptive Brain, I: Cognition, Learning, Reinforcement, and Rhythm, and II: Vision, Speech, Language, and Motor Control.* Amsterdam: North-Holland.

GROSSBERG, S., & E. MINGOLLA. (1985a). "Neural Dynamics of Form Perception: Boundary Completion, Illusory Figures, and Neon Color Spreading." *Psychological Review,* 92: 173–211. Reprinted in Grossberg, [1988].

GROSSBERG, S., & E. MINGOLLA. (1985b). "Neural Dynamics of Perceptual Grouping: Textures, Boundaries, and Emergent Segmentations." *Perception and Psychophysics,* 38: 141–171. Reprinted in Grossberg, [1988].

HAINES, K., & R. HECHT-NIELSEN. (1988). "A BAM with Increased Information Storage

Capacity.'' *IEEE International Conference on Neural Networks, San Diego, CA,* I:181–190.

HARSTON, C. T. (1990). ''Business with Neural Networks.'' In A. J. Maren, C. T. Harston, & R. M. Pap, eds., *Handbook of Neural Computing Applications.* San Diego: Academic Press, pp. 391–400.

HEBB, D. O. (1949). *The Organization of Behavior.* New York: John Wiley & Sons. Introduction and Chapter 4 reprinted in Anderson & Rosenfeld [1988], pp. 45–56.

HECHT-NIELSEN, R. (1987a). ''Counterpropagation Networks.'' *Applied Optics,* 26(23): 4979–4984.

HECHT-NIELSEN, R. (1987b). ''Counterpropagation Networks.'' *IEEE First International Conference on Neural Networks, San Diego, CA,* II:19–32.

HECHT-NIELSEN, R. (1987c). ''Kolmogorov's Mapping Neural Network Existence Theorem.'' *IEEE First International Conference on Neural Networks, San Diego, CA,* III: 11–14.

HECHT-NIELSEN, R. (1988). ''Applications of Counterpropagation Networks.'' *Neural Networks,* 1(2):131–139.

HECHT-NIELSEN, R. (1989). ''Theory of the Backpropagation Neural Network.'' *International Joint Conference on Neural Networks, Washington, DC,* I:593–605.

HECHT-NIELSEN, R. (1990). *Neurocomputing.* Reading, MA: Addison-Wesley.

HERTZ, J., A. KROGH, & R. G. PALMER. (1991). *Introduction to the Theory of Neural Computation.* Redwood City, CA: Addison-Wesley.

HINTON, G. E., & T. J. SEJNOWSKI. (1983). ''Optimal Perceptual Inference.'' *Proceedings of the IEEE Conference on Computer Vision and Pattern Recognition, Washington DC,* pp., 448–453.

HOPFIELD, J. J. (1982). ''Neural Networks and Physical Systems with Emergent Collective Computational Abilities.'' *Proceeding of the National Academy of Scientists,* 79:2554–2558. Reprinted in Anderson & Rosenfeld [1988], pp. 460–464.

HOPFIELD, J. J. (1984). ''Neurons with Graded Response Have Collective Computational Properties like Those of Two-state Neurons.'' *Proceedings of the National Academy of Sciences,* 81:3088–3092. Reprinted in Anderson & Rosenfeld [1988], pp. 579–584.

HOPFIELD, J. J., & D. W. TANK. (1985). ''Neural Computation of Decisions in Optimization Problems.'' *Biological Cybernetics,* 52:141–152.

HOPFIELD, J. J., & D. W. TANK. (1986). ''Computing with Neural Circuits.'' *Science,* 233: 625–633.

HORNIK, K., M. STINCHCOMBE, & H. WHITE. (1989). ''Multilayer Feedforward Networks Are Universal Approximators.'' *Neural Networks,* 2(5): 359–366.

HORNIK, K., M. STINCHCOMBE, & H. WHITE. (1990). ''Universal Approximation of an Unknown Mapping and Its Derivatives Using Multilayer Feedforward Networks.'' *Neural Networks.* 3(5): 551–560.

HUBEL, D. H., & T. N. WIESEL. (1962). ''Receptive Fields, Binocular Interaction and Functional Architecture in Cat's Visual Cortex.'' *Journal of Physiology* (London) 160: 106–154.

JACOBS, R. A. (1988). ''Increased Rates of Convergence Through Learning Rate Adaptation.'' *Neural Networks,* 1(4):295–307.

JAIN, A. K., & R. C. DUBES. (1988). *Algorithms for Clustering Data*. Englewood Cliffs, NJ: Prentice-Hall.

JEONG, H., & J. H. PARK. (1989). "Lower Bounds of Annealing Schedule for Boltzmann and Cauchy Machines." *International Joint Conference on Neural Networks, Washingtin, DC*, I:581–586.

JOHNSON, R. C., & C. BROWN. (1988). *Cognizers: Neural Networks and Machines that Think*. New York: John Wiley & Sons.

JORDAN, M. (1989). "Generic Constraints on Underspecified Target Trajectories." *International Joint Conference on Neural Networks, Washington, DC*, I:217–225.

KESTEN, H. (1958). "Accelerated Stochastic Approximation." *Annals of Mathematical Statistics*, 29:41–59.

KIRKPATRICK, S., C. D. GELATT, JR., & M. P. VECCHI. (1983). "Optimization by Simulated Annealing." *Science*, 220:671–680. Reprinted in Anderson & Rosenfeld [1988], pp. 554–568.

KLIMASAUSKAS, C. C., ed. (1989). *The 1989 Neuro-computing Bibliography*. Cambridge, MA: MIT Press.

KOHONEN, T. (1972). "Correlation Matrix Memories." *IEEE Transactions on Computers*, C-21:353–359. Reprinted in Anderson & Rosenfeld [1988], pp. 174–180.

KOHONEN, T. (1982). "Self-organized Formation of Topologically Correct Feature Maps." *Biological Cybernetics*, 43:59–69. Reprinted in Anderson & Rosenfeld [1988], pp. 511–521.

KOHONEN, T. (1987). *Content-Addressable Memories* (2nd ed.), Berlin: Springer-Verlag.

KOHONEN, T. (1988). "The 'Neural' Phonetic Typewriter." *Computer*, 21(3):11–22.

KOHONEN, T. (1989a). *Self-organization and Associative Memory* (3rd ed.), Berlin: Springer-Verlag.

KOHONEN, T. (1989b). "A Self-learning Musical Grammar, or 'Associative Memory of the Second Kind'." *International Joint Conference on Neural Networks, Washington, DC*, I:1–5.

KOHONEN, T. (1990a). "Improved Versions of Learning Vector Quantization." *International Joint Conference on Neural Networks, San Diego, CA*, I:545–550.

KOHONEN, T. (1990b). "The Self-Organizing Map." *Proceedings of the IEEE*, 78(9):1464–1480.

KOHONEN, T., K. TORKKOLA, M. SHOZAKAI, J. KANGAS, & O. VENTA. (1987). "Microprocessor Implementation of a Large Vocabulary Speech Recognizer and Phonetic Typewriter for Finnish and Japanese." *European Conference on Speech Technology, Edinburgh, September 1987*. Volume 2, pp. 377–380.

KOLMOGOROV, A. N. (1963). "On the Representation of Continuous Functions of Many Variables by Superposition of Continuous Functions of One Variable and Addition." *Doklady Akademii Nauk SSSR*, 144:679–681. (*American Mathematical Society Translation*, 28:55–59).

KOSKO, B. (1987). "Competitive Adaptive Bidirectional Associative Memories." *IEEE First International Conference on Neural Networks, San Diego, CA*, II:759–766.

KOSKO, B. (1988). "Bidirectional associative memories." *IEEE Transactions on Systems, Man, and Cybernetics*, 18:49–60. Reprinted in Anderson, Pellionisz, & Rosenfeld [1990], pp. 165–176.

Kosko, B. (1992a). *Neural Networks and Fuzzy Systems: A Dynamical Systems Approach to Machine Intelligence.* Englewood Cliffs, NJ: Prentice-Hall.

Kosko, B., ed. (1992b). *Neural Networks for Signal Processing.* Englewood Cliffs, NJ: Prentice-Hall.

Kreinovich, V. Y. (1991). "Arbitrary Nonlinearity Is Sufficient to Represent All Functions by Neural Networks: A Theorem." *Neural Networks,* 4(3):381–383.

Lawler, E. L., J. K. Lenstra, A. H. G. Rinnooy Kan, & D. B. Shmoys, eds. (1985). *The Traveling Salesman Problem: A Guided Tour of Combinatorial Optimization.* New York: John Wiley & Sons.

Lawrence, J. (1993). "Data Preparation for a Neural Network." *Neural Network Special Report, AI Expert,* pp. 15–21.

Le Cun, Y. (1986). "Learning Processes in an Asymmetric Threshold Network." In E. Bienenstock, F. Fogelman-Souli, & G. Weisbuch, eds. *Disordered Systems and Biological Organization.* NATO ASI Series, F20, Berlin: Springer-Verlag.

Le Cun, Y., B. Boser, J. S. Denker, D. Henderson, R. E. Howard, W. Hubbard, & L. D. Jackel. (1990). "Handwritten Digit Recognition with a Backpropagation Network." In D. S. Touretsky, ed., *Advances in Neural Information Processing Systems 2.* San Mateo, CA: Morgan Kaufman, pp. 396–404.

Lee, S. (1992). "Supervised Learning with Gaussian Potentials." In B. Kosko (ed.), *Neural Networks for Signal Processing.* Englewood Cliffs, NJ: Prentice-Hall, pp. 189–228.

Lee, S., & R. M. Kil. (1991). "A Gaussian Potential Function Network with Hierarchically Self-Organizing Learning." *Neural Networks,* 4(2):207–224.

Leonard, J. A., M. A. Kramer, & L. H. Ungar. (1992). "Using Radial Basis Functions to Approximate a Function and Its Error Bounds." *IEEE Transactions on Neural Networks,* 3(4):624–627.

Levine, D. S. (1991). *Introduction to Neural and Cognitive Modeling.* Hillsdale, NJ: Lawrence Erlbaum Associates.

Lippmann, R. P. (1987). "An Introduction to Computing with Neural Nets." *IEEE ASSP Magazine,* 4:4–22.

Lippmann, R. P. (1989). "Review of Neural Networks for Speech Recognition." *Neural Computation,* 1:1–38.

MacGregor, R. J. (1987). *Neural and Brain Modeling.* San Diego: Academic Press.

Makhoul, J., S. Roucos, & H. Gish. (1985). "Vector Quantization in Speech Coding." *Proceedings of the IEEE,* 73(11):1551–1588.

Mazaika, P. K. (1987). "A Mathematical Model of the Boltzmann Machine." *IEEE First International Conference on Neural Networks, San Diego, CA,* III:157–163.

McClelland, J. L., & D. E. Rumelhart. (1988). *Explorations in Parallel Distributed Processing.* Cambridge, MA: MIT Press.

McCulloch, W. S. (1988). *Embodiments of Mind.* Cambridge, MA: MIT Press.

McCulloch, W. S., & W. Pitts. (1943). "A Logical Calculus of the Ideas Immanent in Nervous Activity." *Bulletin of Mathematical Biophysics,* 5:115–133. Reprinted in Anderson & Rosenfeld [1988], pp. 18–28.

McEliece, R. J., E. C. Posner, E. R. Rodemich, & S. S. Venkatesh. (1987). "The Capacity of the Hopfield Associative Memory." *IEEE Transactions on Information Theory,* IT-33:461–482.

MEHROTRA, K. G., C. K. MOHAN, & S. RANKA. (1991). "Bounds on the Number of Samples Needed for Neural Learning." *IEEE Transactions on Neural Networks*, 2(6):548–558.

METROPOLIS, N. A., W. ROSENBLUTH, M. N. ROSENBLUTH, A. H. TELLER, & E. TELLER. (1953). "Equations of State Calculations by Fast Computing Machines." *Journal of Chemical Physics*, 21:1087–1091.

MILLER, W. T., R. S. SUTTON, & P. J. WERBOS, eds. (1990). *Neural Networks for Control*. Cambridge, MA: MIT Press.

MINSKY, M. L., & S. A. PAPERT. (1988). *Perceptrons, Expanded Edition*. Cambridge, MA: MIT Press. Original edition, 1969.

MOODY, J., & C. J. DARKEN. (1989). "Fast Learning in Networks of Locally Tuned Processing Units." *Neural Computation*, 1:281–294.

NGUYEN, D., & B. WIDROW. (1989). "The Truck Backer-Upper: An Example of Self-Learning in Neural Networks." *International Joint Conference on Neural Networks, Washington, DC*, II:357–363.

NGUYEN, D., & B. WIDROW. (1990). "Improving the Learning Speed of Two-Layer Neural Networks by Choosing Initial Values of the Adaptive Weights." *International Joint Conference on Neural Networks, San Diego, CA*, III:21–26.

OJA, E. (1982). "A Simplified Neuron Model as a Principal Components Analyzer." *Journal of Mathematical Biology*, 15:267–273.

OJA, E. (1989). "Neural Networks, Principal Components, and Subspaces." *International Journal of Neural Systems*, 1:61–68.

OJA, E. (1992). "Principal Components, Minor Components, and Linear Neural Networks." *Neural Networks*, 5(6):927–935.

PAO, Y-H. (1989). *Adaptive Pattern Recognition and Neural Networks*. New York: Addison Wesley.

PARK, J., & I. W. SANDBERG. (1991). "Universal Approximation Using Radial-basis Function Networks." *Neural Computation*, 3:246–257.

PARKER, D. (1985). *Learning Logic*. Technical Report TR-87, Cambridge, MA: Center for Computational Research in Economics and Management Science, MIT.

PINEDA, F. J. (1987). "Generalization of Back-Propagation to Recurrent Neural Networks." *Physical Review Letters*, 59:2229–2232.

PINEDA, F. J. (1988). "Dynamics and Architecture for Neural Computation." *Journal of Complexity*, 4:216–245.

PINEDA, F. J. (1989). "Recurrent Back-Propagation and the Dynamical Approach to Adaptive Neural Computation." *Neural Computation*, 1:161–172.

PITTS, W., & W. S. MCCULLOCH. (1947). "How We Know Universals: The Perception of Auditory and Visual Forms." *Bulletin of Mathematical Biophyscis*, 9:127–147. Reprinted in Anderson & Rosenfeld [1988], pp. 32–42.

POGGIO, T. (1990). "Networks for Approximation and Learning." *Proceedings of the IEEE*, 78:1481–1497.

QIAN, N., & T. J. SEJNOWSKI. (1988). "Learning to Solve Random-Dot Stereograms of Dense Transparent Surfaces with Recurrent Back-Propagation." In D. Touretzky, G. Hinton, & T. Sejnowski, eds., *Proceedings of the 1988 Connectionist Models Summer School*. San Mateo, CA: Morgan Kaufmann, pp. 435–443.

REBER, A. S. (1967). "Implicit Learning of Artificial Grammars." *Journal of Verbal Learning and Verbal Behavior,* 5:855–863.

ROCHESTER, N., J. H. HOLLAND, L. H. HAIBT, & W. L. DUDA. (1956). "Tests on a Cell Assembly Theory of the Action of the Brain, Using a Large Digital Computer." *IRE Transactions on Information Theory,* IT-2:80–93. Reprinted in Anderson & Rosenfeld [1988], pp. 68–80.

ROGERS, S. K., & M. KABRISKY. (1989). *Introduction to Artificial Neural Networks for Pattern Recognition.* SPIE Short Course, Orlando, FL, Bellington, WA: Society of Photo-Optical Instrumentation Engineers.

ROSENBLATT, F. (1958). "The Perceptron: a Probabilistic Model for Information Storage and Organization in the Brain." *Psychological Review,* 65:386–408. Reprinted in Anderson & Rosenfeld [1988], pp. 92–114.

ROSENBLATT, F. (1959). "Two Theorems of Statistical Separability in the Perceptron." *Mechanization of Thought Processes: Proceeding of a Symposium Held at the National Physical Laboratory, November 1958.* London: HM Stationery Office, pp. 421–456.

ROSENBLATT, F. (1962). *Principles of Neurodynamics.* New York: Spartan.

RUMELHART, D. E., G. E. HINTON, & R. J. WILLIAMS. (1986a). "Learning Internal Representations by Error Propagation." In D. E. Rumelhart & J. L. McClelland, eds., *Parallel Distributed Processing,* vol. 1 chapter 8. reprinted in Anderson & Rosenfeld [1988], pp. 675–695.

RUMELHART, D. E., G. E. HINTON, & R. J. WILLIAMS. (1986b). "Learning Representations by Back-Propagating Error." *Nature,* 323:533–536. Reprinted in Anderson & Rosenfeld [1988], pp. 696–699.

RUMELHART, D. E., J. L. MCCLELLAND, & THE PDP RESEARCH GROUP. (1986). *Parallel Distributed Processing, Explorations in the Microstructure of Cognition; Vol. 1: Foundations.* Cambridge, MA: MIT Press.

SANGER, T. D. (1989). "Optimal Unsupervised Learning in a Single-Layer Linear Feedforward Neural Network." *Neural Networks,* 2(6):459–473.

SARIDIS, G. N. (1970). "Learning Applied to Successive Approximation Algorithms." *IEEE Transactions on Systems Science and Cybernetics,* SCC-6:97–103.

SCOFIELD, C. L. (1988). "Learning Internal Representations in the Coulomb Energy Network." *IEEE International Conference on Neural Networks, San Diego, CA,* I:271–276.

SEJNOWSKI, T. J., & C. R. ROSENBERG. (1986). *NETtalk: A Parallel Network That Learns to Read Aloud.* The Johns Hopkins University Electrical Engineering and Computer Science Technical Report JHU/EECS-86/01, 32 pp. Reprinted in Anderson & Rosenfeld [1988], pp. 663–672.

SERVAN-SCHREIBER, D., A. CLEEREMANS, & J. L. MCCLELLAND. (1989). "Learning Sequential Structure in Simple Recurrent Networks." In D. S. Touretzky, ed., *Advances in Neural Information Processing Systems 1.* San Mateo, CA: Morgan Kaufmann, pp. 643–652.

SILVA, F. M., & L. B. ALMEIDA. (1990). "Acceleration Techniques for the Backpropagation Algorithm." *Lecture Notes in Computer Science,* 412:110–119.

SIMMONS, G. F. (1972). *Differential Equations with Applications and Historical Notes.* New York: McGraw-Hill.

SIVILATTI, M. A., M. A. MAHOWALD, & C. A. MEAD. (1987). "Real-time Visual Computations Using Analog CMOS Processing Arrays." In P. Losleben, ed., *Advanced Research in VLSI: Proceedings of the 1987 Stanford Conference*. Cambridge, MA: MIT Press, pp. 295–312. Reprinted in Anderson & Rosenfeld [1988], pp. 703–712.

SONEHARA, N., M. KAWATO, S. MIYAKE, & K. NAKANE. (1989). "Image Data Compression Using a Neural Network Model." *International Joint Conference on Neural Networks, Washington, DC,* II:35–41.

SPECHT, D. F. (1967). "Vectorcardiographic Diagnosis Using the Polynomial Discriminant Method of Pattern Recognition." *IEEE Transactions on Bio-Medical Engineering,* BME-14:90–95.

SPECHT, D. F. (1988). "Probabilistic Neural Networks for Classification, Mapping, or Associative Memory." *IEEE International Conference on Neural Networks, San Diego, CA,* I:525–532.

SPECHT, D. F. (1990). "Probabilistic Neural Networks." *Neural Networks,* 3(1):109–118.

SPRECHER, D. A. (1965). "On the Structure of Continuous Functions of Several Variables." *Transactions of the American Mathematical Society,* 115:340–355.

SUTTON, R. S. (1986). "Two Problems with Backpropagation and Other Steepest-descent Learning Procedures for Networks." *Proceedings of the Eighth Annual Conference of the Cognitive Science Society,* pp. 823–831.

SZU, H. H. (1986). "Fast Simulated Annealing." In J. S. Denker, ed., *AIP Conference Proceedings 151, Neural Networks for Computing, Snowbird, UT*. New York: American Institute of Physics.

SZU, H. H. (1988). "Fast TSP Algorithm Based on Binary Neuron Output and Analog Neuron Input Using the Zero-Diagonal Interconnect Matrix and Necessary and Sufficient Constraints of the Permutation Matrix." *IEEE International Conference on Neural Networks, San Diego, CA,* II:259–265.

SZU, H. H. (1989). *Neural Networks: Theory, Applications and Computing*. Lecture Notes for UCLA Engineering Short Course, Engineering 819.185, March 20–23, 1989.

SZU, H. H. (1990a). "Colored Noise Annealing Benchmark by Exhaustive Solutions of TSP." *International Joint Conference on Neural Networks, Washington, DC,* I:317–320.

SZU, H. H. (1990b). "Optical Neuro-Computing." In A. J. Maren, C. T. Harston, & R. M. Pap, eds., *Handbook of Neural Computing Applications*. San Diego: Academic Press.

SZU, H. H., & R. HARTLEY. (1987). "Fast Simulated Annealing." *Physics Letters A,* 122(3,4):157–162.

SZU, H. H., & A. J. MAREN. (1990) "Multilayer Feedforward Neural Networks II: Optimizing Learning Methods." In A. J. Maren, C. T. Harston, & R. M. Pap, eds., *Handbook of Neural Computing Applications*. San Diego: Academic Press.

TAKEFUJI, Y. (1992). *Neural Network Parallel Computer*. Boston: Kluwer Academic Publishers.

TAKEFUJI, Y., & H. H. SZU. (1989). "Design of Parallel Distributed Cauchy Machines." *International Joint Conference on Neural Networks, Washington, DC,* I:529–532.

TANK, D. W., & J. J. HOPFIELD. (1987). "Collective Computation in Neuronlike Circuits." *Scientific American,* 257:104–114.

TEPEDELENLIOGLU, N., A. REZGUI, R. SCALERO, & R. ROSARIO. (1991). "Fast Algorithms for Training Multilayer Perceptrons." In B. Soucek & the IRIS Group, *Neural and Intelligent Systems Integration*. New York: John Wiley & Sons.

TOLAT, V. V., & B. WIDROW. (1988). "An Adaptive 'Broom Balancer' with Visual Inputs." *IEEE International Conference on Neural Networks, San Diego, CA,* II:641–647.

VEMURI, V., ed. (1988). *Artificial Neural Networks: Theoretical Concepts*. Washington, DC: IEEE Computer Society Press.

VON NEUMANN, J. (1958). *The Computer and the Brain*. New Haven: Yale University Press. Pages 66–82 are reprinted in Anderson & Rosenfeld [1988], pp. 83–89.

WEIR, M. (1991). "A Method for Self-Determination of Adaptive Learning Rates in Back Propagation." *Neural Networks,* 4:371–379.

WERBOS, P. (1974). *Beyond Regression: New Tools For Prediction and Analysis in the Behavioral Sciences* (Ph.D. thesis). Cambridge, MA: Harvard U. Committee on Applied Mathematics.

WHITE, H. (1990). "Connectionist Nonparametric Regression: Multilayer Feedforward Networks Can Learn Arbitrary Mappings." *Neural Networks*, 3(5): 535–549.

WIDROW, B., & M. E. HOFF, JR. (1960). "Adaptive Switching Circuits." *IRE WESCON Convention Record,* part 4, pp. 96–104. Reprinted in Anderson & Rosenfeld [1988], pp. 126–134.

WIDROW, B. (1987). "The Original Adaptive Neural Net Broom-balancer." *International Symposium on Circuits and Systems*. New York: IEEE. pp. 351–357.

WIDROW, B., & M. A. LEHR. (1990). "30 Years of Adaptive Neural Networks: Perceptron, Madaline, and Backpropagation." *Proceedings of the IEEE,* 78(9):1415–1442.

WIDROW, B., P. E. MANTEY, L. J. GRIFFITHS, & B. B. GOODE. (1967). "Adaptive Antenna Systems." *Proceedings of the IEEE,* 55:2143–2159.

WIDROW, B., & S. D. STEARNS. (1985). *Adaptive Signal Processing*. Englewood Cliffs, NJ: Prentice-Hall.

WIDROW, B., R. G. WINTER, & R. A. BAXTER. (1987). "Learning Phenomena in Layered Neural Networks." *IEEE First International Conference on Neural Networks,* II:411–429.

WILSON, G. V., & G. S. PAWLEY. (1988). "On the Stability of the Travelling Salesman Problem Algorithm of Hopfield and Tank." *Biological Cybernetics,* 58:63–70.

XU, L., E. OJA, & C. Y. SUEN. (1992). "Modified Hebbian Learning for Curve and Surface fitting." *Neural Networks,* 5(3):441–457.

Index

A

Aarts, E., 37, 337, 338, 339, 424, 437
Abu-Mostafa, Y. S., 140, 437
Acceptance probability, 339, 342, 347, 348, 358–59, 362 (*see also* Metropolis condition)
Accretion, 201, 422
Ackley, D. H., 26, 334, 337, 346, 367, 369, 371, 372, 424, 437
Activation, 3, 4, 20, 422
Activation function, 3, 17–19, 422
  ART2, 250
  brain-state-in-a-box, 132
  Gaussian, 315, 388, 427
  Heaviside function, 17, 434 (*see* Step function)
  identity, 17, 314
  log, 314
  sigmoid, 17–19
    adaptive slope, 312–13
    arctangent, 313, 433
    binary or logistic, 4, 17–18, 143, 293, 423, 433
    bipolar, 18–19, 293–94, 424, 434
    customized, 309–12
    hyperbolic tangent (tanh), 17, 19, 293, 298, 350, 351, 358, 434

step function, 17–18, 27, 41, 109, 434
threshold, 17, 435 (*see* Step function)
Activity level, 3, 337, 358
ADALINE, 7, 23–24, 40, 41, 80–88, 97, 422
  algorithm, 81–82
    derivation, 86–88
  applications, 82–86
    AND function, 83–84
    AND NOT function, 85
    OR function, 85–86
  architecture, 81
  exercises and project, 99, 100
Adaptive antenna systems, 24, 447
Adaptive architectures, 385–98 (*see also* Probabilistic neural nets; Cascade correlation)
Adaptive resonance theory (ART), 16, 25, 157, 218–87, 422 (*see also* ART1; ART2)
  algorithm, 220–22
  architecture, 219–20
Ahmad, S., 299, 437
Aiso, H., 357, 427, 437
Akiyama, Y., 357, 427, 437
Algorithm, 3, 15–16, 423 (*see also* Learning rule; Supervised training; Unsupervised training)
  Adaline (delta rule), 81–82
  adaptive resonance theory, 220–21, 225–29, 250–57

Algorithm (*cont.*)
  anti-Hebbian (modified), 365
  autoassociator with threshold, 133
  backpropagation, 290–300, 321–23
  backpropagation for fully recurrent nets,
    305
  backpropagation in time, 383–84
  bidirectional associative memory (BAM),
    143
  Boltzmann machine with learning, 369–72
  Boltzmann machine without learning,
    340–42
  brain-state-in-a-box (BSB), 132
  cascade correlation, 394–97
  counterpropagation:
    forward, 208–9
    full, 200–201
  Hamming net, 166–67
  Hebb rule, 49, 103–5, 122
  Hopfield, continuous, 350–51
  Hopfield, discrete, 136–37
  learning vector quantization (LVQ),
    188–89
  Madaline, 89–92
  Maxnet, 159
  Mexican hat, 161–63
  modified Hebb, 365–66
  neocognitron, 407–11
  perceptron, 61
  probabilistic neural net, 388
  QuickProp, 397–98
  self-organizing map (SOM), 170, 172
  simple recurrent net, 375–76
  Widrow-Hoff (*see* Adaline)
Allman, W. F., 36, 437
Almeida, L. B., 307, 384, 385, 437, 445
AND function, 29, 44–45, 50–55, 62–69,
    83–84
AND NOT function, 30, 85
Anderson, J. A., 9, 22, 23, 24, 35, 102, 104,
    121, 130, 131, 132, 149, 282, 398, 424,
    437, 438
Angeniol, B., 24, 172, 184, 211, 438
Annealing (*see* Simulated annealing)
Annealing schedule, 371, 423, 442 (*see also*
    Cooling schedule)
Anti-Hebbian learning, 365–66
Applications: (*see also* Business; Character
    recognition; Control; Data
    compression; Encoder problem;
    Logic functions; Medicine, Pattern
    recognition; Signal Processing;
    Speech Production; Speech
    Recognition; Traveling salesman
    problem)
  adaptive antenna systems, 24, 447
  broom balancer, 24, 447
  character strings (context-sensitive
    grammar), 373–77, 445

curve fitting, 365–66
  handwritten character recognition, 8–9,
    398–417, 428, 443
  heart abnormalities, 39, 389, 446
  instant physician, 9
  mortgage risk assessment, 11, 439
  multiplication tables, 24
  musical composition, 24, 442
  NETtalk, 10, 289, 445
  phonetic typewriter, 10, 432, 442
  shift register, 380–81
  sinc function (damped sinusoid), 381–84
  telephone noise cancellation (echo
    cancellation), 7, 426
  time varying patterns, 372
  truck-backer-upper, 8, 24, 436, 444
Arbib, M. A., 76, 438
Architecture, 3, 4, 12–15, 423
  adaptive, 335, 385–98
  competitive layer, 14–15
  feedforward, 12
  multilayer, 12–14
  partially recurrent, 372
  recurrent, 12
  single layer, 12–14
  traveling salesman problem, 338
Arctangent function, 313, 433
Aristotle, 101
Arozullah, M., 304, 438
ART1, 25, 218, 222–46, 282, 422
  algorithm, 225–29
  analysis, 243–46
  applications, 229–42
    character recognition, 236–42
    simple examples, 229–36
  architecture, 222–25
    computational units, 223
    supplemental units, 224–25
  exercises and projects, 283, 286
ART2, 25, 218, 246–82, 422, 439 (*see also*
    Noise suppression)
  activation function, 250
  algorithm, 250–57
    parameters, 255–56
  analysis, 275–82
    differential equations, 277–79
    initial weights, 281–82
    instant ART2, 275–76
    reset, 279–81
  applications, 257–74
    character recognition, 273–74
    simple examples, 257–68
    spanning tree, 268–72
  architecture, 247–50
  exercises and projects, 284–86
ART3, 439
Artificial neural network, 3 (*see also* Neural
    network)

Associative memory, 16, 25, 101–55, 211, 423 (*see also* Autoassociative memory; Bidirectional associative memory; Heteroassociative memory; Hopfield net)
Assumptions:
  artificial neural network, 3
  McCulloch-Pitts neuron, 26–27
Asynchronous updates, 135, 138, 148, 347, 423
Autoassociative memory, 9, 16, 24, 102, 121–40, 423, 442
  exercises and project, 151, 153
  feedforward, 16, 121–29
    algorithm, 122
    application, 122–25
    architecture, 121
    storage capacity, 125–29
  recurrent, 16, 129–40
    brain-state-in-a-box, 131–32
    discrete Hopfield, 135–40 (*see also* Hopfield net)
    linear autoassociator, 130–31
    with threshold, 132–35
Autoassociator, 423
Axon, 5, 6, 423

B

Bachmann, C. M., 426, 438
Backpropagation for fully recurrent nets, 384–85, 437
Backpropagation in time, 8, 377–84
  algorithm, 379, 383–84
  application, 380–84
  architecture, 377–79
Backpropagation net, 9, 23, 289–333, 441
  activation functions:
    adaptive slope, 312–13
    arctangent, 313
    binary (logistic) sigmoid, 293
    bipolar sigmoid, 293–94
    customized sigmoid, 309–12
    non-saturating, 314
    non-sigmoid, 315
  algorithm, 290–300, 321–23
    adaptive learning rates, 306–9
    batch weight updating, 306
    momentum, 305
    standard, 294–96
    two hidden layers, 321–23
    weight initialization, 296–98, 444
  analysis, 305–28
    derivations, 324–28
    variations, 305–23
  applications, 300–305, 314–15
    data compression, 302–4

Xor function, 300–302, 306, 314
      Nguyen-Widrow initialization, 302
      product of sines, 314–15
  architecture, 290–91, 320–21
    standard, 290–91
    two hidden layers, 320–21
  exercises and projects, 330–33
Backpropagation training, 25, 289, 294–96, 321–28, 423, 445, 447 (*see also* Backpropagation for fully recurrent nets; Backpropagation in time; Backpropagation net; Simple recurrent net)
BAM (*see* Bidirectional associative memory)
Barhen, J., 385, 438
Barto, A. G., 307, 438
Batch updating, 87, 306, 397
Baum, E. B., 298, 438
Baxter, R. A., 82, 89, 91, 447
Bayes decision theory, 26, 187, 335, 385–87, 432
Beck, H., 306, 439
Bias, 20, 21, 41, 81, 165, 290, 423
  adaptation, 20, 21, 49, 61, 90–91, 295–96
  relation to threshold, 41–43
Bidirectional associative memory (BAM), 140–49, 423, 440, 442
  continuous, 143–44
  discrete, 140–43, 144–49
    algorithm, 141–43
    analysis, 148–49
    application, 144–48
    architecture, 141
  exercises and projects, 152, 154–55
Bidirectional connections, 338
Binary, 17, 423
Binary data:
  And function, 29, 50–51
  And Not function, 30
  character recognition, 236–42, 273–74
  Or function, 29
  pattern association, 110–15
  Xor function, 30–31, 300–301
Binary sigmoid, 4, 17–18, 143, 293, 423, 433
Binary/bipolar hybrid data:
  And function, 51–52, 62–68, 83–84
  pattern association, 116
Biological neuron, 6–7
Biological neural network, 5–7, 37
Bipolar, 17, 423
Bipolar data:
  And function, 44–45, 52–55, 69, 84
  And Not function, 85
  character recognition, 55–56, 71–76, 119–20, 144–47
  Or function, 46, 85–86
  pattern association, 116–19, 123–27, 133–34
  Xor function, 47, 92–95, 301–2

Bipolar sigmoid, 18–19, 293–94, 424, 434
Bivalent, 15, 424, 429
Block, H. D., 23, 59, 91, 432, 438
Boltzmann machine, 26, 37, 437, 443 (*see also* Boltzmann machine with learning; Boltzmann machine without learning)
Boltzmann machine with learning, 334, 367–72, 424, 437, 442
  algorithm, 369–70
  application, 371–72
  architecture, 368
Boltzmann machine without learning, 16, 334, 338–48, 424
  algorithm, 340–42
  analysis, 346–48
  application, 342–46
  architecture, 340
  exercises and projects, 418–21
Boolean functions (*see* Logic functions; AND function; AND NOT function; OR function; XOR function)
Boser, B., 443
Bottom-up signals, 221, 246
Bottom-up weights, 219, 222–23, 225, 256, 424
  initial values, 246, 282
Boundary (*see* Decision boundary)
Boundary contour system (BCS), 424
Brain-state-in-a-box (BSB), 9, 24, 131–32, 424
Broom balancing, 24, 447
Brown, C., 6, 23, 26, 36, 442
Brown, J., 306, 439
Bryson, A. E., 25, 438
BSB (*see* Brain-state-in-a-box)
Business, applications of neural networks, 11, 441
Butler, C., 39, 439

C

Capacity, 125–29, 140, 149, 424, 437, 440, 441, 443
Carpenter, G. A., 25, 218, 222, 224, 229, 243, 247, 248, 251, 252, 275, 277, 282, 422, 438, 439
Cascade correlation, 335, 390–98, 424, 439
  algorithm, 394–97
  architecture, 391–94
Cater, J. P., 307, 439
Cauchy machine, 334, 359–362, 424, 442, 446
Cauchy (or colored) noise, 359
Caudill, M., 39, 431, 439
Cell (*see* Unit; Neuron)

Chain rule, 87–88, 107–8, 324–25, 326–27
Character recognition, 25 (*see also* Handwritten character recognition)
  ART, 236–42, 273–74
  bidirectional associative memory, 144–48
  counterpropagation, 215–16
  Hebb net, 55–56
  heteroassociative memory, 119–21
  perceptron, 71–76
  self-organizing map, 176–78
Chen, S., 330, 439
Clamped, 367, 369, 372, 424
Classification, 424 (*see* Pattern classification)
Cleeremans, A., 372, 373, 445
Cluster unit, 16, 218, 219, 220, 202, 207, 222, 223, 425
Clustering, 425
Clustering neural networks, 157 (*see also* Self-organizing map; Adaptive resonance theory)
Code-book (or code) vector, 157, 187, 218, 429
Cognitron, 25, 440
Cohen, M. A., 144, 384, 439
Collins, E., 11, 433, 439
Combinatorial optimization, 437 (*see also* Constrained optimization problems; Constrained optimization nets)
Competition, 156 (*see also* Adaptive resonance theory; Competitive neural net; Counterpropagation; Learning vector quantization; Self-organizing map; Winner-take-all)
Competitive layer, 14–15, 219, 223
Competitive learning, 425 (*see also* Kohonen learning)
Competitive neural net, 158–69, 425 (*see also* Hamming net; Maxnet; Mexican hat)
Components:
  minor, 365–66, 444
  principal, 363–65, 444
Connections, 3
  bidirectional, 140–41, 158, 338, 348
  excitatory, 27
  inhibitory, 27
Conscience, 425
Consensus function, 339, 425
Constrained optimization nets, 16, 135, 335–62
  Boltzmann machine, 338–48, 424
  Cauchy machine, 359–62, 424
  Gaussian machine, 357–59, 427
  Hopfield net, 348–57, 428
Constrained optimization problems, 335, 425
Content addressable memory, 102, 135, 425, 442
Context unit, 372–74, 425

Contrast enhancement, 160
Control, applications of neural networks, 8, 36
Controller module, 8 (*see* Truck-backer-upper)
Convergence, 425
Cooling schedule:
  Boltzmann machine, 342
  Cauchy machine, 359–60, 362
Cooper, L. N., 26, 426, 438
Correlation:
  among input vectors, 104, 106
  encoding, 106, 148, 425
  matrix, 22, 82, 363, 442
Cortical units, 316–19
Cottrell, G. W., 302, 304, 439
Coulomb energy net, 425, 445 (*see also* Reduced coulomb energy network)
Counterpropagation, 16, 26, 157, 195–211, 426, 441
  exercises and projects, 213–14, 215–17
  forward only, 206–11
    algorithm, 206–9
    application, 209–11
    architecture, 206–7
  full, 196–206
    algorithm, 199–201
    application, 201–6
    architecture, 196–98
Counting layers in a net, 12
Covariance matrix, 363, 365, 395
Cowan, C. F. N., 330, 439
Cross talk, 104–5, 110, 426
Curve fitting, 365–66, 447

D

Darken, C. J., 316, 444
DARPA, 26, 36, 164, 439
Data compression, 195, 302–4, 368, 438, 439, 446
Data representation, 48, 102, 115–19, 298–99, 443 (*see also* Binary data; Bipolar data)
  binary vs bipolar, 48, 118–19, 298–99
  continuous vs discrete, 298–99
Dayhoff, J. E., 199, 286, 439
Decay term:
  Hopfield net, 349, 357
  Oja rule, 363
Decision boundary (or region), 42–46, 51–55, 57–58, 63–66, 68, 84–86, 93–95, 386, 389, 426
Delta rule, 23, 40, 86–88, 106–8, 121, 199, 426

extended (arbitrary activation function), 106, 107–8, 427
generalized (backpropagation), 106, 289, 294–96
one output unit, 86–87
several output units, 87–88, 106–7
Delta-bar-delta, 307–9, 426
Dembo, A., 426, 438
Dendrites, 5, 426
Denker, J. S., 443
DeRouin, E., 306, 439
Differential equations:
  ART1, 243–46
  ART2, 277–79
  Cohen-Grossberg, 144, 384
  Fully recurrent net, 384
  Hopfield, 349, 357, 384
Dot product, 129, 426
  metric, 158, 169, 196, 201, 209
  net input, 21, 114
Dubes, R. C., 442
Duda, W. L., 22, 445
Durbin, R., 354, 439

E

Echo cancellation, 7, 426
Eigenvalue, 82, 127, 128, 131, 363, 365, 432
Eigenvector, 127, 128, 131, 363, 364, 365, 432
Elman, J. L., 372, 375, 425, 439
Emulator, 8 (*see also* Truck-backer-upper)
Encoder problem, 334, 367, 368, 371–72, 426
Energy function, 135, 148–49, 337, 426 (*see also* Lyapunov function)
  BAM, 148–49
  Boltzmann machine, 346–47 (*see also* Consensus function)
  Hopfield net, 135, 138, 139–40, 349, 354, 357
Epoch, 221, 252, 255, 296, 341, 342, 350, 369, 426, 436
Equilibrium:
  activations, 245–46, 277–78
  thermal, 346
  weights, 221, 229, 243–45, 247, 262–63, 265–68, 279
Erasing an association, 148
Error propagation (*see* Backpropagation)
Error (*see* Total squared error; Squared error; Mean squared error)
Euclidean distance, 158, 169, 188, 196, 201, 209, 268, 427
Excitatory connection, 27–28, 427
Excitatory signal (or input), 225, 245, 247, 407

Exclusive or (*see* XOR function)
Exemplar, 16, 157, 158, 164–66, 169, 218, 427
Extended delta rule, 106, 107–8, 427

F

Fahlman, S. E., 307, 390, 391, 394, 397, 398, 424, 433, 439
Farhat, N. H., 26, 440
Fast learning, 221, 225, 229, 247, 251, 255, 256, 264–67, 270–72, 427 (*see also* ART1; ART2; Slow learning)
Fast optimization, 355–56
Fast simulated annealing 359, 446 (*see* Cauchy machine)
Fault tolerance, 6, 427
Fausett, D. W., 7, 316, 435, 440
Fausett, L. V., 306, 439, 440
Feedforward network, 12, 102, 427 (*see also* Adaline; Backpropagation net; Cascade correlation; Hebb net; Madaline; Perceptron; Probabilistic neural net)
Feedforward phase, 290–91 (*see also* Backpropagation training)
Fixed-point cycle, 135
Fixed weight nets, 16, 157, 158–69, 335–62, 427 (*see also* Boltzmann machine without learning; Hamming net; Hopfield net; MAXNET; Mexican hat; McCulloch-Pitts neuron)
Fukushima, K., 25, 398, 431, 440
Funahashi, K., 328, 440
Function, activation (*see* Activation function)
Function approximation, 5, 195, 202–6, 210–11, 314–15, 381–84, 440–43, 447

G

Gain, 427
Gain control units, 224
Gaussian distribution or function, 315, 316, 348, 358, 362, 386–88, 427, 443
Gaussian machine, 334, 357–59, 427, 437
Gelatt, C. D., 26, 342, 442
Geman, D., 26, 342, 440
Geman, S., 26, 342, 440
Generalization, 48, 96, 121, 290, 298, 427, 437, 438 (*see also* Memorization)
Generalized delta rule, 428 (*see* Backpropagation)

Geva, S., 330, 440
Ghosh, S., 11, 433, 439
Gish, H., 189, 443
Global minimum, 306
Gluck, M. A., 37, 440
Golden, R. M., 9, 438
Golub, G. H., 365, 440
Goode, B. B., 24, 447
Gradient , 86–87, 107, 296
Gradient descent, 289, 296, 384 (*see also* Backpropagation; Delta rule)
Grandmother cells, 428
Grant, P. M., 330, 439
Griffiths, L. J., 24, 447
Grossberg learning, 199, 428
Grossberg, S., 24, 25, 144, 218, 222, 224, 229, 243, 247, 248, 251, 252, 275, 277, 282, 384, 422, 424, 431, 438, 439, 440
Gulati, S., 385, 438
Gutschow, T., 26

H

Haibt, L. H., 22, 445
Haines, K., 141, 149, 440
Hamming distance, 147, 164, 166, 428
Hamming net, 164–69, 428
  application, 166–69
  architecture, 165–66
Handwritten character recognition, 8–9, 398–417, 428, 443
Hard limiter 354 (*see* Step function)
Hardware, 2, 26
Harston, C. T., 11, 441
Hartley, R., 26, 359, 424, 446
Haussler, D., 298, 438
Heaviside function, 17, 434 (*see also* Step function)
Hebb learning rule, 22, 23, 40, 48, 96, 102–6, 149, 428
  exercises, 150
  modified, 362–66, 447
  normalization, 105–6, 365
  outer products, 104, 111–13
  setting the weights 103–4
Hebb net, 48–58, 428
  algorithm, 49
  application, 50–58
    AND function, 50–55
    character recognition, 55–56
    limitations, 56–58
  exercises, 97–98
Hebb, D. O., 22, 48, 96, 149, 441
Hecht-Nielsen, R., 9, 26, 82, 88, 132, 138, 141, 149, 195, 199, 201, 208, 211, 298, 314, 328, 330, 377, 426, 440, 441

Hecht-Nielsen Theorem, 329
Henderson, D., 443
Hertz, J., 76, 105, 316, 363, 364, 365, 372, 377, 384, 385, 441
Heteroassociative memory, 16, 102, 108–21, 428 (*see also* Bidirectional associative memory)
  application, 108–21
    character recognition, 119–21
    procedure, 108–9
    simple examples, 110–19
  architecture, 108
  exercises and projects, 150–51, 152–53
Hidden layers:
  backpropagation with two, 320–23
  number of, 96, 290, 299, 320
Hidden units, 4, 14, 428
  number needed, 303
Hinton, G. E., 25, 26, 289, 330, 334, 337, 338, 346, 367, 369, 371, 372, 377, 380, 424, 437, 439, 441, 445
History of neural networks, 22–26, 37
Ho, Y-C., 25, 438
Hoff, M. E., 23, 24, 80, 89, 97, 106, 422, 447
Holland, J. H., 22, 445
Hopfield, J. J., 25, 121, 130, 135, 136, 137, 139, 149, 349, 350, 351, 352, 353, 357, 384, 428, 441, 446
Hopfield net, 12, 25, 135–40, 149, 348–57, 428, 437
  capacity, 140
  continuous, 16, 334, 348–57
    algorithm, 350–51
    analysis, 352–57
    application, 351–52
    architecture, 350
  discrete, 135–40
    algorithm, 136–38
    analysis, 139–40
    application, 138–39
    architecture, 136
Hornik, K., 329, 441
Howard, R. E., 443
Hubbard, W., 443
Hubel, D. H., 398, 441
Hybrid learning (*see* Counterpropagation)
Hyperbolic tangent function (tanh), 17, 19, 293, 298, 350, 351, 358, 434
Hyperplanes, 43

I

Image compression, 302–4 (*see also* Data compression)
Inhibited cluster unit, 219–20, 224, 251
Inhibitory connection, 27, 428

Inhibitory signal (or input), 27, 225, 245, 247, 407
Inner product (*see* Dot product)
Input units, 4, 428
Input vector, 16, 20
Instant physician, 9
Iteration, 428–29
Iterative net, 12, 102 (*see also* Recurrent network)
Ito, T., 25, 398, 431, 440

J

Jackel, D., 443
Jacobs, R. A., 306, 307, 308, 426, 441
Jain, A. K., 442
Jeong, H., 348, 359, 361, 362, 442
Johnson, R. C., 6, 23, 26, 36, 442
Jones, R. S., 24, 438
Jordan, M., 329, 442

K

Kabrisky, M., 7, 445
Kajiura, M., 357, 427, 437
Kangas, J., 24, 442
Kawato, M., 304, 446
Kesten, H., 307, 442
Kil, R. M., 316, 330, 443
Kirkpatrick, S., 26, 342, 442
Klimasauskas, C. C., 24, 26, 442
Kohonen learning, 157, 199, 429 (*see also* Self-organizing map)
Kohonen, T., 7, 10, 22, 24, 101, 102, 104, 149, 157, 160, 161, 169, 172, 178, 179, 182, 187, 189, 192, 194, 195, 211, 268, 429, 430, 432, 434, 442
Kohonen unit, 206
Kolmogorov, A. N., 328, 442
Kolmogorov theorem, 328–29, 441
Korst, J., 37, 337, 338, 339, 424, 437
Kosko, B., 36, 140, 141, 143, 144, 148, 149, 423, 442, 443
Kramer, M. A., 316, 443
Kreinovich, V. Y., 330, 443
Krogh, A., 76, 105, 316, 363, 364, 365, 372, 377, 384, 385, 441

L

Lawler, E. L., 335, 443
Lawrence, J., 299, 443

Layer, 12, 429
  cluster, 196
  competitive, 14
  hidden, 12
  output, 12
Learning algorithm, 429 (*see* Algorithm)
Learning rate, 21, 429
  Adaline, 82
  ART2, 256
  backpropagation, 292, 305, 306–9
  perceptron, 59, 61
  self-organizing map, 172
Learning rate adjustment:
  backpropagation, 306–9, 439, 441, 447
  self-organizing map, 172
Learning rules: (*see also* Algorithm;
      Supervised training; Unsupervised
      training)
  delta rule, 107
  Grossberg, 199, 208
  Kohonen, 170, 199, 207
  Oja, 363, 364
  outstar, 199
  Sanger, 364
Learning trial, 220, 221, 250, 255, 429
Learning vector quantization (LVQ), 16,
      157, 187–95, 211, 212–13, 215, 429,
      442
  algorithm, 188–89
  application, 189–93
    geometric example, 190–93
    simple example, 189–90
  architecture, 187–88
  exercises and projects, 212–13, 215
  variations, 192–95
    LVQ2, 192–94
    LVQ2.1, 194–95
    LVQ3, 195
Least-mean-square (LMS), 23, 80, 426
Lebiere, C., 390, 391, 394, 424, 433, 439
Le Cun, Y., 9, 25, 443
Lee, S., 316, 330, 386, 443
Lehr, M. A., 24, 88, 91, 97, 430, 447
Lenstra, J. K., 335, 443
Leonard, J. A., 316, 443
Le Texier, J-Y., 24, 172, 185, 211, 438
Levine, D. S., 37, 443
Linear autoassociator, 429
Linear independence, 106
Linear separability, 43–47, 429
  AND function, 44–45
  OR function, 46–47
  XOR function, (non-separability of) 47
Linear threshold units, 429
Linear units, 363
Lippmann, R. P., 10, 15, 158, 164, 211, 229,
      425, 428, 443
LMS rule (*see* Delta rule; Least-mean-
      square)

Local minima, 306
Logic functions, 22, 28, 429 (*see* AND
      function; AND NOT function; OR
      function; XOR function)
Logistic sigmoid function (*see* Sigmoid)
Long term memory, 430
Look-up table, 195, 196, 202
LVQ (*see* Learning vector quantization)
Lyapunov function, 135, 138, 139, 148–49,
      430 (*see also* Energy function)
Lyapunov, A. M., 135

M

McClelland, J. L., 22, 25, 48, 105, 106, 121,
      130, 132, 289, 303, 330, 372, 373, 423,
      432, 443, 445
McCulloch, W. S., 22, 23, 26, 31, 37, 430,
      443, 444
McCulloch-Pitts input/output function, 354
      (*see* Step function)
McCulloch-Pitts neuron, 2, 22, 26–35, 430
  activation function, 27
  applications, 29–35
    AND function, 29
    AND NOT function, 30
    Hot and Cold, 31–35
    OR function, 29
    XOR function, 30
  architecture, 27–28
  assumptions, 26–27
  exercises, 37–38
McEliece, R. J., 140, 443
MacGregor, R. J., 37, 443
MADALINE, 24, 40, 41, 48, 81, 88–96, 97,
      430, 447
  algorithm, 89–92
    MRI, 89–91
    MRII, 91–92
  application, 92–96
    XOR function, 92–95
  architecture, 88–89
  exercises and projects, 99–100
Mahowald, M. A., 26, 446
Makhoul, J., 189, 443
Mantley, P. E., 24, 447
Mapping networks (*see* Backpropagation;
      Counterpropagation; Pattern
      association; Pattern classification)
Maren, A. J., 360, 424, 446
Markov chain:
  Boltzmann machine, 347–48
  Cauchy machine, 361–62
Matrix-multiplication notation, 21
MAXNET, 15, 156, 158–60

application, 159–60
architecture, 159
Mazaika, P. K., 443
Mead, C. A., 26, 446
Mean squared error, 80, 430
Medicine (applications of neural networks):
　heart abnormalities, 39, 389, 446
　instant physician, 9
Mehrotra, K. G., 320, 444
Memory (*see* Content addressable memory;
　　Long term memory; Short term
　　memory)
Memorization, 289, 298, 430 (*see also*
　　Generalization)
Metropolis, N. A., 347, 444
Metropolis condition, 347
Mexican hat, 156, 158, 160–64, 430
　algorithm, 161–63
　application, 163–64
　architecture, 161
Miller, W. T., 8, 36, 444
Mingolla, E., 424, 440
Minsky, M. L., 23, 24, 37, 39, 43, 48, 76,
　　79, 80, 97, 432, 444
Missing data, 118–19, 124, 133–34, 430
Mistakes in data, 115, 118–20, 123–24, 134,
　　138–39, 430
Miyake, S., 25, 304, 398, 431, 440, 446
Mohan, C. K., 320, 444
Momentum, 430
Moody, J., 316, 444
Mortgage assessment, 11, 439
Multilayer network 13–14 (*see*
　　Backpropagation; Backpropagation in
　　time; Counterpropagation; Cascade
　　correlation; MADALINE;
　　Neocognitron; Probabilistic neural
　　net; Simple recurrent net)
Multilayer perceptron, 430 (*see*
　　Backpropagation net)
Munro, P., 302, 304, 439
Murphy, G. L., 9, 438
Musical composition, 24, 442

N

Nakane, K., 304, 446
Namphol, A., 304, 438
Neighborhood, 10, 169–71 (*see also* Self-
　　organizing map)
　hexagonal, 171, 178, 181
　linear, 170, 176, 202, 208, 210
　rectangular, 171, 178, 180–81
Neocognitron, 9, 25, 335, 398–417, 431, 440
　algorithm, 407–17
　　training patterns, 411–17

training process, 408–11
　architecture, 399–406
Net input, 3, 20, 21, 431
NETtalk, 10, 289, 445
Network (*see* Applications; Architectures;
　　Feedforward network; Multilayer
　　network; Recurrent network; Single-
　　layer neural net)
Neural nets, 431 (*see* Neural networks)
Neural networks, 431
　artificial, 3–5
　　characteristics, 3, 6
　　history, 22–26, 36
　biological, 5–7, 37
　interest:
　　interdisciplinary, 2, 7, 22, 23
　　renewed, 2
　optical, 26
Neurocomputing, 2, 35, 431
Neuron, 3, 431
　artificial, 4
　biological, 6–7
Nguyen, D., 8, 24, 297, 330, 436, 444
Nguyen-Widrow initialization, 297–98
Node, 3, 431 (*see also* Neuron)
Noise, 431 (*see also* Missing data; Mistakes
　　in data)
　colored (Cauchy), 359
　suppression in ART, 246, 250, 251, 255,
　　260–61
　used to generate training data, 367
　white (Gaussian), 359
Norm of vector, 20, 225 (*see also* Euclidean
　　distance; Dot product metric)
Normalization:
　adaptive resonance theory, 226–27,
　　246–47, 250
　dot product metric, 158, 196
　Hebb rule, 105–6, 365
　neocognitron, 407
　probabilistic neural net, 388
Notation, 20–21

O

Oja, E., 334, 363, 364, 365, 366, 444, 447
Oja learning rule:
　1-unit, 363–64
　M-units, 364–65 (*see also* Sanger rule)
Optical neural networks, 26, 440, 446
Optimization (*see* Combinatorial
　　optimization; Constrained
　　optimization)
OR function, 29, 46–47, 85–86
Orthogonal vectors (or patterns), 105, 122,
　　126, 129, 431

Outer product, 104, 431
Output, 431
    function (*see* Activation function)
    layer, 12
    unit, 4, 431
Outstar, 431

P

Paek, E., 26, 440
Palmer, R. G., 76, 105, 316, 363, 364, 365,
    372, 377, 384, 385, 441
Pao, Y-H., 37, 440
Papert, S. A., 23, 24, 37, 39, 43, 59, 76, 79,
    97, 432, 444
Parallel distributed processing, 25, 443, 445
Park, J., 316, 444
Park, J. H., 348, 359, 361, 362, 442
Parker, D., 25, 444
Pattern, 431
Pattern association, 14, 15–16, 101–55, 195,
    432 (*see also* Autoassociative
    memory; Bidirectional associative
    memory; Heteroassociative memory;
    Hopfield net)
Pattern associator, 432
Pattern classification, 12, 15, 39–100, 432
    (*see also* ADALINE; Backpropagation;
    Cascade correlation; Hebb net;
    Learning vector quantization;
    Neocognitron; Perceptron;
    Probabilistic neural net)
Pattern classifier, 432
Pattern recognition, 8–9, 37 (*see also*
    Character recognition)
Pawley, G. S., 336, 352, 353, 447
Pellionisz, A., 24, 35, 282, 438
Perceptron, 23, 37, 59–80, 96–97, 432, 438,
    444, 445, 447
    activation function, 59
    algorithm, 23, 61
    application, 62–76
        AND function, 62–69
        character recognition, 71–76
        simple example, 69–71
    architecture, 60
    convergence theorem, 76–80
    exercises and projects, 98–99, 100
    learning rule, 23, 40, 59, 432
Phonetic typewriter, 10, 432, 442
Pineda, F. J., 384, 444
Pitts, W., 22, 23, 26, 31, 430, 443, 444
Plasticity, 219, 432
Poggio, T., 316, 444
Posner, E. C., 140, 443

Prata, A., 26, 440
Principal components, 363–65, 432
Probabilistic neural net, 316, 335, 446,
    385–89, 432
    algorithm, 388
    analysis, 389
    application, 388–89
    architecture, 387
Problems solved by neural nets, 3, 15
    clustering, 16 (*see* Adaptive resonance
        theory; Self-organizing map)
    constrained optimization, 16, 25 (*see*
        Boltzmann machine without learning;
        Hopfield net, continuous)
    mapping, 16 (*see* Backpropagation;
        Counterpropagation; Pattern
        association; Pattern classification)
Probabilistic state transition (*see* Acceptance
    probability)
Processing element, 432 (*see also* Neuron;
    Unit)
Psaltis, D., 26, 440

Q

Qian, N., 385, 444
QuickProp, 390, 397–98, 433

R

Radial basis function, 330, 427, 433, 443
Radial basis function network, 316, 439, 444
Ranka, S., 320, 444
Reber, A. S., 374, 445
Recurrent network, 12, 102, 433 (*see also*
    Bidirectional associative memory;
    Boltzmann machine; Hopfield net;
    MAXNET; Recurrent nets trained by
    backpropagation)
Recurrent nets trained by backpropagation,
    8, 335, 372–85, 444 (*see also*
    Backpropagation for fully recurrent
    net; Backpropagation in time; Simple
    recurrent net)
Reduced (or restricted) coulomb energy
    network, 26, 433
Regularization network, 316
Relaxation, 433
Reset unit (or mechanism), 219, 222, 223,
    224–25, 251, 279–81
Resonance, 221, 433
Rezgui, A., 313, 447
Rinnooy Kan, A. H. G., 335, 443

Ritz, S. A., 24, 438
Rochester, N., 22, 445
Rodemich, E. R., 140, 443
Rogers, S. K., 7, 445
Rosario, R., 313, 447
Rosenberg, C. R., 10, 445
Rosenblatt, F., 23, 59, 97, 432, 434, 445
Rosenbluth, A. W., 347, 444
Rosenbluth, M. N., 347, 444
Rosenfeld, E., 22, 23, 24, 35, 149, 282, 398, 438
Roucos, S., 189, 443
Rumelhart, D. E., 22, 25, 37, 48, 105, 106, 121, 130, 132, 289, 303, 330, 377, 380, 423, 432, 440, 443, 445

S

St Jacques, J.-M., 140, 437
Sandberg, I. W., 316, 444
Sanger, T. D., 364, 365, 445
Sanger learning rule, 364
Saridis, G. N., 307, 445
Saturate, 293, 433
Scalero, R., 313, 447
Schneider, M., 306, 439
Scofield, C. L., 11, 426, 433, 439, 445
Sejnowski, T. J., 10, 26, 334, 337, 338, 346, 367, 368, 369, 371, 372, 385, 424, 437, 439, 441, 444, 445
Self-organization, 149, 433, 442 (see also ART1; ART2; Self-organizing map)
Self-organizing map (SOM), 7, 10, 16, 24, 157, 169–87, 211, 433, 438 (see also Neighborhood)
  algorithm, 170–72
  application, 172–87
    character recognition, 176–78
    geometric, 178–85
    simple example, 172–75
    spanning tree, 178–81
    traveling salesman problem, 182–87
  architecture, 169–71
  exercises and projects, 211–12, 214–15
Self-supervised learning, 15, 365
Sequential updates, 135, 140, 339, 348 (see also Asynchronous updates; Synchronous updates)
Servan-Schreiber, D., 372, 373, 374, 375, 445
Shmoys, D. B., 335, 443
Short term memory, 433
Shozakai, M., 24, 442
Sigmoid functions, 17–19, 433
  arctangent, 313, 433

binary range (logistic), 4, 17–18, 143, 293, 423, 433
  bipolar range, 18–19, 293–94, 424, 434
  customizing, 309–12
  hyperbolic tangent (tanh), 17, 19, 293, 298, 350, 351, 358, 434
Signal processing, 7, 36, 37, 97, 443, 447
Signals, 3, 5, 434 (see also Bottom-up signals; Excitatory signals; Inhibitory signals; Top-down signals)
Silva, F. M., 307, 445
Silverstein, J. W., 24, 438
Simmons, G. F., 135, 445
Simple recurrent net, 372–77, 445
  algorithm, 373
  application, 373–77
  architecture, 372–73
Simulated annealing, 26, 37, 339, 434, 437, 442 (see also Cooling schedule)
Sinc (damped sine) function, 381–84
Single-layer neural net, 12–14, 434 (see also Pattern association; Pattern classification)
Single-layer perceptron, 434
Sitte, J., 330, 440
Sivilatti, M. A., 26, 446
Slab, 434
Slope (or steepness) parameter, 18, 434
Slow learning, 221, 247, 251, 255, 256–57, 268, 270–72 (see also ART1; ART2; Fast learning)
SOM (see Self-organizing map)
Soma, 5, 434
Sonehara, N., 304, 446
Spanning tree:
  clustering using ART2, 270–72
  clustering using self-organizing maps, 178–81
  data, 179, 269, 287, 434
Specht, D. F., 39, 385, 389, 432, 446
Speech production, 9–10
Speech recognition, 10, 24, 443
Squared error, 86, 87, 107, 434
Squashing function, 329
Sprecher, D. A., 328, 446
Sprecher theorem, 329
Stable state, 434
  spurious, 138, 434
Stability, 218–19, 251, 434
Stearns, S. D., 7, 37, 97, 306, 426, 447
Steepness parameter (see Slope parameter)
Step function, 17–18, 27, 41, 109, 434
Stinchcombe, M., 329, 441
Storage capacity (see Capacity)
Storing vectors, 121, 125–29
Strength, 434 (see Weight)
Strictly local backpropagation, 316–19, 435, 440

Suen, C. Y., 365, 366, 447
Supervised training, 15–16, 157, 289, 435
    (*see also* Hebb rule; Delta rule;
    Backpropagation)
Sutton, R. S., 8, 36, 307, 438, 444, 446
Synapse, 435
Synaptic units, 316–19
Synchronous processing, 435
Synchronous updates, 131, 148, 435
Szu, H. H., 26, 121, 127, 135, 336, 352, 354,
    356, 359, 360, 361, 424, 446

T

Takefuji, Y., 335, 348, 354, 357, 359, 430,
    446
Tank, D. W., 25, 135, 149, 349, 350, 351,
    352, 353, 441, 446
Target (or target vector), 289, 291, 292, 435
Teller, A. H., 347, 444
Teller, E., 347, 444
Telephone noise suppression, 7, 426
Temperature, 339, 358
Tepedelenlioglu, N., 313, 447
Tesauro, G., 299, 437
Thalamic units, 316–19
Thermal equilibrium, 346
Threshold, 17, 20, 27, 41, 59, 61, 62, 435
Threshold function, 435 (*see* Step function)
Tolat, V. V., 24, 447
Tolerance, 303, 435
Top-down signals, 221, 225, 246
Top-down weights, 219, 222–23, 225, 247,
    256, 435
    initial values, 246, 282
Topology preserving map, 169 (*see* Self-
    organizing map)
Topological neighborhood, 169–171, 435 (*see*
    Neighborhood)
Torkkola, K., 24, 442
Total least squares, 365
Total squared error, 83, 84, 85, 91, 289, 298,
    435
Training (*see also* Supervised training;
    Unsupervised training)
    algorithm, 436 (*see also* Algorithm)
    epoch, 436 (*see* Epoch)
    input vector, 20
    set (how many patterns), 298, 444
    output vector, 20 (*see also* Target vector)
Transfer function, 436 (*see* Activation
    function)
Traveling salesman problem, 24, 25, 172,
    182, 211, 335–37, 436, 438, 439, 447
    architecture, 338, 343–44, 350
    results:
        Boltzmann machine, 344–46
        Boltzmann/Cauchy hybrid, 360
        Cauchy machine, 360

fast optimization, 354–56
Hopfield net, 351–52
self-organizing map, 182–187
Wilson and Pawley, 352–54
setting the weights, 340–41, 344, 350
Willshaw initialization, 354–55
Truck-backer-upper, 8, 24, 436, 444
Two-thirds rule, 225

U

Underrepresented classes, 306, 436
Ungar, L. H., 316, 443
Unit, 3, 436 (*see also* Cluster unit; Context
    unit; Cortical unit; Hidden unit; Input
    unit; Kohonen unit; Linear unit;
    Output unit; Synaptic unit; Thalamic
    unit)
Unlearning, 148
Unsupervised learning, 16, 157, 170–72, 218,
    221–22, 225–29, 250–57, 362–366, 436
Updating (*see* Asynchronous updating;
    Batch updating; Synchronous
    updating)

V

Van Loan, C. F., 365, 440
Vaubois, G., 24, 172, 185, 211, 438
Vecchi, M. P., 26, 342, 442
Vector, 436
Vector quantization, 436 (*see also* Learning
    vector quantization)
Vemuri, V., 398, 447
Venkatesh, S. S., 140, 443
Venta, O., 24, 442
Vigilance parameter, 220, 225–29, 256, 281,
    422, 436
    effect on cluster formation, 230–42,
    270–72
VLSI implementations, 26
Von Neumann, J., 23, 447

W

Weight, 3, 20, 436
Weight change, 21 (*see also* Fixed weight
    nets)
    backpropagation, 295–96, 325
    backpropagation, 295–96, 325
    Boltzmann machine, 367
    cascade correlation (QuickProp), 397

counterpropagation, 199, 207–8
delta rule (Adaline), 86, 87, 107
extended delta rule, 107
Hebb rule, 49
Kohonen learning, 157
modified Hebb, 363, 364, 365
perceptron, 59
Weight matrix, 20
symmetric, 130, 132, 135, 140, 149, 158, 348, 357
Weight vector, 20
Weir, M., 307, 447
Werbos, P. J., 8, 25, 36, 444, 447
White, H., 329, 330, 441, 447
Widrow, B., 7, 8, 23, 24, 37, 39, 40, 80, 82, 88, 89, 91, 97, 106, 297, 306, 330, 422, 426, 430, 436, 444, 447
Widrow-Hoff learning rule, 23, 80, 422, 426, 436 (*see also* Delta rule)
Wiesel, T. N., 398, 441
Williams, R. J., 25, 289, 330, 377, 380, 445
Willshaw, D. J., 354, 439
Wilson, G. V., 336, 352, 353, 447
Winner-take-all, 15, 156, 158–60, 206, 247, 251, 425, 436
Winter, R. G., 82, 89, 91, 447

X

Xor function, 30–31, 47, 92–95, 300–302, 306, 314
Xu, L., 365, 366, 447

Y

Yamashita, A., 357, 366, 447

Z

Zak, M., 385, 438
Zeitouni, O., 426, 438
Zeroing out the diagonal, 121, 124–25, 135, 140, 149
Zipser, D., 302, 304, 439